The Complex Mind

The Complex Mind

An Interdisciplinary Approach

Edited by

David McFarland
University of Oxford, UK

Keith Stenning
University of Edinburgh, UK

and

Maggie McGonigle-Chalmers
University of Edinburgh, UK

Selection and editorial matters © David McFarland, Keith Stenning, Maggie McGonigle-Chalmers 2012
Chapters © their individual authors 2012
Epilogue © David McFarland 2012
Obituary for Ulrich Nehmzow © Mark Lee 2012

All rights reserved. No reproduction, copy or transmission of this publication may be made without written permission.

No portion of this publication may be reproduced, copied or transmitted save with written permission or in accordance with the provisions of the Copyright, Designs and Patents Act 1988, or under the terms of any licence permitting limited copying issued by the Copyright Licensing Agency, Saffron House, 6–10 Kirby Street, London EC1N 8TS.

Any person who does any unauthorized act in relation to this publication may be liable to criminal prosecution and civil claims for damages.

The authors have asserted their rights to be identified as the authors of this work in accordance with the Copyright, Designs and Patents Act 1988.

First published 2012 by
PALGRAVE MACMILLAN

Palgrave Macmillan in the UK is an imprint of Macmillan Publishers Limited, registered in England, company number 785998, of Houndmills, Basingstoke, Hampshire RG21 6XS.

Palgrave Macmillan in the US is a division of St Martin's Press LLC, 175 Fifth Avenue, New York, NY 10010.

Palgrave Macmillan is the global academic imprint of the above companies and has companies and representatives throughout the world.

Palgrave® and Macmillan® are registered trademarks in the United States, the United Kingdom, Europe and other countries

ISBN: 978–0–230–24757–4

This book is printed on paper suitable for recycling and made from fully managed and sustained forest sources. Logging, pulping and manufacturing processes are expected to conform to the environmental regulations of the country of origin.

A catalogue record for this book is available from the British Library.

A catalog record for this book is available from the Library of Congress.

10 9 8 7 6 5 4 3 2 1
21 20 19 18 17 16 15 14 13 12

Printed and bound in the United States of America

The proceeds from this book will go to a fund held by the School of Philosophy, Psychology and Language Sciences, University of Edinburgh, under the name of Brendan McGonigle in support of post-graduate research and scholarship of an inter-disciplinary nature.

Contents

List of Illustrations ix

Preface xii

Notes on Contributors xiv

Part I Complexity and the Animal Mind

Introduction 3
Maggie McGonigle-Chalmers

1 Relational and Absolute Discrimination Learning by Squirrel Monkeys: Establishing a Common Ground with Human Cognition 12
Barry T. Jones

2 Serial List Retention by Non-Human Primates: Complexity and Cognitive Continuity 25
F. Robert Treichler

3 The Use of Spatial Structure in Working Memory: A Comparative Standpoint 38
Carlo De Lillo

4 The Emergence of Linear Sequencing in Children: A Continuity Account and a Formal Model 55
Maggie McGonigle and Iain Kusel

5 Sensitivity to Quantity: What Counts across Species? 80
Sarah T. Boysen and Anna M. Yocom

Part II Complexity in Robots

Introduction 99
David McFarland

6 Towards Cognitive Robotics: Robotics, Biology and Developmental Psychology 103
Mark Lee, Ulrich Nehmzow and Marcos Rodriguez

7	Structuring Intelligence: The Role of Hierarchy, Modularity and Learning in Generating Intelligent Behaviour *Joanna J. Bryson*	126
8	Epistemology, Access and Computational Models *George Luger*	144
9	Reasoning about Representations in Autonomous Systems: What Pólya and Lakatos Have to Say *Alan Bundy*	167

Part III Language, Evolution and the Complex Mind

	Introduction *Keith Stenning*	187
10	How to Qualify for a Cognitive Upgrade: Executive Control, Glass Ceilings and the Limits of Simian Success *Andy Clark*	197
11	Private Codes and Public Structures *Colin Allen*	223
12	The Emergence of Complex Language *Wolfram Hinzen*	243
13	Language Evolution: Enlarging the Picture *Keith Stenning and Michael van Lambalgen*	264

Epilogue: Brendan McGonigle 283

Index 289

Illustrations

Tables

4.1a	Example sequence of binary comparisons for Game A in which size difference is combined with the mean and variance of the urgency variables	68
4.1b	An example stream of bet outcome data in terms of stimulus and referent pairs within one monotonic trial	69
6.1	Development of competencies	108

Figures

1.1	Pairwise training given to squirrel monkeys, and an example of a monkey extending a relational choice ('larger') in the context of triadic testing	18
2.1	Representation of three different 5-item lists combined to form a 15-item list	33
2.2	Mean proportion choice scores of 4 monkeys in a within-subject design conforming to a 15-item list when combining three earlier-learned 5-item lists	34
3.1	Average number of moves (visits to containers) required by Rats (blue line); Monkeys (dotted line) and children (green line) in order to complete an exhaustive search of a set of nine containers	42
3.2	Micro-development of search organisation in monkeys (*Cebus apella*) exploring a linear configuration of baited containers	44
3.3	A static pattern of the animations presented to human participants who were asked to write a description and indicate the perceived complexity of the most representative search path used by rats (left), monkeys (middle) and children (right)	46
3.4	A) Number of words required by human participants to describe the most representative search path of rats, monkeys and children. B) Complexity ratings reported for the same paths. Error bars = 1 SE	47
4.1	The touchscreen seriation task	57
4.2	The trial number at which each item within a five item monotonic size sequence reached criterion	63
4.3	Summarised frequency distribution of forwards and backwards errors during five-item size sequence learning	64

4.4a	The model representation of stimuli as objects held within a task lattice	66
4.4b	The representation of the size property of the stimuli for a monotonic set [12345] as Gaussian probability distributions	67
4.5	Model architecture and information flow within a single binary comparison between stimuli 1 and 2	70
4.6	Flow of information (Steps 1 to 3) within a simulation trial through the components illustrated in Figure 4.5	71
4.7	Model architecture and information flow (Steps 1 to 3) within a single binary comparison between stimuli 1 and 2.	72
4.8a	The distribution of incorrect selections across set [12345] for the simulations	74
4.8b	The distribution of incorrect selections across set [ABCDE] for the simulations	74
4.9	Model representation of the urgency (u) variables for monotonic set [12345] as Gaussian probability distributions for a typical simulation	75
5.1	Introduction of Arabic numerals as choice stimuli in the Quantity Judgment Reverse Contingency Task (RCT) compared with performance when candy arrays were presented.	86
5.2	Quantity judgment reverse contingency task (RCT) by chimpanzees using candy and rock arrays.	87
6.1	A typical modern mobile robot	107
6.2	Sandwich assembly	109
6.3	Four snapshots of two trajectories	111
6.4	Phase space example	112
6.5	Trajectories of wall-following and obstacle-avoidance as recorded from an overhead camera	114
6.6	Robot under manual control... and under model control	116
6.7	Left: when 2D constraints are satisfied for face and eye tracking, structured light is projected. Right: the captured 2D image is processed into 3D and texture mapping can be changed by 3D post-processing operations	121
6.8	Automatic pose alignment and feature extraction	122
7.1	A gross abstraction of neural control	133
8.1	A representation of the numbers of people having a symptom, e, and a disease, h	151
8.2	The set of evidence, E, is partitioned by the set of possible hypotheses, h_i	152
8.3a.b	Examples of two different failures of discrete component semiconductors	155

8.4	A Bayesian belief network representing the causal relationships and data points implicit in the discrete component semiconductor domain	157
8.5	Real-time data from the transmission system of a helicopter's rotor	158
8.6	The data of Figure 8.5 is processed using an auto-regressive hidden Markov model as in Figure 8.6	159
8.7	Cognitive model use and failure: Above) A model-calibration algorithm; Below) For assimilation and accommodation of new data	160
8.8	A set of hypotheses h_i are linked to multiple (supporting) perceptions, evidence e_j	162
9.1	A worked example of the proof	174
9.2	The hollow cube	175
9.3	Joined tetrahedra	176
9.4	Kepler's star-polyhedron	177

Preface

Questions about the minds of humans, other animals and artificial agents are now the subject of intensive research in a number of academic disciplines. This is perhaps one of the first books to combine studies of animal minds, artificial minds and human evolution within a common framework.

The book is in three parts, each focusing on a key aspect of this interdisciplinary agenda. The first part – Complexity and the Animal Mind – focuses on the advances made by comparative psychologists in breaking the mould of traditional learning theory and explaining the intelligent behaviour of primates in terms of information gain and serial search. The second part – Complexity in Robots – focuses on the importance of utilising such ideas in the overall design of artificial autonomous systems. The third part – Language, Evolution and the Complex Mind – brings into view the cognitive products of language evolution in the light of what we now know about what the non-human can achieve through non-verbal means.

How does complex behaviour arise from simple elements? In an ongoing process? In the development of a process? In the evolution of processes? The twentieth century saw this question taken up under various banners, in a range of fields, as counterpoint to the ever-dominant current of reductionism in science. Still today many identify science with the explanation of phenomena by their reduction to causal processes of smaller units. Reduction is undoubtedly one of science's major modes. In the nineteenth century, reductionism served spectacularly to relate disparate sciences – physics to chemistry being the *pièce de résistance*. But reduction is not without its discontents. We may be able to explain, post hoc, how a given macro-phenomenon reduces to a microprocess. But in many cases, prediction still defeats us at the lower level. We may be satisfied that the weather reduces to a few physical principles, but we do not turn to calculations of these principles to predict the weather. Instead we talk of depressions, fronts, isobars – and rain. Or imagine Darwin's chances of coming to his theory of natural selection by working up from what was known about biochemistry. Chemists too are still in business. Reduction can reassure us that we have not missed ingredients for explanation, or left spirits lurking in the data, but to predict and control and theorise we have to deal directly with high-level complexity – and complexity often turns out not to reduce the way that we assumed. In the young behavioural sciences, premature reduction can threaten to throw out the baby we need to focus on with the reductionist bathwater. And even in contemporary physics, high-level theories are required to deal with complex systems – thermodynamics is perhaps the best known example.

This need has produced several prominent general mathematical and computational approaches to complexity. The Wikipedia entry http://en.wikipedia.org/wiki/File: Complexity-map-overview.png nicely illustrates the complexity of twenty-first century concepts of complexity. When psychology had to establish its very scientific possibility, behaviourist reduction was the order of the day. But now that this battle is won, psychology also needs to be able to address complexity.

A recurring issue in this book is whether 'associations' between events can serve for the construction of complex mental structures and processes. For the most part this book is not concerned with particular mathematical treatments of complexity, but with qualitative analyses of intuitively complex processes. But it is certainly concerned with the counterbalance to reduction to which these mathematical frameworks contribute.

The unusual thread that unites the authors of this book is the inspiration their work has drawn from the research and teaching of the late Brendan McGonigle. The Epilogue records a taste of these personal connections. But it is testimony to the coherence of his diverse work that these samples of his influence could each contribute to papers on current thinking and constitute a well-founded book.

Contributors

Editors

David McFarland is Emeritus Fellow at Balliol College, University of Oxford, UK. He is the author and editor of many books on behaviour; his most recent publications include *Artificial Ethology, Oxford Dictionary of Animal Behaviour* and *Guilty Robots, Happy Dogs: The Question of Alien Minds*.

Maggie McGonigle-Chalmers is Senior Lecturer in the Department of Psychology at the University of Edinburgh, UK. Her research interests include developmental cognition, executive functioning in childhood autism and Fragile X syndrome and cognitive neuroscience.

Keith Stenning is Professor of Human Communication at the University of Edinburgh, UK. His research interests include comparison of modalities of representation, learning/teaching of formal knowledge.

Contributors

Colin Allen is Provost Professor of Cognitive Science and History and Philosophy of Science, College of Arts and Sciences, Indiana University, USA.

Sarah T. Boysen is Associate Professor of Psychology, The Ohio State University, USA.

Joanna J. Bryson is Reader in the Department of Computer Science, University of Bath, UK, where she is also the research group leader for Bath Artificial Intelligence.

Alan Bundy is Professor, School of Informatics, The University of Edinburgh Informatics Forum, Scotland.

Andy Clark is Professor of Logic and Metaphysics Philosophy, Psychology and Language Sciences, University of Edinburgh, Scotland.

Wolfram Hinzen is Professor and Director of PG Studies in the Department of Philosophy, Durham University, UK.

Barry T. Jones is Emeritus Professor and Honorary Senior Research Fellow, School of Psychology, University of Glasgow, Scotland.

Iain Kusel is a doctoral student at The Design Group, Faculty of Maths, Computing and Technology, The Open University, Milton Keynes, UK.

Michael van Lambalgen is Professor and Chair of Logic and Cognitive Science, Department of Philosophy, University of Amsterdam, The Netherlands.

Mark Lee is Professor of Intelligent Systems in the Department of Computer Science, at the University of Wales, Aberystwyth, UK.

Carlo De Lillo is Senior Lecturer, University of Leicester School of Psychology, UK.

George Luger is Professor of Computer Science, Linguistics and Psychology, University of New Mexico, USA.

Ulrich Nehmzow was Professor of Cognitive Robotics, School of Computing and Intelligent Systems, University of Ulster, Northern Ireland, UK.

Marcos Rodriguez is Professor of Computer Science, Geometric Modelling and Pattern Recognition Research Group, Sheffield Hallam University, UK.

F. Robert Treichler is Emeritus Professor, Department of Psychology, Kent State University, USA.

Anna M. Yocom works in the Department of Psychology, The Ohio State University, USA.

Part I
Complexity and the Animal Mind

Introduction
Maggie McGonigle-Chalmers

'Contemporary study of the animal mind is perhaps the most powerful integrative movement in biological science since Darwin.'
– Hendry (2008)

Humans are well practiced at lauding their achievements. A recent example from Penn, Holyoak and Povinelli (2008) notes, for example, that human animals 'build fires and wheels, diagnose each other's illnesses, communicate using symbols, navigate with maps, risk their lives for ideals, collaborate with each other, explain the world in terms of hypothetical causes, punish strangers for breaking rules, imagine possible scenarios and teach each other how to do all of the above' (Penn, Holyoak and Povinelli, 2008), a list which they proceed to augment in particular and considered detail with respect to analogical reasoning, rule-based learning, reasoning about spatial relations, transitive inference, hierarchical reasoning, causal reasoning and theory of mind. All of these achievements deserve the epithet 'complex', insofar as this refers to emergent higher-order properties of the biological system. Whether this is represented conceptually via the interacting elements of mental representations, or more literally via the large pathways connecting sensory, motor, memorial and planning areas of the brain, it is not difficult to argue the case for the complexity of the human mind.

It is ironic that such a capability, which has used its powers to penetrate the mysteries of the physical universe for several millennia, has so very belatedly come to look at how it itself may have emerged from a cognitive substrate that we share with other species. It is nothing short of astonishing that, having pronounced so shrilly on our 'unique' linguistic capabilities down the centuries, we have so very recently only started to actually 'listen' for differentiation of meaning and communicative intentionality in other species (Cheney and Seyfarth, 2005; Deecke, Ford and Slater, 2005; Hollen and Manser, 2005; Marler and Evans, 1997). But perhaps the greatest misfortune that befell comparative psychology in its inception was the force of

Morgan's canon when he presaged decades of the tyranny of behaviourism with the edict that:

> In no case may we interpret an action as the outcome of the exercise of a higher psychical faculty, if it can be interpreted as the outcome of one which stands lower on the psychological scale. (Morgan, 1894)

The problem was that in the reductionism that followed (albeit arguably based on a misinterpretation of Morgan; see Thomas, 1998), the tools of measurement were perhaps paradoxically drawn from our own invented culture of science and engineering. As Jones describes in Chapter 1, the criteria by which animal intelligence was judged were based on dependable indices such as energic values measured in actual size, brightness or pitch. And although the (much later) birth of a new comparative approach in the form of animal 'cognition' eschewed much of the old reductionism, it introduced criteria that drew substantially on abilities that appear relatively late within human development, with little regard for the extent to which they owed their emergence to enculturation and formal schooling. From this tradition there have been some 'no-win' debates about whether animals make transitive 'inferences' using criteria based on deductive mechanisms that are not particularly in evidence even in late childhood (Clark, 1969; Hunter, 1957) or show a readiness to adopt counting mechanisms that in human development are patently not only the product of formal tuition but actually work 'against the grain' of the human tendency to quantify by more primitive means (Fuson, 1988; Nunes and Bryant, 1996).

It is therefore only relatively recently that animal cognition has started to embrace the study of core cognitive skills that are not end-state defined, but instead are highly plausible candidates for how 'higher-level' abilities get off the ground. In doing so they have had to re-establish what that ground actually consists of. One place to start in every sense is with the basic connectives in the visual world without which complex behaviours and the internal representations that guide them would have no roots or primitives from which to grow. These connectives are argued by Fodor and Pylyshyn (1988) to provide the core indivisible semantics for human cognition and include causal, temporal and spatial relations and relations of equality, similarity and difference.

One of the first triumphs of research in animal cognition, therefore, was to start reclaiming this 'relational' ground from the grip of behaviourist accounts that had attempted to replace the concept of relational perception through reduction. Denying animals the ability to 'comprehend' a relation between one object and another, theories were built instead around the energic properties of individual stimuli, rendering 'relational learning' in animals as a mere mechanical outcome of stimulus/response strengths. In Chapter 1 of this volume, Jones details a probing archival study that was one of the first to challenge the veracity of this claim. In the spirit of simply 'opening the horse's mouth', the ease and stability of size-relational learning in monkeys was compared explicitly with the proposed alternative of absolute stimulus learning, concluding not only that the former is a perceptual primitive in non-human primates, but that the absolute response is also fundamentally relational. Drawing, as they appear to, on ad hoc relationships within the stimulus environment, such absolute codes are shown to be highly unstable when that environment changes. These conclusions are virtually identical in every respect to those drawn by the developmental psychologist, Peter Bryant, during his pioneering studies of size-relational learning in young children (Lawrenson and Bryant, 1972).

With a common ground based on fundamental relational connectives firmly reinstated by such research, we might ask, 'Where next?' One route is to see how far relational comprehension can extend beyond the binary case, encompassing relations between relations such as 'middle-sized' and 'second smallest' (see McGonigle and Chalmers, 2002). The monkeys were successful in all of this and patterns of acquisition and relative difficulty were almost astonishingly similar to those obtained with preschool children. While gratifying to obtain yet another analogue with human performance, was this capability as evinced within the lab really informing us as to how 'complex' knowledge grows from simple primitives in evolution and development? Certainly it had been demonstrated that by making the multiple response requirements explicit, we could indeed sharpen the fine-tuning of the relational learning (and also of transitivity of choice; see McGonigle and Chalmers, 1992), but by what means, and with what exact implications for natural knowledge growth? Is it plausible that the heavily supervised designs of single-choice discrimination tasks are remotely analogous to the environmental conditions that have provoked complex decision-making in nature? The real world of dominance hierarchies, foraging and mating may be one within which binary choice may still effectively prevail ('this one or that one', 'fight or flee', 'aggressor or appeaser', etc.), but those decision-making capabilities appear to emerge in the form of a ranking process that can unite piecemeal input and thereby govern rational choice. In our search for cognitive continuities across species, is it not this ability to create overall order from partial sampling of the data that holds a clue to what may lie at the roots of higher-order logico-mathematical abilities in humans? Discrimination learning was not the tool for exploring this possibility. It was time to look at ranking behaviours more explicitly. And it was time to do so in the laboratory.

The second chapter of this volume reviews one of the most fundamental of all ranking mechanisms; the order of events in time. Despite Lashley's injunction to reject simple associative chaining accounts in a seminal paper on the problem of serial order and to view it as 'the most complex type of behaviour', that is, 'the logical arrangement of thought and action' (Lashley,

1951), seriality has become synonymous with a low-level 'explanation' of a solution to a cognitive task. Within the long tradition of assessing cognition in animals through training and test, the issue of whether the animal has simply conserved the order of events during training rather than their 'deeper' structure has plagued areas such as transitivity research. Here evidence is sought that binary relations of, for example, size (red is bigger than green; green bigger than blue) can be integrated into a linear structure affording 'deductions' based on combining these relations into a connected order (red>green>blue). To assess this, the original pairings have to be trained. In young children, responses to novel test pairings do seem to be initially dependent on a substructure formed by the temporal ordering of the relations in a monotonic linear sequence, AB, BC, CD and so on, and any training that violates this temporal sequencing is likely to result in failure (Kallio, 1982). Despite the fact that this might afford a clue as to how children subsequently become aware of transitive relations based on object properties alone, there has been nothing short of an obsession with eliminating temporal chaining when studying transitive choice mechanisms in animals. Without the end-state evidence (as with children) that at some point a higher-level solution will emerge based on the relational connectives alone (Halford, 1984), this is perhaps not too surprising. More surprising is how long it has taken to see that temporal relations can themselves be integrated into linear structures in precisely the same way as object relations. If A comes before B and before C, can the relationship between A and C be resolved by some mechanism of integration and ranking? And, if so, what are the properties of the resulting representations; what long-term dependencies can be supported by learning several adjacent relations? How precise and fine-tuned are the resulting linear orders? Given how fundamental temporality is to all organised behaviour, these were questions that were long overdue for investigation in the lab.

In Chapter 2, Robert Treichler reviews the two main types of experimental paradigm designed to assess temporal-ordering mechanisms. The first, the 'concurrent conditional' method, is largely inherited from binary discrimination techniques: the transitivity training paradigm (after Bryant and Trabasso, 1971) that involves presenting a set of linked comparisons in which reward contingencies are switched (A>B; B>C; C>D; D>E) such that novel comparisons can only be resolved if the pairs are united into an overall structure (A>B>C>D>E) (Bryant and Trabasso, 1971). With numerous theories as to how associative mechanisms could explain the patterns of transitive choices on novel pairs (such as B>D) in monkeys without implying overall ranking or serial integration, some disambiguation came in the form of the simultaneous chaining method pioneered at Columbia University by Herb Terrace and his group based on the method of rewarding serial choices such as select A then B, B then C and so forth. The switch to looking directly at ordering operations rather than inferring them from single choices has led to a wealth of new and impressive evidence on ranking capabilities in monkeys that Treichler reviews. He also provides a persuasive argument that these sorts of serial behaviours offer a platform for converging neuro-imaging evidence on brain organisation across human and non-human primates, bringing with it a new objectivity into the continuity/discontinuity debate. Finally, Treichler returns to the current conditional paradigm exemplified by some of his own research, but adding a new twist to the story of serial-order memory in monkeys. Here we learn how macaques could combine two separate serial 'lists' such as A<B<C<D<E and F<G<H<I<J simply by dint of subsequently learning one connecting pair (E<F). Latency measures and tests for ordinal position all combine to present a convincing case that serial memory is both highly organised and independent of test procedure in at least some non-human primate species.

If temporal order is a fundamental, then surely spatial order is too? Also once considered to be an 'artefact' getting in the way of the measurement of transitive reasoning abilities in children and monkeys (McGonigle and Chalmers, 2001), the inescapable fact (and the very reason for this concern) is that humans use spatial location as an aid to memory and spatial vectors as an aid to thinking about relations (Huttenlocher, 1968; Trabasso, 1977). But viewed as an important cognitive device, rather than an uninteresting basis for solving a problem, the deployment of spatial location in the control behaviour is now receiving the attention it deserves as a fundamental property of the complex mind. Clearly what is at issue here is not so much the spatial coding of the environment that permits instinctive foraging, homing and migrational behaviour, but rather the active and flexible use of spatial information to manage novel environments in a way that is maximally adaptive. In Chapter 3, Carlo De Lillo addresses the important issue of how the spatial properties of an environment may be used to reduce Working Memory (WM) demands. A first important strand in the evidence he reviews was the move towards allowing space to operate in an ecologically valid way in experiments on animals' memory for food locations. Thus instead of artificially engineering a spatially neutral environment (like the radial maze), De Lillo and others have now manipulated natural spatial properties (such as clustering) and measured the extent to which different species can exploit spatial structure to remember food locations and to avoid revisiting locations that have already been searched. De Lillo reports clear and interesting evidence of large differences across taxa in terms of this index of WM, from mice, tree shrews, capuchin monkeys and human children. An obvious question is whether this is a cognitive achievement of the individual or whether it is an adaptation engineered by evolution over which the monkey (say) has no control. For example, when foraging efficiency improves in monkeys when clusters rather than random layouts are provided, is this simply because the clusters, like fruit distributions on trees, represent their natural 'patchy' foraging niche? De Lillo argues that this not the case, as primates also seem able to benefit from circular and linear layouts. Crucially, however, his research has also shown that monkeys actively learn how to exploit spatial layout in novel environments. Spatial clusters when exhaustively searched can be represented hierarchically, each as a region that can be abandoned before moving on, thus reducing the need to remember every single location within it. Monkeys did not start with exhaustive searches within clusters but gradually acquired expertise in doing so. De Lillo then raises an important observation regarding the emergence of such efficient searches. Not only do they reduce the cognitive (and motoric) costs of inefficient search, they also produce trajectories that we would describe as 'simple' (i.e. linear, circular etc.). He describes intriguing results based on adults' Likert scale ratings of the simulated travel paths of rats, monkeys and children as well as actual verbal descriptions in support of this view. Towards the end of his chapter, De Lillo reports on his recent and novel variant of the Corsi tapping task in which he confirms the hypothesis that memory for serial spatial information is enabled by hierarchical and other types of spatial organisation, suggesting the representation of space and time in a coherent and memory-sparing manner. Although some of the studies reviewed in this chapter were carried out with human adults, the substance of the research reviewed by De Lillo strongly suggests that what emerges in human evolution is the cognitive cost-reducing aspects of principled spatial search, precursors of which are likely to be found in nonhuman primates.

Emergence of efficient search is also the theme of Chapter 4 by McGonigle-Chalmers and Kusel, but now in the context of the perhaps more contentious territory of how one moves from trial-and-error search to logico-mathematical understanding. A long-standing index of this in human development is size seriation (Kingma, 1984; Piaget and Szeminska, 1941). Analogous to the linear spatial searches documented by De Lillo, rods or blocks of different sizes must be placed in a linear arrangement from biggest to smallest or vice versa, true 'success' being the ability to do so using an errorless strategy of selection and 'end-to-end' placement. That this is a skill that is not 'readily available' (to borrow a phrase from De Lillo) until well into the school years is beyond dispute. But how and why it becomes available in the form of spontaneous, non-trained behaviour of older children and adults has never been satisfactorily explained. One good reason is that the experiences provoking such change almost certainly occur beyond the capture of laboratory situations. However, the seeds of these changes are likely to be found in the early trial-and-error routes to success. If this is true, is it possible to characterise the mechanisms of change in such a way that they could characterise in principle the nature of the learning in human and animal that might lead to such a powerful device? The objective that lies at the heart of this chapter is to capture how simple binary relational rules (as described in Chapter 1 with regard to squirrel monkeys' 'natural' mode of computing size differences) can generate, by means of self-regulated learning, the fine-tuned 'ranking' that seriation requires. This is tackled by an attempt to computationally model the acquisition of five-item seriation using minimalist assumptions, while incorporating known psychological constraints on how size relations are apprehended by young children. The rationale for the latter is that it is these constraints that provide a non-arbitrary rule for searching and selecting items and in that respect make size seriation very different from association learning based on arbitrary list learning. The simulation successfully models the obtained learning data and does indeed map onto the very specific profile obtained in size seriation as compared with an arbitrary sequence learning task (a colour string). The changes that occur to allow spontaneous seriation are not captured (as yet) by the model, and it is acknowledged that we may have to look to the role of a symbol system (or other cultural devices) in supporting this change (see Clark, Chapter 11). However, the force of this chapter is in showing how the effort to control potentially explosive amounts of information can be supported by a cycle of exchange between simple binary perceptual judgements and their actionable consequences. With sufficient experience, principled ranking and its associated characteristics of self-regulated learning and apparent 'simplicity' of a linear monotonic solution (as also seen in the spatial searches of De Lillo's primate subjects) is what emerges, and is surely where to look for the origins of logico-mathematical structure.

In Chapter 5, Boysen and Yocom take us still further into the underpinnings of logico-mathematical structure: the apprehension of numerosities. Reminding us at the start that discrimination of different numerosities has now been studied (and found) across vertebrate classes including fish, birds and mammals, they focus in particular on what can be learned from relative difficulty with different number judgements made and its implications for how numerosity is represented by different animals. We first learn of accumulating evidence across all species that smaller numerosities are easier to discriminate than larger ones, leading to a proposal that the fundamental (and perhaps universal) representation of number is logarithmic and that numerosity comparisons are ratiomorphic, a proposal that leads in turn to the speculation that species taxonomically closer to humans can discriminate amounts at smaller ratios. In the child, number judgements become inextricably entangled with the count alphabet and formal tuition in how to enumerate (Fuson, 1988). A truly fascinating insight into how support through the symbol system can indeed radically alter the level at which number judgements are made forms the core of this chapter. For in Boysen's research with chimpanzees (Sarah, famous for being one of the pioneer language-trained chimps of David Premack, and Sheba), a plan to study social deception was gradually hijacked by an even more compelling issue: why could the chimps obey a reverse contingency training regime (RCT) and select a smaller amount of food (rewarded by giving them the larger amount) when the amount was signalled by the Arabic numerals (on which both chimps had been trained) but not by the food itself? Boysen and colleagues went on to show that this was not a function of competition with another chimp, nor even of some overriding compulsion from the appetitive system, as it also occurred when the amounts to be compared were represented by inedible items. They review subsequent studies with monkeys and other species of apes (in which the 'symbolic' cue was the colour of the food container), all of which found the same basic effect. Boysen and Yocom conclude their chapter by drawing a compelling analogy between the 'biological imperative' operating with apes in the presence of a real food reward and a similar effect found with three-year-children and children with autism in deception tasks. This leads them to the plausible speculation that the representational symbols that allow apes to overcome this imperative alter their level of functioning in ways that are directly comparable to the maturational changes that take place in children between the ages of three and four.

In this last chapter, therefore, we cross a threshold from natural adaptive skills of perception and judgement to the deployment of symbols that have been explicitly taught and can be used to control behaviour in new ways. It is worth noting that this final theme of Section 1 is taken up by Andy Clark in Chapter 11 of this book, where he comments on how the symbol system might operate to overcome the intrinsic 'limits' of simian success, a thesis which he himself notes was partly prompted by Boysen's discoveries with Sarah and Sheba; in other words, with a species that is denied access to the infinite store of symbols provided by speech, but, as this section attests, perhaps less so to the coherent structuring of thought and action on which the symbol system is predicated.

References

Bryant, P. E. and Trabasso, T. (1971). Transitive inferences and memory in young children. *Nature*, 232: 456–8.

Cheney, D. L. and Seyfarth, R. M. (2005). Constraints and preadapatations in the earliest stages of language evolution. *Linguistic Review*, 22: 135–59.

Clark, H. H. (1969). Influence of language on solving three term series problems. *Journal of Experimental Psychology*, 82(2): 205–215.

Deecke, V. B., Ford, J. K. B. and Slater, P. J. B. (2005). The vocal behaviour of mammal-eating killer whales: Communicating with costly calls. *Animal Behaviour*, 69(2): 395–405.

Fodor, J. and Pylyshyn, Z. (1988). Connectionism and cognitive architecture. *Cognition*, 28: 3–71.

Fuson, K. C. (1988). *Children's counting and concepts of number*. New York: Springer.

Halford, G. S. (1984). Can young children integrate premises in transitivity and serial order tasks? *Cognitive Psychology*, 16:65–93.

Hollen, L. and Manser, M. (2005). Studying alarm call communication in meerkats. *Cognitie Creier Comportament*, 9(3): 525–37.

Hunter, I. M. L. (1957). The solving of three-term series problems. *British Journal of Psychology,* 48: 286–98.
Huttenlocher, J. (1968). Constructing spatial images: a strategy in reasoning. *Psychological Review,* 75: 550–60.
Kallio, K. D. (1982). Developmental change on a five-term transitive inference task. *Journal of Experimental Child Psychology,* 33: 142–164.
Kingma, J. (1984). The sequence of development of transitivity, correspondence and seriation. *Journal of Genetic Psychology,* 144(2): 271–84.
Lashley, K. S. (1951). The problem of serial order in behaviour. In L. A. Jeffries (ed.), *Cerebral mechanisms in behaviour.* New York: John Wiley.
Lawrenson, W. and Bryant, P. E. (1972). Absolute and relative codes in young children. *Journal of Child Psychology and Psychiatry,* 13: 25–35.
Marler, P. and Evans, C. S. (1997). Communication signals of animals: Contributions of emotions and reference. In U. C. Segestrale and P. Molnar (eds), *Nonverbal communication: Where nature meets culture.* Hillsdale, NJ: Lawrence Erlbaum.
McGonigle, B. and Chalmers, M. (2001). Spatial representation as cause and effect: Circular causality comes to cognition. In M. Gattis (ed.), *Spatial schemas and abstract thought.* London: MIT Press.
McGonigle, B. and Chalmers, M. (2002). The growth of cognitive structure in monkeys and men. In S. B. Fountain, M. D. Bunsey, J. H. Danks and M. K. McBeath (eds), *Animal cognition and sequential behaviour: Behavioral, biological and computational perspectives.* Boston: Kluwer Academic, pp. 269–314.
Morgan, J. L. (1894). *An introduction to comparative psychology.* London: W. Scott.
Nunes, T. and Bryant, P. (1996). *Children doing mathematics.* Oxford: Blackwell.
Penn, D. C., Holyoak, K. J. and Povinelli, D. J. (2008). Darwin's mistake: Explaining the discontinuity between human and nonhuman minds. *Behavioral and brain sciences,* 31: 109–30.
Piaget, J. and Szeminska, A. (1941). *La genese du nombre chez l'enfant.* Neuchatel: Delachaux and Niestle.
Trabasso, T. (1977). The role of memory as a system in making transitive inferences. In R. V. Kail and J. W. Hagen (eds), *Perspectives on the development of memory and cognition.* Hillsdale, NJ: Lawrence Erlbaum.

1
Relational and Absolute Discrimination Learning by Squirrel Monkeys: Establishing a Common Ground with Human Cognition

Barry T. Jones

> It is far better to ask important questions than to answer trivial ones with great precision.
> – Warren and McGonigle (1969)

1.1 Introduction

In scientific enquiry, simple explanations of phenomena are preferred to less simple (complex) explanations. Part of the rationale appears to be that simple explanations are thought to be predicated on fewer assumptions than complex alternatives. In this respect, the simple explanations are often described as being more 'elegant', more 'beautiful', 'nicer', 'safer' or simply 'better' than the more complex, competing positions. Both in its strong form as an axiom or in its weak form as a guiding rule of thumb, this process (Occam's razor) has been a driving force close to the centre of the development of explanations in most domains of systematised thought. Unfortunately, what is often poorly recognised is that the productive application of this explanatory strategy predicates on having already made appropriate decisions on what is simple and what is complex and having assessed and accepted as defensible the assumptions that prop up these decisions.

Nowhere in psychology was this difficulty more pronounced than in the first half of the twentieth century with competing explanations of what animals might 'see' and 'learn about' when making discriminations in their worlds. In the laboratory the principal paradigm used to explore seeing and learning was the two-stimulus simultaneous discrimination learning paradigm and the target phenomenon was 'transposition'. Typically animals were rewarded for responding to one of two simultaneously presented stimuli that differed along a single dimension such as size or brightness, and by carrying out post-acquisition transfer testing trials that included stimuli from other points of the stimulus dimension. The way in which an animal's learned responses generalised to new stimulus pairs would be measured and conclusions would be drawn about what had been seen and learned about in the original discrimination.

One view was that in these circumstances the stimuli were *'comprehended in relation to one another'* (Wheeler, 1940, cited by Reese, 1968: 5), the relationship learned about in discrimination training and the knowledge of the relationship subsequently applied in transfer test trials – demonstrating the transposition of the relationship and thereby accounting for the stimulus equivalence. Somewhat loosely, either the relationship itself or the responses learned to the relationship were said to have been transposed. This was called the 'relational' hypothesis and relational seeing and learning such as this was at the centre of Gestalt psychology (e.g. Kohler, 1929).

The 'absolute' hypothesis of seeing and learning was a competing view, and one which preserved an evolutionary discontinuity thesis, by arguing that only humans were capable of the higher-order ability of comprehending relations (see Kendler, 1995, for a review). Accordingly, twentieth-century versions of the older philosophies of associationism were developed within a strongly formulaic framework – as classical conditioning (Pavlov, 1927) and as instrumental conditioning (Thorndike, 1911) – and formed the basis of most subsequent learning explanations in comparative psychology (although considerably extended by others, e.g., Hull, 1952; Spence, 1956; and Sutherland and Mackintosh, 1971). Rather than learn about the relationship between the *two* simultaneously presented stimuli in acquiring the discrimination, it was claimed that animals learned about the *single* stimulus that they were rewarded for responding to (and also, independently, about the *single* stimulus they were not rewarded for responding to). In subsequent transfer testing trials, advocates of the absolute hypothesis held that the single stimulus that would be chosen would be the one most resembling (on the dimension used in the original discrimination) the stimulus that was originally rewarded. Stimulus equivalence such as this was explained by the gradients of excitation and inhibition that both accumulated on and diffused along this stimulus dimension as discrimination training progressed, with their respective peaks at the loci of the rewarded and unrewarded stimuli. The algebraic effect of the positive and negative gradient interaction explained which stimulus presented in transfer test trials should be responded to and which should not. Although these hypothetical processes provide support for the absolute hypothesis in discrimination learning, with a range of additional assumptions about gradient growth rate, shape and shape change with growth, coupled with the original distance apart of the gradient peaks, it was claimed that the absolute hypothesis also could explain apparent relational learning and responding.

One might think that through the application of Occam's razor, the apparent simplicity of the relational hypothesis with its few and relatively straightforward assumptions might attract the majority support. However, the scientific culture underpinning the successes of industrial revolution in the late nineteenth and early twentieth centuries was built on the application of simple scientific and engineering principles captured in easily understood formulae and applied on a massive scale to produce towering buildings, bridges, factories and a range of industrial processes. As a consequence, the formulaic representations underpinning the absolute hypothesis of seeing and learning in the early years of the twentieth century were leaning against a door of acceptance that was already ajar and a momentum was built up in favour of the absolute rather than the relational explanation. Indeed, to illustrate the extent to which interests in the absolute hypothesis had obscured interests in the relational hypothesis by the end of the 1960s, one of the most influential books on animal discrimination learning (Sutherland and Mackintosh, 1971) is cited as 'devoting so little space (*two* out of some 500 pages) to "relational learning"' (McGonigle and Jones, 1978: 636; italics added). Although the Sutherland and Mackintosh review spends so little time on studies that demonstrate relational learning and responding with the two-stimulus discrimination paradigm, they do concede that it appears that *'at least in certain circumstances* [our emphasis] animals learn to respond to the relationship between two stimuli rather than to their absolute values' (cited by McGonigle and Jones, 1978: 636). Reese (1968) in a comprehensive review of relational and transpositional studies also concluded that there is support for the relational hypothesis in some studies and for the absolute hypothesis in others. With respect to the relational hypothesis he writes, 'even if relational perception is undeniably *possible, it* is still important to ask whether it is usual. ...' (cited by McGonigle and Jones, 1978: 636).

Curiously, Reese did not attempt to provide any general synthesis of the different studies that he reviewed in support of either absolute or relational learning – not even that one hypothesis was generally more likely than the other, or generally more readily achieved. Indeed, by the end of the 1960s, changes were occurring in comparative psychology that would have made any attempt that he might have made at a general synthesis quite inappropriate anyway, for two related reasons.

First, Seligman (1970) and Seligman and Hager (1972) reviewed data on and clearly showed that the 'equivalence of associability' (Seligman, 1970: 407) or the 'equipotentiality premise' (Seligman and Hager, 1972: 2), which had been at the core of comparative psychology, was wholly invalid. Within this framework it was held that a 'pure' measure of any species' learning could be got by measuring, for example, whether and if so how quickly any species-*neutral* stimulus could be associated with any species-*neutral* response. Reviewing from across the widest range of species, Seligman showed that when this path was followed an arbitrary and not a pure measure of learning would be derived and therefore would be of no value in representing a species' abilities nor in comparing them. Consequently, when this path was followed, the psychology of seeing and learning developed would also be arbitrary, not pure. Typical of the many types of experiment Seligman reviewed in his rejection of the equivalence of associability (or the equipotentiality premise) was Garcia and Koelling's (1966) so-called bright-noisy-tasty water experiment in which rats were exposed to concurrent audiovisual and gustatory stimuli on licking a water nozzle. When paired with electric floor shocks, the learned avoidance was through the audiovisual, not gustatory, cue; when paired with illness-inducing toxins or X-rays, the reverse was the case. Consequently, any effort to synthesise experiments which in these terms produced non-comparable data (i.e. from single-species experiments using different S-R combinations or from cross-species experiments using any single S-R combination) would be a wasted effort.

Second, Hodos and Campbell (1969) had shown that the phylogenetic *scale* – hitherto the backbone of comparative psychology and the device through which the measures of different species' pure learning would normally have been compared (see comparative psychology texts such as Ratner and Denny, 1964; Waters, 1960) – was bogus, an imaginary linear sequence wrongly masquerading as a representation of the evolution of, say, learning. As Hodos and Campbell assert, the phylogenetic *tree* represents how each species alive today has evolved and how they will have adapted to their own ecological niche. The temporal proximity of two different species' common ancestor as well as the physical constraints of their respective ecologies might well predict communalities in form and function but there would be few grounds for predicting that either absolute of relative seeing or learning might be paramount. Of course, in the same vein, there would be no grounds for predicting that both would not be present and active. Presence or absence, by the end of the 1960s it remained an unanswered empirical question.

At the end of the 1960s, it was clear that a new start was required in exploring the roles of the absolute and relative hypotheses in discrimination learning. It was clear that a start within a single species and with the use of a single stimulus dimension was initially indicated and that following 2-stimulus discrimination training on the single dimension, more extensive and imaginative transfer testing would be required than had hitherto been used. Accordingly, and as an introduction to the outcomes of a series of size discrimination experiments with squirrel monkeys reported in the next part of this chapter, McGonigle and Jones (1978: 636) write:

> One reason [why the absolute-relational issue has not been resolved in spite of the many reports of experiments designed to bear on them] seems to lie in the overdependence of investigators on the simple

transfer model which has become the hallmark of transposition research. Conventionally, in experiments based on this paradigm, animals are first trained to discriminate a single pair of stimuli differing usually in size and brightness: they are then tested for preferences when for example the training 'positive' stimulus is re-paired with a novel one differing from it along the same 'dimension' of change. Given the many possible bases for discrimination provided in any single discrimination task, this method seems far too limited to permit any detailed evaluation of the performance characteristics of those subjects who might respond primarily to 'a relational set of cues' as distinct from others who would discriminate on the basis of 'absolute' stimulus properties.

In the experiments we describe here, we have adopted, by contrast, a much more extensive ('replication via the consequences') strategy comprising many training and testing episodes. Just as importantly, we have varied the task requirements and consequently *the criteria of evaluation* from subject to subject. Some subjects have been required to make relational judgements, others to conserve a response to specific size or brightness stimuli over a wide variety of stimulus collections and their transformation(s). From the performance profile(s) which have resulted from such tests, we now feel able to draw at least some preliminary conclusions concerning both the primacy of relative over absolute stimulus perception by monkeys and the information processing demands made by each sort of task.

Three concatenating themes or questions bind together McGonigle and Jones' series of size experiments that are reported in this chapter. They are briefly introduced here. First, 'ability and stability' – how well do monkeys learn to conserve a size relationship as compared with an absolute/specific stimulus value and how relatively stable are the two types of learning? Second, 'widening horizons' – how well does conserving a size relationship and an absolute/specific stimulus value learned in 2-stimulus discriminations extend to 3-stimulus discriminations and to an extended stimulus set? This theme also includes whether other relationships can be learned, such as middle, and how does it compare with learning larger. Third, 'size building blocks' – what features of the size stimuli and the immediate environment are recruited in making size discriminations?

In McGonigle and Jones' (1978) size discrimination experiments reported in this chapter, the same feral-born, sub-adult experimentally naive male squirrel monkeys (*Saimiri scuireus*) were used; they were housed in large communal cages with permanent access to water and to MRC Diet four-hours per day. They were transferred to individual cages for testing in a matte grey Wisconsin General Testing Apparatus using orthodox testing procedures with shelled half peanuts as reward. A pool of five cuboid-size stimuli was made from polystyrene covered with polished alabaster and painted matte white – A (6.25 cm), B (5.08), C (3.81), D (2.54) and E (1.27).

1.2 Ability and stability: conserving a relationship versus a specific stimulus value

The first experiment (1a in McGonigle and Jones) compared the performance of subjects encouraged to learn the relation 'larger' with subjects encouraged to conserve a single 'absolute' stimulus size (see Figure 1.1). The encouragement came through using a stimulus pool of three (B, C, D) but presented in randomly-alternating B-C and C-D pairs for 2-stimulus discrimination learning. Group R was rewarded for responding to B of B-C and C of C-D while Group SS was rewarded for responding to C of B-C and C of C-D. Experiment 1a was different from any other previously published experiment using randomly training alternating pairs (e.g. Campbell and Kral, 1958 [birds]; McCulloch, 1935 [rats]; and Schwaartzbaum and Pribram, 1960 [monkeys]): first, *both* relative and absolute learning was encouraged in the same experiment in the two groups (R and SS) and, second, *identical* stimulus, methodological and procedural set-ups were used for both groups (reward contingency differences notwithstanding). This novel within-experiment feature was critical, for the absolute-relational comparisons that the experiments were designed to permit. Groups R and SS met the initial conventional acquisition criterion (18 correct from 20 consecutive trials) in just over 100 trials. Group R was quicker, with fewer errors, but not significantly so. Overtraining was carried out to a much more strict criterion of no more than one error per session for three consecutive sessions during which Group SS made significantly more errors than did group R. It appears that conserving a relationship and a specific stimulus value are both possible under these identical conditions of test but that the former learning is significantly more stable than the latter. Indeed, Wertheimer (1959) made this very prediction: that it would be more difficult for subjects to remember a specific stimulus value than a directional relationship. The monkey's overtraining data in experiment 1a are consistent with his view.

McGonigle and Jones' second experiment (1b) was designed to extend the test of Wertheimer's hypothesis by pre-exposing each group to potential *distracter* stimuli just out of reach for 30 seconds prior to being allowed to respond to the usual B-C and C-D pairs. Four different novel stimuli sets were used as distracters: the largest (A) and smallest (E) stimulus and the largest configuration (A-B-C) and the smallest (C-D-E). Group R's normal responding was barely disrupted by the distracters whereas Group SS's normal responding was severely disrupted to the extent that it reverted to chance on the B-C and C-D pairs. One can only speculate on what is driving Group SS's disruption in these trials. It is clearly only (very) temporary. What is also clear is that an analysis shows that the smallest distracter stimulus (E) is significantly more disruptive than the other three distracters, which would be predicted by any (psycho)metric that represents the relative positions of the different stimuli on a size dimension. This is also consistent

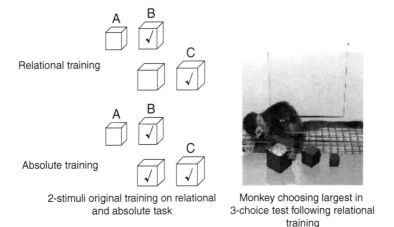

Figure 1.1 Pairwise training given to squirrel monkeys, and an example of a monkey extending a relational choice ('larger') in the context of triadic testing

with Helson's (1964) theoretical position in which a 'running average' (or adaptation level) of the different stimuli in very recent use is used as a reference point from which specific stimuli are subsequently and temporarily judged. Accordingly, once the original training trials are reintroduced in 1b, the adaptation level quickly returns to the pre-disruption level. In terms of the issues of 'ability and stability' – it appears that there is good evidence from 1a to support both the absolute and relational hypothesis. The overtraining in experiment 1a and the distracter trials of 1b, however, strongly suggest that whatever Group SS has seen and learned about in discrimination training is most unstable as compared with Group R. The relative stability issue was further explored below.

1.3 Widening horizons: from 2-stimulus to 3-stimulus size discriminations

McGonigle and Jones' third experiment (2a) extended 2-stimulus testing with the B-C and C-D pairs to 3-stimulus testing with a B-C-D triad (see Figure 1.1 for an example of triadic testing). The initial trials of the day were with the triad and were differentially rewarded (Group R responding to B; Group SS to C); the remaining trials of the day were with the normal training pairs. Group R reached the conventional criterion on the B-C-D triad and a stricter one without error, whereas Group SS made significantly more errors in reaching both – adding to the evidence that relational learning is much more stable than absolute stimulus learning. Neither group's normal 2-stimulus training was disrupted by the new triad training, however. McGonigle and Jones' fourth experiment (2b) extended 3-stimulus discrimination training further to include triads A-B-C and C-D-E having components not yet used in training trials. With this extended stimulus set, Group R and SS conserved the relationship and the specific stimulus value almost perfectly.

Their fifth experiment (2c) was a continuation of 2a and 2b, extending 3-stimulus training even further using a learning set design and adding a set of novel triads. Group R and SS were retained but a subset of the latter group were used to create a new group, SSI, rewarded for responding to the middle-sized stimulus of whatever training triad was used. New stimuli were added to the ABCDE pool to increase the number of novel triads available: A' (5.72 cm), B' (4.45), C' (3.18), D' (1.91). From the newly extended pool, Group R and SSI were given four triads (AA'B, BB'C, CC'D, DD'E) in a counterbalanced order and Group SS were given four equivalent triads but containing stimulus C (AA'C, BB'C, CC'D, CD'E). Forty-eight trials per triad were given; one triad per day. Group R conserved the relationship, larger, almost perfectly over the four different triads; Group SS performed less well but still reached the conventional criterion with all triads, but Group SSI did not perform much better than chance on any of the four triad learning tasks. When given a subsequent set of easier-to-discriminate triads however (Group R and SSI were given A'CD, AB'D, BD'E; Group SS were given ACD', BCE, CC'E), even Group SSI reached the conventional training criterion. Moreover, following completion of the easy set, training returned to the initial harder set and this time round SSI reached criterion on each of the triads. An examination of errors on the first ten trials of each of these sets suggests that in common with the results of the earlier experiments, Group R applies whatever they had learned previously with almost immediate success (conserving the relationship, larger). Group SS is also successful (conserving stimulus C) but with significantly less stability than Group R, but Group SSI performs almost at chance. This profile strongly suggests that an intermediate relationship has not been learned and transferred by this group but instead they seem to treat each new triad as a new single stimulus problem – middle, therefore, is not the same type of relationship as larger in the squirrel monkeys under these conditions of test.

1.4 The building blocks of size: what stimulus and environmental features are used?

The foregoing shows that monkeys can discriminate and learn about different sizes, which raises the question, how? For example, do they carry out an irreducible and automatic volume computation on the cuboids or do they use a rule of thumb (such as height or width) as a proxy for it? McGonigle and Jones began to address this question with Group R and SSI in their sixth experiment (3a) and Group R and SS in their seventh (3b). First in 3a and in common with earlier training, Group R were rewarded for choosing stimulus A from the training triad A-B-C and C from C-D-E, while Group SSI were rewarded for choosing B and D, respectively. Planometric (2D) versions of the training stimuli were mounted on red plaques (the same colour as the WGTA testing tray) and presented in transfer trials in either an upright or flat position. Equivalent stereometric (3D) versions were also made with plaques and also presented in upright and flat positions in equivalent transfer trials. Neither Group R nor SSI's responding was disrupted during the training trials, and transfer to the stereometric versions of the training stimuli was good for both groups – but there was little or no transfer to the planometric versions by either group. Consequently, it appears that there is something about the original training stimuli that is not captured in the planometric versions and the success with stereometric versions demonstrate that it is not merely the presence of the plaques that is responsible for the poor performance with the planometrics.

Experiment 3b investigates this possibility further by manipulating volume and the linear size dimensions in three different ways. Groups R and SS were used in 3b with the training pairs B-C and C-D. First, in transfer trials in which all transfer pairs (B-C, C-D) were the same height as stimulus C but retained the volume of their parent stimulus, Group R appeared to respond to the largest volume and ignored the equal heights. In the same vein, Group SS appeared to respond to the volume of specific stimulus C ignoring the equal heights. It appears that volume information has been retained from the original discriminations by both groups and is used in these transfer trials. Second, in transfer trials in which all stimuli had the volume of C but retained the heights of the parent stimulus, Group R responded reliably to the highest stimulus but Group SS's responses were mixed in that they appeared to respond to height not volume with C-D but not with B-C. Here it appears that when differential volume information is absent, Group R (and for some of the trials) Group SS are able to use in transfer trials height information from previous training. Third, in trials in which all transfer stimuli retained the volume of the parent stimulus but had heights negatively correlated with their volumes (e.g. tallest stimulus had the smallest volume), consistent responding was seen in neither group. Earlier trials suggested that both volume and height information might have been used in discrimination training but when one is pitched directly against the other, neither appear to be used effectively.

From these two somewhat limited but nonetheless instructive experiments, McGonigle and Jones (1978: 646) conclude, 'It is clear from the results that the monkey does not depend exclusively on any one dimension of change which may normally be correlated with volume (variation). In our view, the data may be best interpreted as showing a multidimensional specification of volume by the monkey, which is weighted by the context of alternatives provided for inspection.' This adds further support to Garner's (1974: 10) thesis that 'the perceived properties of any single stimulus change as that stimulus is variously paired with other stimuli, since the dimension which meaningfully differentiates a pair of stimuli depends on a particular pair of stimuli involved.' Indeed, and in support of this point, McGonigle and colleagues (McGonigle and Jones, 1975, and Jones, Rana and McGonigle, 1980) have shown the extent to which other animals' perception of pattern in a single stimulus is governed by the contrastive stimulus with which it is temporarily paired.

An inter-stimulus comparison is only one type of context within which a stimulus might be evaluated, however. The immediate environment provides another. Accordingly, one might speculate that the conservation of an absolute stimulus value (Groups SS and SSI) might be made more difficult if the immediate environment is visually 'removed' whereas the same difficulty might not apply to learning that relies on inter-stimulus comparisons (Group R). In McGonigle and Jones' eighth experiment (3c) self-luminous transfer stimuli were used in the dark for the testing phase (jewellers' paste was used to coat the stimuli and they were kept charged in a light box). A day's testing began with Groups R, SS and SSI being trained on the triads A-B-C and C-D-E, followed by a switch to total darkness and to self-luminous versions of the triads for some of the day's trial blocks over several days. In the dark, Group R's responding was almost perfect in their conservation of the relationship, larger, whereas Groups SS and SSI's responses fell to chance. These data strongly support the view that seeing and learning about stimuli of an absolute size recruits information from the immediate environment whereas the conservation of a relationship is much less dependent on such a process.

McGonigle and Jones' ninth and tenth experiments (3d, 3e) took place in the light and focused on extending the question of what aspects of the immediate environment might be used in (particularly) conserving absolute stimulus values. In experiment 3d, the red stimulus tray (the closest environmental feature to the stimuli) was made to look apparently smaller in size by putting a matte grey border around it of the same colour as the WGTA and training and testing was carried out with the usual white triads A-B-C and C-D-E. The change in apparent size of the tray did not disrupt responding in any of the three groups which suggests that either the immediately adjacent visual environment (the red tray) has no role in specifying absolute size, or if it does it can be overshadowed by less immediately adjacent features.

In experiment 3e, the relationship between the stimuli and the environment was explored further. Groups R, SSI and SS took part with training trials with A-B-C and C-D-E but new transfer stimuli were made by retaining the white ones and painting a red border on all their faces, or painting the stimuli red with a white border on all their faces. All groups transferred their usual responding to the red-with-white-border stimuli almost perfectly, suggesting that the outside dimensions of the stimuli (the white borders) were sufficient to specify whatever had been learned about size. On the other hand, while it also appears that the white-with-red-borders stimuli were responded to somewhat systematically, the groups responded differently. Group R showed no errors, transferring their learning perfectly. Groups SS and SSI appeared to respond to the white surfaces enclosed by the red borders as though the white surfaces represented the total surface size of each transfer cube – each transfer stimuli 'looked smaller' because of the red borders. The somewhat contradictory results of the 'border' experiments with groups SS and SSI suggest that a number of different yardsticks might be used in absolute size judgements. They also raise the question of whether it is even possible to learn to conserve a single stimulus value when these different yardsticks are removed – as in the self-luminous transfer testing introduced in 3c.

McGonigle and Jones' eleventh and twelfth experiments (3f, 3g) were designed to help answer this question. Experiment 3f was carried out entirely in the dark with self-luminous stimulus triads (the same seven different triads used in 2c with the easiest three given before the harder four). The usual reward contingencies were employed. Once again, Group R learned all seven problems with hardly any errors, whereas although Groups SS and SSI reached the conventional criterion on all problems, it was after making a considerable number of errors and with very little evidence of a carry-over from one problem to the next. Experiment 3g was designed to find out whether what was learned in the dark with respect to size discriminations could be transferred to the light and vice versa. Groups SS and SSI were recombined since it was clear that both groups were conserving absolute stimulus values. Group R was not included in this experiment since it is equally clear that they are impervious to almost any stimulus change in their application of the relationship, larger.

A single triad (A'B'D) was used throughout 3g with normal white and self-luminous versions. There were eight testing sessions on consecutive days. A single testing session was for 70 trials divided into first and second halves and in half of the sessions the rewarded stimulus was changed between halves and the other half was kept the same. A single testing session could either be carried out totally in the light or in the dark (with appropriate stimuli) or with the first half in the light and the second in the dark or vice versa. With such a crossed design permitting appropriate controls, the conservation of absolute stimulus values was found to be poorer in the dark than the light. A striking result, however, was that whereas learning in the dark appeared to transfer readily to the light, the reverse was not the case, indicating that there must be more than one way of encoding absolute values and the one used depends on the task requirements.

1.5 Size discrimination conclusions

For the first time, the conservation of the size relationship 'larger' can be *properly* compared with the conservation of the size relationship 'intermediate' and of an 'absolute' stimulus value in a coordinated series of size experiments with monkeys (McGonigle and Jones, 1978). They write, 'Insofar... as the relational code "larger/largest" was by far the most difficult one to disrupt, we can reasonably infer it to be the most "primitive"... and the most likely form of encoding to be at work *as a general rule* (at least in simultaneous discriminations) unless task requirements demand more of the subject by way of a "deeper" or at least different form of processing' (McGonigle and Jones, 1978: 658). The inability of monkeys to learn the relationship 'intermediate' in spite of extensive opportunities to do so – appearing to solve each successive intermediate problem through new absolute stimulus learning – indicates the danger of treating relationships generically in the absolute versus relational debate. Indeed, and with reference to the absolute–relational conundrum introduced at the beginning of this chapter, the data from this series of size experiments (coupled with an equivalent series of brightness experiments also reported by McGonigle and Jones, 1978) strongly suggest that a full understanding of simultaneous discrimination learning is tantamount to identifying not *whether* relationships might be used in discrimination learning (the traditional approach to the absolute–relational controversy) but *which* relationships are used. For Group R are clearly using relationships *between* the discriminanda, but Group SS and SSI, on the other hand, far from conserving absolute stimulus values in some mystical way that is qualitatively different from Group R's solution, also appear to be *using relationships to encode the so-called absolute value*; a range of features from the visible environment appear to be indicated in this series of experiments and are worthy of further investigation.

Indeed, identifying the different relationships used in the squirrel monkeys' discrimination learning has been shown to be important not only in size and brightness discriminations (McGonigle and Jones, 1978) but also in pattern discriminations (McGonigle and Jones, 1975). Whatever conclusions can be drawn from this corpus of data, a clear one (using the traditional nomenclature) is that for squirrel monkeys to conserve the relationship larger is much 'easier' than to conserve an absolute stimulus value; this suggests that the former has been selected for in evolutionary terms over the latter and in this sense the conservation of relationships is more basic or primitive than is the conservation of absolute values. In the context of evolutionary continuity, relational perception is surely more of a common ground shared across primate species than an exclusive property of the human mind.

References

Campbell, D. T. and Kral, T. P. (1958). Transposition away from a rewarded stimulus card to a nonrewarded one as a function of a shift in background. *Journal of Comparative and Physiological Psychology,* 51: 592–5.

Garcia, J. and Koelling, R. A. (1966). Relation of cue to consequence in avoidance learning. *Psychonomic Science,* 4: 123–4.

Garner, W. R. (1974). *The processing of information and structure.* New York: Wiley.

Helson, H. (1964). *Adaptation-level theory.* New York: Harper and Row.

Hodos, W. and Campbell, C. B. G. (1969). Scala Naturae: Why there is no theory in comparative psychology. *Psychological Review,* 76: 337–50.

Hull, C. L. (1952). *A behavior system.* Newhaven: Yale University Press.

Jones, B. T., Rana, R. K. and McGonigle, B. O. (1980). The role of context in the perception of orientation by the rat. *Perception,* 9: 591–8.

Kendler, T. S. (1995). *Levels of cognitive development.* Mahwah, NJ: Lawrence Erlbaum.

McCulloch, T. L. (1935). The selection of the intermediate of a series of weights by the white rat. *Journal of Comparative Psychology,* 20: 1–11.

McGonigle, B. O. and Jones, B. T. (1975). The perception of linear gestalten by rat and monkey: Sensory sensitivity or the perception of structure. *Perception,* 4: 419–29.

McGonigle, B. O. and Jones, B. T. (1978). Levels of stimulus process by the squirrel monkey: relative and absolute judgements compared. *Perception,* 7: 635–59.

Pavlov, I. P. (1927). *Conditioned reflexes.* London: Oxford University Press.

Ratner, S. C. and Denny, M. R. (1964). Comparative psychology. Homewood, IL: Dorsey.

Reese, H. W. (1968). *The perception of stimulus relations: Discrimination learning and transposition.* London: Academic Press.

Schwaartzbaum, J. S. and Pribram, K. H. (1960). The effects of amygdalectomy in monkeys on transposition along a brightness continuum. *Journal of Comparative and Physiological Psychology,* 53: 369–99.

Seligman, M. E. P. (1970). On the generality of the laws of learning. *Psychological Review,* 77: 406–20.

Seligman, M. E. P. and Hager, J. L. (1972). *Biological boundaries of learning.* New York: Meredith.

Spence, K. W. (1956). *Behavior theory and conditioning.* New Haven: Yale University Press.

Sutherland, N. S. and Mackintosh, N. J. (1971). *Mechanisms of animal discrimination learning.* New York: Academic Press.

Thorndike, E. L. (1911). *Animal intelligence.* New York: Macmillan.

Waters, R. H. (1960). The nature of comparative psychology. In R. H. Waters, D. A. Rethlingshafter and W. E. Caldwell (eds), *Principles of comparative psychology.* New York: McGraw-Hill.

Warren, J. M. and McGonigle B. O. (1969). Attention theory and discrimination learning. In R. M. Gilbert and N. S. Sutherland (eds), *Animal discrimination learning.* London: Academic Press: 113–36.

Wertheimer, M. (1959). On discrimination experiments; I. Two logical structures. *Psychological Review,* 66: 252–66.

2
Serial List Retention by Non-Human Primates: Complexity and Cognitive Continuity

F. Robert Treichler

> Increasingly, people seem to misinterpret complexity as sophistication, which is baffling – the incomprehensible should cause suspicion rather than admiration. Possibly this trend results from a mistaken belief that using a somewhat mysterious device confers an aura of power on the user.
>
> *Niklaus Wirth, computer scientist; inventor of Pascal language*
>
> Out of intense complexities, intense simplicities emerge.
>
> *Winston Churchill*

2.1 Introduction

One implication of such quotations is that understanding complex issues is enhanced by preliminary knowledge of simpler component processes. In the present instance, research on memory organisation in animals is encouraged as a tactic to allow insights into integrative operations underlying cognitive processes in other species. It should be noted that here the appeal to simplicity derives, not from reductionist rationale, but from the prospect that animal research provides advantages in experimental control and precision of measurement. In his comparative treatments, Charles Darwin (1859, 1988) used the terms 'organs' and 'instincts' when referring to anatomy and behaviour, respectively. He stated that: 'I can see no difficulty in natural selection preserving and continually accumulating variations of instinct to any extent that may be profitable. It is thus, that I believe, that all the most complex and wonderful instincts have originated.' Accordingly, it seems just as appropriate to seek animal analogues of complex behaviours as to study the evolution of morphology.

2.2 Serial memory in animals

In this chapter, one specific kind of behaviour, serial memory, is explored to determine whether or not it entails associative learning exclusively or, alternatively, might involve internal (and covert) organisation. To qualify for inclusion here, the retained information must be acquired as an arbitrarily generated ordinal series with no systematic physical differences that might serve as cues to an item's location in the order. Further, this information should be retained as reference memory, that is, accessible over a relatively long (multiple session) term, in contrast to immediately post-event, short-term memory. It seems especially appropriate to consider serial reference memory in a volume dedicated to Brendan McGonigle for it was he and Margaret Chalmers who pioneered its measurement in non-human primates. Their adaptation of Bryant and Trabasso's (1971) developmental test provided the first systematic study of ordered-list memory in monkeys. McGonigle and Chalmers' 1977 report served as the precursor to many research undertakings that utilise the now familiar 'transitive inference' paradigm. Preliminarily, their version of the test required learning four overlapping two-choice ('premise') pairs representing a five-item ordered list (usually designated by the consecutive letters, ABCDE, and trained as pairs A-B, B-C, C-D and D-E with reward contingencies conforming to either the ascending or descending serial order). Training involves progressive acquisition of each problem-pair until all four appear within the same session. Accordingly, correct choices vary with the specific paired combination presented on any trial. After acquisition, all other possible (but heretofore unseen) pairings (especially B versus D) are displayed as tests of retention characteristics. Since publication of the original McGonigle and Chalmers (1977) work, the procedure has been extended to a wide variety of species ranging from pigeons (von Fersen et al., 1991) to chimpanzees (Gillan, 1981). Typically problems are presented in simultaneous two-choice format with some accommodation to species characteristics. Accordingly, pigeons choose alternative pecking keys, rats press levers or dig in sand and non-human primates touch one of two display buttons or displace alternative objects on a tray (as in the Wisconsin General Test Apparatus). Retention characteristics revealed in 'transitive' tests serve as indicants of serial memory when human and animal performances are compared on analogous tasks, and such outcomes are often cited when the issue of continuity of cognitive function (similarity/dissimilarity of processing across species) is considered. Transitivity performances have been reviewed, either as one among a variety of behavioural criteria (Premack, 2007; Penn, Holyoak and Povinelli, 2008) or as indicants of processing limitations in animals' serial memory (Wynne, 1998; Vasconcelos, 2008). Reference to the four specific reviews noted in the previous sentence is provided to allow contrast of their general conclusions to the views advanced in this chapter. Both Premack (2007) and Penn et al. (2008) have contended that tests of transitivity (and a variety of other behavioural measures) fail to provide support for the concept of continuity of cognitive function, that is, the Darwinian (1871) view that 'there is no fundamental difference between man and the higher mammals in their mental faculties'. The reviews by Wynne (1998) and Vasconcelos (2008) both advance a similar, but more specific, agenda. Each contends that animals' transitive performances are attributable to associative learning and do not involve integrative operations that qualify as organisational or 'cognitive'.

2.3 Testing primate list memory

The intent of the present chapter is to offer alternative interpretations and qualification of both the non-continuity view and strict associative interpretation of serial memory. Empirical support for this dissent is derived principally from non-human primate (mostly, macaque) behavioural outcomes, although reference to neurophysiological and neuroanatomical evidence from a variety of species is included.

When serial reference memory is evaluated, the two most frequently utilised methodologies are termed the 'concurrent conditional' and the 'simultaneous chain' procedures. The earlier-noted tests of McGonigle and Chalmers (1977) provide a good example of the concurrent conditional method. The label derives from the requirement for intra-session learning of multiple problems (concurrently) with reward contingent (conditional) upon choice of the appropriate alternative within the specific pair displayed on a trial. Under this procedure, the multiple problems allow training of a list of near-unlimited length, but if the list is short (only three or four items), choices on the novel recombinations that constitute retention tests could be determined, not by memory for the serial order, but by appropriately choosing or avoiding the series' end-items. Accordingly, many investigators (e.g., Gillan, 1981; Fersen et al., 1991) have held that serial lists of at least five items are required for valid assessment of retention after concurrent conditional training. However, when the novel pairings available for retention tests are generated, a five-item list yields only one pair (B versus D) comprised of items that do not appear with an always- or never-rewarded item (an end-anchor) and have been both rewarded and non-rewarded during training. However, it seems inefficient to train four premise pairs to yield just one test pair. One remedy is to train more pairs because the number of appropriate test pairs advances incrementally with provision of longer lists (a six-item list yields three test pairs; a seven-item, six test pairs; an eight-item, ten; etc.). For example, Rapp, Kansky and Eichenbaum (1966) used seven-item concurrent conditional lists for a developmental test of monkey serial memory, and list lengths of 15 (Treichler, Raghanti, and Van Tilburg,

2003) and 45 items (Treichler and Raghanti, 2007) have been trained with macaques.

The other typical procedure, 'simultaneous chaining', was first utilised for monkey testing by D'Amato and Columbo (1988). An updated review of simultaneous chaining (Terrace, 2005) notes that it avoids a criticism often voiced by proponents of conditioning explanations. Such critics contend that the pairs in concurrent conditional tasks are learned via conventional reinforcing mechanisms, and the overlapping multiple appearances of discriminanda yield different (and ordered) reward values for the various items. An appropriate example is *value transfer theory* (von Fersen et al., 1991) which posits that value accrues to any list item as a consequence of both its rewarded choice and its appearance with other rewarded items, even though the target item might not be the correct one in that pairing (a conditioned reinforcing explanation). Thus, when an item appears in test pairings, choice probability is determined by the sum of its specific and conditioned (transferred) reinforcing values. An optimal strategy is: always choose the consistently rewarded item, but, for all pairings without this item, select whichever alternative had been nearer to the more-frequently-rewarded ('good') end of the serial order. Simultaneous chaining provided a procedure that avoided this issue of differential reward of the list items (and their postulated 'values') by initiating each trial with simultaneous display of all items in the series. Reward was then made contingent on touching all arrayed items in correct order. Additionally, the trial-to-trial location of items in the display was shuffled to require varied motor responses with different arrays. Initial training typically entails ordered choice on just two stimuli, but the number of items that comprise the list can rapidly be advanced to attain longer lengths. After training, retention is evaluated by presenting test trials comprised of limited portions of the series (e.g., two or more items from one list or items from several separately trained lists). Perhaps the best illustrations of simultaneous chain performance are provided by Terrace, Son and Brannon (2003). In their experiments, macaques began with shorter lists, but subsequently learned seven-item lists with all items present from the onset of training. These lists revealed that the number of sessions to an acquisition criterion decreased systematically with successive lists, indicating development of a facility similar to learning set (Harlow, 1949). Further, when test pairs were comprised of items from different lists, choice was determined by ordinal list position, that is, if an item that had occupied a lower position in one list was paired with one from any higher list position, the higher-ordered item was reliably selected (91 per cent for all such possible pairs). Seemingly, monkeys remembered the ordinal position of an item and applied that information to guide choice on any novel pair. Greater differences between ordinal positions of paired items yielded systematically increased likelihood of choice conforming to the serial order, and this was characteristic of both within- and between-list pairs. The implication was that choosing between items close to one another in ordinal position was difficult. Results from reaction time measures suggested that same characteristic because pairings of nearby items yielded systematically longer choice latencies than were obtained with large differences between ordinal positions. Thus, the outcomes from several dependent measures indicated that monkeys remembered the ordinal positions of items and stored this information as a representation of location along a continuum. Such internal representation of ordered lists accords well with D'Amato and Colombo's (1990) organisational or paralogical view and challenges the contention that serial information is retained as a compendium of associations between adjacent list members (a conditioning view, e.g., Wynne, 1998).

2.4 Behavioural and neurological contrasts

That pigeons retain serial information in a manner different from primates (involving end-anchors) is well documented (Terrace, 1993; Terrace and McGonigle, 1994), but species specificity of transitive performance remains controversial (see Wynne, 1998, for a species-inclusive proposal). It should be noted that pigeon results are not characteristic of all birds because some corvid species have shown serial retention properties strikingly like those of monkeys (Bond, Kamil and Balda, 2003). However, because of apparent differences in memory characteristics among species tested in analogous situations, it seems probable that contrasting outcomes are based on different (less complex?) integrative neuronal mechanisms.

Acknowledgement and recognition that simpler neuronal integrative mechanisms might be present in animals could serve to temper some of the controversy associated with the 'continuity vis-à-vis discontinuity of mind' issue. In broad and polarised terms, that issue asks whether the 'minds' of animals and humans differ qualitatively or merely quantitatively. Certainly it is acknowledged that humans have capabilities that surpass those of animals, but are these differences derived from integrative mechanisms so unique as to bear no evolutionary relationship to those of other organisms? One of the previously noted treatments of this issue (Penn, et al., 2008), reviewed a variety of behavioural indicants and recommended that the 'discontinuity' alternative be favoured. While circumspect and scholarly in advocating that view, some of the performances cited as dichotomous indicants of animal vis-à-vis human capabilities were limited in scope (e.g. no transitive inference investigations of lab primates) or ambivalent in definition (animals show no 'higher-order relations involving mental states'). Among the responsive commentaries on Penn et al.'s 2008 position statement was a rejoinder from McGonigle and Chalmers (2008). A major emphasis of their critique was that Penn et al. (2008) based their conclusion of human uniqueness on an unduly restricted set of (cultural) factors and neglected influential developmental effects. They specifically recommended tests of cognitive development in children as prospective indicators of continuity.

Another previously noted review (Premack, 2007) favoured the discontinuity approach based on evidence for distinctive human neurophysiology and neuroanatomy combined with the premise that such structures or functions reflected specific cognitive operations not shared by animals. However, that rationale assumes a correlation between uniqueness of physical and behavioural components that remains speculative. Although some associations between language and human brain anatomy have been noted (Buxhoeveden, Switala, Roy and Casanova, 2001; Buxhoeveden and Casanova, 2002), evidence that these reflect unique cognitive capabilities is unconfirmed. Further, some neuroanatomical features initially considered species specific now appear more widely distributed. For example, Von Economo neurons once thought restricted to humans and great apes (Nimchinsky et al., 1999) have now been discovered in cetaceans and elephants (Butti et al., 2009; Hakeem et al., 2009). Indeed, some of the evidence cited by reviewers (Penn et al., 2008) as support for functional relationships between language and neuroanatomical characteristics (e.g., enlarged neuropil space, see Sherwood, Rilling, Holloway and Hof, 2009) was not considered an indicant of non-continuity by its authors. Indeed, the first author of the report noted in the previous sentence (along with several colleagues; Sherwood, Sabiaul and Zawidzki, 2008) has recently published an extensive review of structural and functional neurological findings related to cognitive function. On the basis of that comprehensive treatment, those authors found sufficient empirical support for evolutionary cognitive development to conclude their review with this statement: 'We consider it inescapable to recognize the continuity in mentality between the LCA (last common [primate] ancestor) and us, despite the significant disparity in phenotypes'.

Premack's (2007) support for the discontinuity view was based on his hope that it might facilitate 'work linking these (cellular-level) neural changes to cognitive processes'. While that is certainly a commendable (and shared) goal, an alternative, of perhaps more heuristic value, is offered by acknowledging that the evolutionary focus inherent in continuity views provides rationale for broadened research strategies. One such approach is represented by investigation of neural mechanisms related to relatively simple behaviours and, subsequently, seeking their relationship to more complex integrative operations. Note that such endeavours fit with the theme expressed at the beginning of this chapter; that is, treating complexity by comprehending simplicity. Illustrative of this approach is the work of Roitman, Brannon and Platt (2007) who recorded single neuron firing rates from the lateral intraparietal cortex in macaque monkeys. They found that rates from selected cells in this area varied in monotonic relationship to the number of visually displayed elements that served as choice alternatives in a numerical discrimination task. Thus, it seemed that such cortical cells might provide neurological analogs of the 'summation units' postulated in some computational models of number processing (e.g., Dehaene and Changeux, 1993). Another neural mechanism that may bear a more direct relationship to the serial memory concerns of this chapter has been studied by Opstal, Fias, Perigneux and Verguts (2009). These investigators applied neuroimaging techniques to track sites of activation in human brains during the course of training on an arbitrarily ordered sequence. In accord with Eichenbaum's (2004) view of hippocampal and associated structure involvement in sequence memory, activation during the initial stages of training occurred in the hippocampal-angular gyrus, but with further training, the activation extended to the left inferior frontal gyrus. Recent and related research by Bongard and Neider (2010) using fMRI techniques has found similar properties in macaque monkeys. They discovered that specific areas of the monkey's frontal cortex were activated during the performance of tasks that required application of simple rules about serial ordering (e.g., greater or less than). These results, considered in concert with an earlier Neider (2004) finding on the activation of the intraparietal sulcus during numerosity judgement tasks, were interpreted as indicating the existence of a hierarchial 'rule-coding unit'. Note that these neural system investigations are not specifically comparative, but they converge on common issues by adopting a continuity perspective. Further, none of these studies are grounded in rationale derived from the 'anthropomorphism' that has been a focus of criticism by proponents of non-continuity (see Wynne, 2007). With no indication of any 'reference' species or presumed cognitive accomplishment, these investigations explore functionally defined processes and seek the integrative networks that are selectively activated when specific behaviours occur.

2.5 Further behavioural assessment – list linking and more

Just as the elucidation of relevant neural networks is enhanced by the study of analogous systems in several species, so too does comparative investigation of behaviour involving informational organisation aid in understanding performances termed 'cognitive'. Accordingly, convergence of tactical approaches from comparative neurophysiology and the development of enhanced behavioural assessments could aid understanding of integrative processes. Several examples of tests designed to provide finer-grained measures of behaviour related to organisation may be noted.

One such venture reemployed the concurrent conditional procedure to test an implication of D'Amato and Colombo's (1990) contention that monkeys remember lists as internally organised sequences with 'symbolic distances' between items. The rationale was that, if monkeys could retain more than one serial list, perhaps these could be linked by training only one pair that included the highest end-item of one list and the lowest of another. Might that combine both original lists with consequent 'symbolic distances' among all items? The issue seemed especially well suited to the concurrent conditional procedure because all trials in both acquisition and retention are presented as simultaneous, two-choice problems, and several lists could be independently trained. More importantly, isolated training could be provided on a conditional pair that linked between lists. Thus, acquisition, retention and manipulation of the major independent variable (linking) would all be administered in the same format with no indication (to subjects) that different conditions were imposed.

Initially, Treichler and Van Tilburg (1996) trained two different, five-item, serially ordered lists of objects and then gave isolated training on just one linking pair; for example, train an A<B<C<D<E list and an F<G<H<I<J list followed by E-F alone with F rewarded. Subsequent novel-pair retention tests indicated that, within the first block of test trials, macaque monkeys treated the information as an inclusive, ten-item, A to J ordered list. Seemingly, link training integrated lists. A follow-up study (Treichler et al., 2003) linked three lists in similar manner and provided additional control conditions (see Figure 2.1). This study confirmed the original outcome, but also showed that memory for ordinal positions and list linking were coexisting influences. Monkey serial memory appeared to reflect organised, internally maintained knowledge of list positions, but it was subject to revision by new information. Together, these influences supported emergence of integrated knowledge. Previously learned lists were internally reorganised to yield near-immediate adoption of choices consonant with an inclusive and appropriately ordered new list (see Figure 2.2A). Further, in a control condition where linking was not provided, monkeys could combine three five-item lists into a 15-item list, but this required many sessions (720 trials) of training on the pairings comprising the new, longer list. Initial choices after non-linking seemed determined by memory for the ordinal list positions of original learning (note Fig. 2.2B). However, after either linking or extensive training without linking, memory for long lists showed the same pattern of inter-item pairing distance effects that were seen in Terrace et al.'s (2003) successive chaining tests. Paired items from widely separated ordinal positions were rarely missed while ones from nearby locations continued to yield errors (between 30 and 40 per cent) well into a 15-day course of testing. Although choice latencies were not reported in this linking study, another experiment employing the same monkeys tested five different five-item lists (Treichler and Van Tilburg, 2002) and found a pattern of shorter latencies to the more widely separated pairs much like that reported in Terrace's (2003) successive chain tests. So, under both concurrent conditional and simultaneous chaining procedures, inter-item pairing distance and latency outcomes challenged the contention that reinforcement-based associations provide the best explanation of serial memory in animals (e.g., Vasconcelos, 2008). In further disputation of that

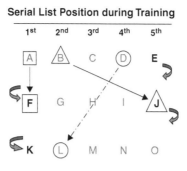

Figure 2.1 Representation of three different 5-item lists combined to form a 15-item list. Serial positions during original acquisition (1st through 5th) are noted, and pairs later trained as isolated links are depicted in bold with flat arrows indicating linking direction. Some examples of position contrasts possible in 15-item test pairings are indicated by pair B-J (triangles) where a correct object is 3 positions ahead of its originally experienced location, pair D-L (circles) where the correct object is 2 positions behind its training location (-2), and pair A-F (squares) representing objects from the same original list position. All between-list test trials can be categorized on a 9-point scale representing location contrasts that vary from 4 ahead, through same, to 4 behind their originally trained position. See Figure 2.2 for number of test trials representing each contrast level and the relationship of this dimension to test performances

view, non-human primates have consistently shown that greater inter-item distances yield progressively less error, and the most difficult discriminations are those of immediately adjacent pairs. Concurrent conditional test results have typically been used to support associative explanations of serial memory (e.g. value transfer). Under this procedure, immediately adjacent (i.e. premise) pairs are the ones most extensively trained and, if frequency of reward determines choice, should be proficiently retained. However, linking results from monkeys, despite being obtained with the favoured procedure of associative proponents, show precisely the opposite outcome. Original training pairs generate the *most* error. Clearly, the outcomes from linking tests provide convincing evidence that contrasts in serial retention performance are related to species differences, not methodological ones.

Another example of precision in assessing serial organisation is provided by the work of Matsuzawa and his associates with chimpanzees (see Biro and Matsuzawa, 1999). Utilising the successive chaining technique, they gave extensive, progressive training on an ordered series of nine visual stimuli and then tested the effects of changes in the number and nature of items displayed. One intent was to assess existence (and characteristics) of response planning, and an overview indicated reliable retention of organised lists with pre-planning of choice over the entire series (but see Beran,

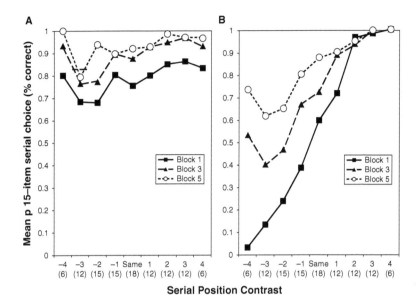

Figure 2.2 Mean (of 4 monkeys in a within-subject design) proportion choice scores conforming to a 15-item list (% correct) when combining three earlier-learned 5-item lists. Scores for each of 9 levels of position contrast (see Figure 2.1 for basis of levels) are shown. Numbers of trials per block at each level are indicated by numbers in parentheses below the label. These scores represent 108 between-list pairs included in 144 trial blocks that also contained 36 within-list pairs. Scores at three stages (Blocks 1, 3, & 5) of testing are displayed. The left panel (A) shows results after list linking and the right panel (B), without linking

Pate, Washburn and Rumbaugh, 2004, for an alternative outcome). The most recent empirical results from the procedure (Inoue and Matsuzawa, 2009) have confirmed earlier demonstrations of memory organisation and are notable in following an admonition made by McGonigle and Chalmers (2008) in their response to Penn et al.'s (2008) advocacy of non-continuity. The suggestion was to consider developmental issues, and, indeed, different aged chimpanzees did show distinctively different organisation on some measures. Several other techniques have assessed informational organising by non-human primates in different contexts. These include studies of ordinal numerical knowledge (Brannon, Cantlon and Terrace, 2006) and transfer of metacognitive skill in serial tasks (Kornell, Son and Terrace, 2007). Tests of list linking have been further employed to aid in defining some limits (Treichler et al., 2007) and test characteristics related to spontaneous serial combinations (Treichler and Raghanti, 2010).

All of these finer-grained investigations of informational organisation share sensitivity to another caution raised by McGonigle and Chalmers

(2008) in their reaction to the Penn et al. position paper (2008). Their concern was that human uniqueness might derive from factors they termed 'cultural' or grounded in extensive training histories. Consequently, integrative capabilities of animals would be underestimated because requisite competencies had not been developed. Notably, each of the aforementioned investigations employed subjects that acquired task sophistication via multiple-phase preliminary training supportive of their later performances. Gaining information from research of this kind follows on this chapter's theme of considering simplicity as a prelude to understanding complexity. Further, the sophistication requirement encourages long-term use of individual animals, and, at risk of being labelled 'primatocentric' (a term coined by Emery and Clayton, 2004), that strategy commends using non-human primates as subjects.

2.6 Breadth of the comparative agenda

Although this chapter advocates continuity (vis-a-vis non-continuity) when considering animal cognition, note that this is truly an 'academic' difference. Competent scholars representing a spectrum of opinion on this issue all seek understanding of mechanisms that influence memory, both as behavioural and physiological processes. So long as the intent of dissention is not to demean or discourage comparative research, contributions derived from any perspective or discipline are worthwhile. A recent overview of comparative cognition (Shettleworth, 2009) notes its extensive interactive relationships with other biological and behavioural sciences (see her fig. 4, p. 215). She also provides a cogent reminder about comparative study wherein she advocates the following:

> Instead of setting up one-off critical tests that members of a species either pass or fail, more progress may be made by breaking a broadly defined capability down into components, asking which are shared by species, and under what conditions and why.

That seems to represent a meaningful and inclusive approach to unravelling some of the complexity encountered in comparative study of serial memory mechanisms.

References

Biro, D. and Matsuzawa, T. (1999). Numerical ordering in a chimpanzee *(Pan troglodytes)*: Planning, executing and monitoring. *Journal of Comparative Psychology*, 113: 178–85.

Bonard, S. and Neider, A. (2010). Basic mathematical rules are encoded by primate prefrontal cortex neurons. *Proceedings of the National Academy of Science, USA*, doi: 10. 1073/pnas/0909180107.

Brannon, E. M., Cantlon, J. F. and Terrace, H. M. (2006). The role of reference points in ordinal numerical comparisons by rhesus monkeys *(Macaca mulatta)*. *Journal of Experimental Psychology: Animal Behavior Processes.* 32: 124–30.

Bryant, P. E. and Trabasso, T. R. (1971). Transitive inferences and memory in young children. *Nature,* 232: 456–8.

Butti, C., Sherwood, C. C., Hakeem, A. Y., Allman, J. M. and Hof, P. R. (2009). Total number and volume of Von Economo neurons in the cerebral cortex of cetaceans. *Journal of Comparative Neurology,* 10 July 2009, 515(2): 243–59.

Buxhoeveden, D. P., Switala, A. E., Roy, E., Litaker, M. and Casanova, M. F. (2001). Morphological differences between minicolumns in human and nonhuman primate cortex. *American Journal of Physical Anthropology* 11: 361–71.

Buxhoeveden, D. P. and Casanova, M. F. (2002). The minicolumn and evolution of the brain. *Brain, Behavior and Evolution,* 60: 125–51.

D'Amato, M. R. and Colombo, M. (1988). Representation of serial order in monkeys *(Cebus apella)*. *Journal of Experimental Psychology: Animal Behavior Processes,* 14: 131–9.

D'Amato, M. R., and Colombo, M. (1990). The symbolic distance effect in monkeys *(Cebus apella)*. *Animal Learning & Behavior,* 18: 133–40.

Darwin, C. (1859). *The origin of species.* London: John Murray; New York: Viking Penguin (1988 reprint).

Darwin, C. (1871). *The descent of man, and selection in relation to sex.* London: John Murray.

Dehaene, S. and Changeux, J-P. (1993). Development of elementary numerical abilities: A neuronal model. *Journal of Cognitive Neuroscience,* 5: 390–407.

Emery, N. J. and Clayton, N. S. (2004). Comparing the complex cognition of birds and primates. In L. J. Rodgers and G. Kaplan (eds), *Comparative vertebrate cognition: Are primates superior to non-primates.* New York: Kluwer/Plenum, pp. 3–48.

Gillan, D. J. (1981). Reasoning in the chimpanzee: II. Transitive inference. *Journal of Experimental Psychology: Animal Behavior Processes,* 7: 150–64.

Hakeem, A. Y., Sherwood C. C., Bonar C. J., Butti, C., Hof, P. R. and Allman, J. M. (2009). Von Economo neurons in the elephant brain. *Anatomical Record (Hoboken)* 292, 242–8.

Inoue, S. and Matsuzawa, T. (2009). Acquisition and memory of sequence order in young and adult chimpanzees *(Pan troglodytes)*. *Animal Cognition,* 12: 159–69.

Kornell, N., Son, L. K. and Terrace, H. S. (2007). Transfer of metacognitive skills and hint seeking in monkeys. *Psychological Science,* 18: 64.

McGonigle, B. O. and Chalmers, M. (1977). Are monkeys logical? *Nature,* 267: 694–6.

McGonigle, B. and Chalmers, M. (2008). Putting Descartes before the horse (again). *Behavioral and Brain Sciences,* 31: 142–3.

Neider, A. (2004). A parieto-frontal network for visual numerical information in the monkey. *Proceedings of the National Academy of Science, USA,* 101: 7457–62.

Nimchinsky, E. A., Gilissen, E., Allman, J. M., Perl, D. P., Erwin. J. M. and Hof, P. R. (1999). A neuronal morphologic type unique to humans and great apes. *Proceedings of the National Academy of Science, USA,* 96: 5268–73.

Opstal, F. V., Fias, W., Peigneux, P. and Verguts, T. (2009). The neural representation of extensively trained ordered sequences. *NeuroImage,* 47: 367–75.

Penn, D. C., Holyoak, K. J. and Povinelli, D. J. (2008). Darwin's mistake: Explaining the discontinuity between human and nonhuman minds. *Behavioral and Brain Sciences,* 31: 109–30.

Premack, D. (2007). Human and animal cognition: Continuity and discontinuity. *Proceedings of the National Academy of Sciences, USA,* 104: 13861–7.
Rapp, P. R., Kansky, M. T. and Eichenbaum, H. (1996). Learning and memory for hierarchical relationships in the monkey: Effects of aging. *Behavioral Neuroscience,* 110: 887–97.
Roitman, J. D., Brannon, E. M. and Platt, M. L. (2007). Monotonic coding of numerosity in Macaque lateral intraparietal area. *Public Library of Science, Biology,* 5(8): e208.
Sherwood, C. C., Rilling, J. K., Holloway, R. I. and Hof, P. R. (2009). Evolution of the brain in humans – specializations in a comparative perspective. In M. D. Binder, N. Hirokawa, U. Windhorst, M. C. Hirsch (eds), *Encyclopedia of Neuroscience,* Part 5, New York: Springer-Verlag, pp. 1334–8.
Sherwood, C. C., Subiaul, F. and Zawidzki, T. W. (2008). A natural history of the human mind: Tracing evolutionary changes in brain and cognition. *Journal of Anatomy,* 212: 426–54.
Shettleworth, S. J. (2008). The evolution of comparative cognition: Is the snark still a boojum? *Behavioural Processes,* 80: 210–17.
Terrace, H. S. (1993). The phylogeny and ontogeny of serial memory: List learning by pigeons and monkeys. *Psychological Science,* 4: 162–9.
Terrace, H. S. (2005). The simultaneous chain: A new approach to serial learning. *Trends in Cognitive Sciences,* 9: 202–10.
Terrace, H. S. and McGonigle, B. M. (1994). Memory and representation of serial order by children, monkeys and pigeons. *Current Directions in Psychological Science,* 3: 180–9.
Terrace, H. S., Son, L. K. and Brannon, E. M. (2003). Serial expertise of rhesus macaques. *Psychological Science,* 14: 66–73.
Treichler, F. R. and Raghanti, M. A. (2010). Serial list combination by monkeys *(Macaca mulatta)*: Test cues and linking. *Animal Cognition,* 13: 121–32.
Treichler, F. R., Raghanti, M. A. and Van Tilburg, D. (2003). Linking of serially ordered lists by macaque monkeys: List position influences. *Journal of Experimental Psychology: Animal Behavior Processes,* 29, 211–21.
Treichler, F. R., Raghanti, M. A. and Van Tilburg, D. (2007). Serial list linking by macaque monkeys: List property limitations. *Journal of Comparative Psychology,* 121: 250–9.
Treichler, F. R. and Van Tilburg, D. (1996). Concurrent conditional discrimination tests of transitive inference by macaque monkeys: List linking. *Journal of Experimental Psychology: Animal Behavior Processes,* 22: 105–17.
Vasconcelos, M. (2008). Transitive inference in non-human animals: An empirical and theoretical analysis. *Behavioural Processes,* 78: 313–34.
Verguts, T. and Fias, W. (2004). Representation of number in animals and humans: A neural model. *Journal of Cognitive Neuroscience,* 16: 1493–1504.
Von Fersen, L., Wynne, C. D. L., Delius, J. D. and Staddon, J. E. R. (1991). Transitive inference formation in pigeons. *Journal of Experimental Psychology: Animal Behavior Processes,* 17: 334–41.
Wynne, C. D. L. (1998). A minimal model of transitive inference. In C. D. L. Wynne and J. E. R. Staddon (eds). *Models for Action.* Hillsdale, NJ: Erlbaum, pp. 296–307.
Wynne, C. D. L. (2007). What are animals? Why anthropomorphism is still not a scientific approach to behavior. *Comparative Cognition and Behavior Reviews,* 2: 125–35.

3
The Use of Spatial Structure in Working Memory: A Comparative Standpoint

Carlo De Lillo

3.1 The comparative study of Working Memory

The comparative method is a powerful tool for the cognitive scientist. It can provide unique information about viable architectures for complex cognitive systems by showing those competences which are always linked together and those which can be dissociated in different species. Moreover, the assessment of the cognitive abilities of living species at different taxonomic distances from humans allows us to make informed guesses about the presence of specific cognitive skills in common ancestors at different points of our phylogeny.

Ultimately the comparative method could help determine what is unique about human cognition. However, in order to do so it must be applied to mental abilities that are closely related to higher cognitive skills in humans but can nevertheless be meaningfully studied in different non-human species. Working Memory (WM) is a key mental construct which might be particularly well suited to this purpose.

In human cognitive neuroscience, WM refers to the temporary storage of information needed to support cognitive functions such as thinking and provides the necessary buffer for the interfacing of perception, long-term memory and action (Baddeley, 2003). The notion of WM has a close relationship with that of general mental capacity (see Cowan, 2005) and it has been linked to attention and the executive skills which enable the ordering and monitoring of temporal information, the integration of cross-modal information and multitasking (Baddeley, 1996). As such, the systematic study of WM within a comparative framework would provide important insights regarding the evolutionary origins of advanced cognitive skills.

3.2 Working Memory and search organisation in radial maze studies

In both humans and animals, WM studies have mainly focused on its span (conceived as the upper limit of the mental capacity of individuals of a particular age or of a particular animal species), as measured by the number of (allegedly) unrelated items that can be temporarily held in memory. In animals, WM span is typically tested with search tasks where items of food have to be retrieved from a set of target locations. In these tasks, the number of visits paid to the targets can be considered an index of how proficient the animal is in monitoring its past actions. More visits than necessary are indicative of reiterative errors and as such of poor temporary memory for locations already explored within a particular trial (see De Lillo, 1996, for a review).

The most common apparatus for the study of animal WM is the radial maze. Its original version (Olton and Samuelson, 1976) was designed for use with small rodents and consists of a central platform with multiple (typically eight) arms departing from it in radial fashion. Variations of the radial maze, featuring simple circular arrangements of containers, have been used as an analogue of the radial maze with birds (Spetch and Edwards, 1986) and children (Foreman, Arber and Savage, 1984). More recently, virtual reality versions of the task have been used for brain imaging studies with adult humans (Astur et al., 2005).

Circular or radial arrangements of food locations have the property of being highly structured and subjects have the opportunity to exploit the constraints afforded by the search space by deploying algorithmic search patterns consisting, for example, in exploring adjacent locations consecutively (see Dubreuil et al., 2003, for a review). As this type of organised responding enables an efficient search without requiring the subject to hold spatial locations in memory, a number of procedures have been designed to break the spatial regularity of the maze in order to encourage animals to rely solely on memory for the monitoring of their past searches. These procedures include, for example, blocking the entrance of some arms or manipulating their angular arrangement (see Schenk et al., 1995).

Although discouraged and generally not considered worthy of a detailed investigation, there are some indications that the deployment of even simple forms of search organisation may tap non-trivial aspects of cognition. For example, a developmental trend can be observed in the spontaneous emergence of algorithmic responding in children tested with the radial maze (Foreman, Arber and Savage, 1984; Aadland, Beatty and Maki, 1985) suggesting that there may be a relationship between cognitive growth and the use of simple spatial strategies in search tasks. Whereas two-year-old children perform only marginally above chance and show little evidence of search organisation, from four years of age children rarely perform reiterative errors and have a tendency to use algorithmic strategies such as entering adjacent arms in succession and avoiding a change in the direction of travel when searching the maze. The relationship between these search strategies and performance is illustrated by the fact that when older children and adults are prevented from using algorithmic strategies their search performance collapses, in contrast with what happens with younger children whose performance remains typically unchanged in conditions designed to hinder the use of principled search patterns (Foreman, Arber and Savage, 1984; Aadland, Beatty and Maki, 1985).

3.3 Memory for locations in clustered, linear and diffuse configurations

Although the radial maze may be considered an extreme and perhaps rather artificial example of a structured search space, it is important to stress that it is extremely difficult to conceive spatial memory tasks with multiple goal locations where the opportunity to exploit their spatial configuration is completely denied to the animals (see also Kirsh, 1995, for a discussion of related issues in Artificial Intelligence). Primate spatial working memory has been typically assessed using irregular arrangements of containers which the animals search in their own home enclosures (see MacDonald and Wilkie, 1990; MacDonald, Pang and Gibeault, 1994), and it is generally implicitly assumed that it is possible to determine memory span for a set of unrelated items independently of their spatial arrangement and the search organisation that it affords. Nevertheless, it is always possible, in principle, for the searching agent to impose structure on the search space so that the memory demand (often used synonymously with cognitive complexity; see Kirsh, 1995) of the search task is reduced. For instance, a primate enclosure can be divided into separate regions, such as the different vertical level of a large cage or different quadrants of the floor or ceiling or different benches, surfaces, among possible others. If explored one after the other in a principled and exhaustive way, such partitions may enable the animal to keep track of where it has been during a particular trial in terms of regions rather than individual locations (see Schenk et al., 1995, for examples of three-dimensional mazes where some of these variables have been considered). Only by using systematic manipulations of the configuration of the search space and observing the search patterns of the animals in relation to their memory performance is it possible to assess the extent to which different animal species and humans may differ in their ability to use the spatial structure afforded by the material they have to remember.

We explored the relationship between search performance and search organisation in capuchin monkeys (*Cebus apella*) following the manipulation of the configuration of the items to be searched (De Lillo, Visalberghi and

Aversano, 1997; De Lillo et al., 1998). In a first study we evaluated the performance of capuchin monkeys when searching sets of containers arranged in spatial clusters on the ceiling of a large enclosure. With this type of configuration we tried to make it evident that the set afforded data-reducing strategies, consisting in searching each cluster exhaustively before moving on to another cluster. If the animals were able to pick up this affordance, they could potentially reduce their memory load from nine items (the entire set of locations) to three items (three locations, when moving within a cluster, and three clusters, when moving between clusters). We assessed the benefits of this particular spatial organisation of containers by comparing the performance observed when animals were searching a more diffuse (a square matrix) arrangement of containers which did not afford clustering and chunking by spatial proximity. As shown in Figure 3.1, the experiment featured an A-B-A design where performance with the square matrix of containers was first assessed in a baseline condition (A), then in an experimental condition featuring the clustered set of containers (B) and then again in a condition identical to the baseline (A) to control for effects of task practice. Thus, with this experiment, rather than neglecting them, we made explicit the affordances which may have been implicitly present in other studies, and systematically manipulated them to observe their effects. The monkeys (see dotted line in Figure 3.1) were able to monitor their searches much more easily in the experimental clustered configuration compared to the baseline and control conditions featuring the square matrix.

Figure 3.1 also shows, for comparison, the results obtained with preschool children and rats (De Lillo, 2004a) tested in a similar task where food items (rats) and toys (children) had to be retrieved from the top of poles arranged according to the different conditions used by De Lillo et al. (1997). From the figure it can be seen that children and monkeys showed the deepest drop in the number of reiterative moves in the control condition compared to baseline and control. Rats' search efficiency also improved in the control compared to the baseline condition but their performance remained relatively stable in the control condition. Of particular interest is the fact that in the clustered condition (but not in the baseline or the control condition) the children show the highest level of search efficiency, followed by that shown by the monkeys, which, in turn, proved more efficient than that shown by rats. Thus, in a condition strongly affording a hierarchical organisation of search, a trend in the level of performance can be observed which increases in parallel with taxonomic relatedness with humans.

Other studies carried out with mice (Valsecchi et al., 2000) and rats (Foti et al., 2007) using a procedure derived from De Lillo et al. (1997) did not find any difference in performance between clustered and non-clustered search spaces, providing further evidence for the notion that the ability to benefit from clustering may not be trivial and it is particularly developed in primates. The results of another study (Bartolomucci, de Biurrun and Fuchs,

42 *Carlo De Lillo*

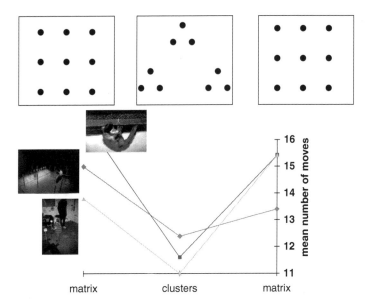

Figure 3.1 Average number of moves (visits to containers) required by Rats (blue line); Monkeys (dotted line) and children (green line) in order to complete an exhaustive search of a set of nine containers. Monkeys (*Cebus apella*) moved on the wire mesh ceiling of a large enclosure and searched for peanuts hidden in containers suspended from the ceiling (De Lillo et al., 1997). Rats for searched for puffed rice placed on a food well at the top of wooden poles and children searched for hidden toys from containers placed at the top of poles (De Lillo, 2004). In order to allow meaningful comparisons between three species, the measurements of the search space were made roughly proportional to the average size of the subjects of each species. According to an ABA design, the set of containers was organised as a diffuse matrix in a baseline and a control condition and as three clusters of three containers each defined by spatial proximity in the experimental condition. A larger number of moves is indicative of a less efficient monitoring of the visits paid to containers in any particular trial

2001) aimed at replicating the results of De Lillo and colleagues (1997) with the tree shrew (*Tupaia belangeri*), a species which shares with primates an enlarged neocortex and visual abilities (Martin, 1990) but is currently classified in a separate order (*Scandentia*), are also consistent with this claim. In fact, even if the tree shrews failed to benefit from the clustered arrangement of containers as capuchin monkeys did (De Lillo et al., 1997), the authors (Bartolomucci, de Biurrun and Fuchs, 2001) claim that they observed more evidence of clustering in this species than in mice (tested by Valsecchi et al., 2000).

It must, however, be pointed out that procedural issues may also account for some of the interspecies differences observed in these studies. For example, mice walked on the floor of an enclosure. Nevertheless, they were required to explore (by standing on their rear legs) containers placed on the ceiling of an enclosure. Presumably this was done to make the set-up as similar as possible to that used with capuchin monkeys (by De Lillo et al., 1997). However, the monkeys, in contrast with mice, moved on the wire-mesh ceiling of the cage itself. For this reason alone rather than because of cognitive differences, mice may have appreciated the affordances of the spatial layout of the set of containers to a lesser extent than monkeys. Also the number of trials used in different studies varied considerably and, as we shall see below, there is reason to believe that task experience is an important factor to take into account for a correct characterisation of the cognitive processes involved in efficient search.

Most comparative approaches emphasise the importance of immediate ecological factors and it is plausible to conjecture that the beneficial effects of clustering just mentioned in capuchin monkeys could be related to the specific adaptations of this particular species, whose diet consists prominently of fruit, to forage on patchy resources such as fruit trees. Indeed, the diet of primates and the cognitive adaptations induced by foraging on patchy resources has been proposed as one of the main causes of the emergence of primate intelligence (Milton, 1981 1993).

One narrow view of the notion that it is the requirement of foraging on patchy resources which causes the emergence of memory skills in fruit-eating species would suggest a specialisation for foraging specifically on search spaces organised in clusters defined by relative spatial proximity. However, studies that followed up the findings of De Lillo et al. (1997) showed that capuchin monkeys benefit from a number of other arrangements of containers, including linearly and circularly arranged sets of goals (De Lillo et al., 1998), just as it did from clustered search spaces (De Lillo et al., 1997) as long as they afford principled and economic search patterns. A more plausible prediction of the notion that cognitive evolution in primates is linked to their diet and foraging niches is that in frugivorous primates 'the ability to remember the exact location of trees that produce desirable fruits and to recall the shortest routes to those trees would enhance foraging efficiency by lowering search and travel costs' (Milton, 1993: 73). The view, expressed in the above sentence, would suggest a general expansion of memory span in primates which would prevent inefficient travel between food patches and would be consistent with widely promoted notions within behavioural ecology and optimal foraging theory (see Krebs and Davies, 1993; MacArthur and Pianka, 1966) and there is indeed recent evidence for an impressive memory capacity in monkeys (Fagot and Cook, 2006).

However, it is also possible that apart from adaptations resulting in a large memory span, which would be expressed in the ability to search efficiently large sets of locations irrespective of their arrangement and of learning, primates (capuchin monkeys in this case) are able to dynamically regulate their

44 Carlo De Lillo

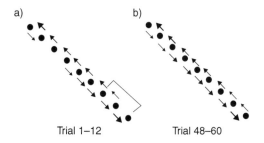

Figure 3.2 Micro-development of search organisation in monkeys (*Cebus apella*) exploring a linear configuration of baited containers. The figure depicts the transitions between successive visits to containers performed by four subjects (combined) in the first (a) and the last (b) blocks of 12 trials in which the data base was divided. Only in the course of task practice monkeys develop a consistent use of end-to-end searches which greatly reduce the memory demand of the task. Redrawn from De Lillo et al. (1998)

behaviour in a way which enables them to learn to take advantage of the structure of the space they have to explore. We have evidence of this from the results of both of our studies which featured different arrangements of search locations (De Lillo et al., 1997, 1998). As mentioned above, a necessary condition for being able to benefit from the affordances of a clustered search space is that each cluster is explored exhaustively before moving on to the next. Only this would allow the animal to benefit from the hierarchical organisation of the search space and remember that a specific area has been explored without having to remember the exact location of each of the locations explored. Instead of deploying this strategy from the outset, in our 1997 study, monkeys gradually acquired expertise with this task. In fact, we observed a negative trend in the number of occasions in which a cluster was abandoned before being exhaustively explored and that this paralleled a trend in search performance as measured by the reduction of visits to containers already explored (De Lillo et al., 1997). This result is important as it indicates that the benefits of structured configurations of goals are not automatically available to the animal, perhaps due to a collusion of the travelling habits of the animal and local constraints of clustered goals.

Similarly, when searching a linear array of locations, only searches which start from one end of the array and move towards the other end by searching adjacent containers consecutively (and without skipping any of them) exploit the spatial arrangement of the containers in the most efficient way by making it possible to search indefinitely numerous locations with minimal (or virtually absent) memory requirements. In our 1998 study (De Lillo et al., 1998), we observed an increase in the number of these 'end-to-end' searches of linear arrays in the course of tasks practice. The trajectories followed at the beginning and at the end of testing are shown in Figure 3.2 below.

However, interestingly in this case, the shift in search pattern occurred in conditions of high search efficiency and was not correlated with performance. This result therefore raises the question of what incentive was there for the animal to induce a behaviour modification. Two possible sources of feedback could in principle be used by the animals: the energetic cost of motor action and the information load imposed by the task. The first of these sources of feedback is consistent with the well-established view that animals strive to minimise the distance travelled, in accordance with well-known accounts of animal behaviour (Milton, 1981, 1993; see also Cramer and Gallistel, 1997, and Menzel, 1973). The second, which has more recently been proposed as potentially important in the context of comparative cognition (see De Lillo, 1994, 1995, 1996; De Lillo and McGonigle, 1997; McGonigle, 1984; McGonigle and Chalmers, 1998; Visalberghi and De Lillo, 1995, for reviews), is that organisms may be striving to minimise the cognitive costs involved in the task. Cognitive costs may therefore (alone or in conjunction with other types of energetic costs) be the currency used for behavioural regulation. Thus, agents may be induced to modify their behaviour if the change produces a less cognitively demanding task solution.

3.3 A comparative analysis of simplicity in search

The notion that the pursuit of cognitive economy could be the driving force for imposing structure over sets of otherwise unconnected items (see also De Lillo, 2004b) could be put in relation to the old and still controversially debated notion of simplicity in perception (von Helmholtz, 1924; Pomencrantz and Kubovy, 1986; Leeuwenberg and Boselie, 1989; Sutherland, 1989) and recently refined and exported to other domains of human cognition (Pothos and Chater, 2002). As search for simplicity has been considered a fundamental human cognitive skill (Chater, 1996, 1997; Pothos and Chater, 2002) the assessment of the extent to which other species conform to it could be important for the understanding of what characterises human cognition and its origins.

In the domain discussed here, it would be of particular interest to be able to show that search performance is related to the type of trajectories that an animal is able to devise as a function of the search space which needs to be explored. In particular, it would be of interest to evaluate the extent to which simpler search patterns sustain higher levels of efficiency and to what extent species taxonomically closer, and possibly cognitively more similar, to humans are better able to devise the least complex search trajectory through the search space. A simple way of measuring complexity of patterns is to look at the length of the description necessary to reproduce that particular pattern. In fact, early experiments on pattern recognition (e.g. Glanzer and Clark, 1964) indicate that the length of English descriptions of visual stimuli is correlated with their perceived complexity (e.g. in ratings

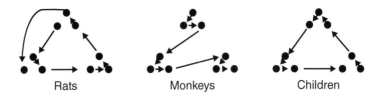

Figure 3.3 A static pattern of the animations presented to human participants who were asked to write a description and indicate the perceived complexity of the most representative search path used by rats (left), monkeys (middle) and children (right)

of pattern complexity). In the study reported below, this measure was used to evaluate the extent to which the trajectories through the search space followed by different species required a description, as written by human participants, of a different length (in terms of word count).

Ninety-six psychology undergraduates at the University of Leicester took part in the study. They were divided into three groups and the participants in each group were tested collectively in a large computer laboratory. Each participant was assigned to a separate PC which was used to administer the task. Following the collection of some information about the participant, the software, developed in-house specifically for this study, allowed the presentation of a diagram consisting of nine dots arranged as three clusters of three dots each (see Figure 3.3) in analogy with the clustered search space used in the search studies with animals and children (De Lillo, 2004a; De Lillo et al., 1997) discussed above.

An animation was then presented where arrows connected the dots, indicating the starting location and the transitions between the different locations which specified a search trajectory of an agent exploring the set. The animation was presented in a loop until the participants submitted their response by clicking on a button on the screen. Participants were instructed to write, in a box provided for this purpose on the screen, a concise but detailed enough description of the path which would have allowed a hypothetical person in possession of the dot pattern (with the arrows missing) to reproduce the path traced in the animation. In order to collect additional information concerning the perceived complexity of the search pattern, a seven-point Likert scale ranging from 'very simple' to 'very complex' was also subsequently provided on the screen. Three patterns were devised, each based on the most representative errorless search path respectively shown by rats, capuchin monkeys and children. The set of search patterns followed by monkeys was taken from the study by De Lillo, Visalberghi and Aversano (1997). The search patterns followed by rats and children were derived from newly collected data (De Lillo, 2004a). The prototypical search path of each species was determined by assigning a different number to each specific transition between two locations in the search space. A Cronbach's alpha

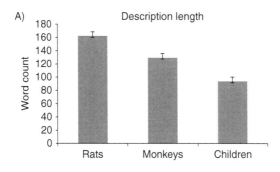
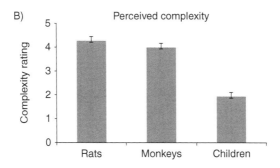

Figure 3.4 A) Number of words required by human participants to describe the most representative search path of rats, monkeys and children. B) Complexity ratings reported for the same paths. Error bars = 1 SE

(Cronbach, 1951) was then calculated to assess which pattern of transitions was the most highly correlated with all the other patterns in the database for a particular species, taking into account where the transitions occurred in the series of moves which defined the search pattern. This was then taken as the most representative search path for the given species. A static representation of the prototypical pattern of each of the species is shown in Figure 3.3. From the figure it can be noticed that the best exemplar of a search path for both children and monkeys is characterised by an exhaustive exploration of each of the clusters before moving into a different cluster. For the children, moreover, the same direction of travel was used in the exploration of each of the clusters. The best example of a rat's trajectory, by contrast, contains some exits before the exhaustive exploration of the subset.

The ninety-six participants were divided into three groups and each group was presented with the typical path of one species only. Thus, a group of participants received only the typical search pattern of rats (N = 31), one group that of monkeys (N = 33), and the third group the exemplar of the search path followed by children (N = 32). The results obtained with the three groups were then compared on the basis of the mean number of words used to describe the pattern that they were presented with. As an additional test, the average points from the Likert scales for the evaluation of perceived complexity was compared between the three groups. The results are presented in Figures 3.4a and 3.4b.

A significant difference ($F(2, 93) = 17.13$, $p < .001$) was observed between the average number of words used to describe the characteristic search path of the three species. The description of the path followed by children required significantly less words than that followed by monkeys (Sheffe test, $p < .01$) which in turn required less words than the one followed by rats (Sheffe test, $p < .05$).

The analysis of the complexity rating also revealed a significant difference in the perceived complexity of the search pattern of the different species ($F(2, 93) = 42.40$, $p < .001$). However, for this variable only the complexity rating for the path followed by the children was found to be less complex than that of both monkeys and rats (Sheffe test, $p < .001$), whereas there was no significant difference in the rating of complexity of the pattern followed by the two animal species.

While caution must be exercised in claiming that the 'less simple' trajectories are cognitively complex, in the sense of claiming that these routes are single entities or action plans that could be recognised, reproduced or remembered, the results presented here provide a first indication that the notion of simplicity as solution to potentially high-working memory demands may be of use in a comparative context – and that it may be better achieved in humans and species taxonomically closer to humans.

3.5   Imposing structure on serial recall

One of the strengths of search tasks implemented in relatively large-scale environments, such as those reviewed above, lies in their ecological plausibility as foraging tasks. However, the use of large-scale environments has a downside as it makes it difficult to determine the relative contribution of cognitive costs as opposed to energetic costs as a motivating factor for the organisms to exploit the spatial affordances of the set of locations they are required to explore. In fact, the simplest paths, which have the property of simplifying the cognitive component of the task by reducing its memory demand, are also the paths which would minimise its energetic costs since they often also feature the shortest route among the locations. Although possible in principle, it has proved very difficult to devise search spaces where the minimisation of cognitive and energetic costs can be dissociated.

The fact that similar functions have been observed in tasks requiring children, monkeys (*Cebus apella*), and pigeons to select items arranged in rows and columns on touch-sensitive computer monitors (see McGonigle, De Lillo and Dickinson, 1992; De Lillo, 1995; and Visalberghi and De Lillo, 1995 for a review) suggests that the role of fatigue in the regulation of behaviour could be minor in these tasks. Nevertheless, the length of the movement required to explore the set is always a spurious variable which makes it impossible to prove unambiguously that cognitive economy plays a role in the organisation of search.

Although of difficult implementation with non-human species, the assessment of serial spatial recall as measured, for example, by the Corsi tapping test (Corsi, 1972; Milner, 1971) allows us to overcome some of the limitations of search tasks where the agents are left free to devise their own searches through the set. The Corsi test has been considered the most important non-verbal test in human neuropsychology (Berch et al., 1998) and consists in the presentation of sequences of flashing icons on a touch screen (or just finger-tapping onto wooden blocks on a tray, in earlier studies) which need to be immediately reproduced by the participants. Traditional versions of the test have (in a way which is not dissimilar to the assumptions mentioned above about animal spatial working memory) ignored the importance of assessing the role of serial spatial structure in determining the memory span of the participants. Typically the test featured an irregular arrangement of items and random sequences of different length were used to assess WM span. However, if appropriately modified (De Lillo, 2004b) it can prove an important complementary tool for the assessment of the relationship between spatial organisation and working memory performance.

In particular, by using an array of clustered spatial locations, analogous to the clustered search space used in the studies by De Lillo and colleagues (1997) and by De Lillo (2004a), it was possible to assess in humans the relationship between search organisation and performance. Three types of screen-based sequences to be reproduced by tapping were designed and referred to as types A, B and C. Type A sequences featured the selection of items within each of the spatial clusters before the selection of items within a different cluster and which had as an added constraint that the direction of movement within each cluster was the same. Such sequences are similar to the typical path shown by children in large-scale search tasks (De Lillo, 2004a). Type B sequences also satisfied the constraint that all the items within each cluster had to be selected before moving to a different cluster. However, in these sequences the pattern of movement within each cluster varied for the different clusters. This type of sequence was considered somehow similar to those spontaneously acquired by monkeys in the study by De Lillo et al. (1997). Finally, Type C sequences were designed so that the serial organisation of the sequences was not consistent with a spatial clustering principle as consecutive items were always selected from different clusters. Although more extreme in their violation of organisational principles, these sequences bear some similarity with the less organised searches deployed by rats (De Lillo, 2004a) as described above. The results clearly indicated that the three types of sequences were reproduced with different levels of accuracy. Specifically, the highest accuracy was recorded for the Type A sequences and the lowest for the Type C sequences. Moreover, an analysis of response times provided the first indication in the literature that spatial clustering can induce hierarchical representation similar to the hierarchical organisation advocated for chunking in non-spatial domains (e.g. Bousfield and Bousfield, 1966).

Although these results do not prove that the advantage observed by monkeys and children in search tasks is necessarily mediated by a form of hierarchical representation akin to chunking, they indicate that spatial grouping could support such sort of representation. Subsequent studies (De Lillo and Lesk, 2009) have clarified better the role of chunking and hierarchical organisation in these types of tasks by showing that the composition of the set in terms of cluster size and number of clusters in the set affects performance. Moreover, by replacing serial recall with a serial recognition task it was possible to show that the benefit emerges in conditions which do not require motor planning, thus indicating that the hierarchical structure pertains to the memory representation of the sequences rather than to the motor plan necessary to reproduce them as a series of actions (De Lillo and Lesk, 2009).

3.6 The organisation of serial recall, prefrontal functions and primate cognitive evolution

fMRI studies which have looked at the use of structure in serial spatial recall (but using slightly different definitions of structure, such as moving along the same line column or diagonal in a display not very different from the layout used by McGonigle, De Lillo and Dickinson (1992), where clear differences emerged between search organisation of birds, non-human primates and children as described in Visalberghi and De Lillo (1995) interestingly found that there is a selective activation of the dorsolateral prefrontal cortex (DLPFC) when human participants reproduce structured sequences which are also easier to retain than unstructured sequences (Bor et al., 2003). As the DLPFC is typically associated with WM function, it seems paradoxical that this area should be more active when WM is less loaded. This negative relationship between DLPFC activity and memory demand, however, makes sense in the light of the issues discussed here. In fact, the activation of the DLPFC could reflect the use of attention and executive resources which may be needed for the detection and use of the structure in the to-be-remembered material. Prefrontal functions thus may mediate the ability to benefit from structure in WM which, as we have seen, is peculiar to primates, and to humans in particular. Thus, the behavioural evidence reviewed in this chapter for a trend in the ability to impose structure on to-be-remembered material in the different species considered would have a neuroanatomical correlate in the progressive expansion of the frontal lobes in living primates (Fuster, 1989).

It has proved more difficult to find a similar pattern of interspecies differences in relation to taxonomic distance from humans by looking at the function of other brain structures such as those which mediate the ability to consolidate a large number of items in Long Term Memory (see MacPhail and Bolhuis, 2001; see also Fagot and Cook 2006, for evidence of extremely large storage capacity for arbitrarily paired stimuli and responses in baboons and pigeons). It is therefore a contention of this chapter to promote the study of organisation in WM within a comparative perspective as it has the potential to provide prime information concerning primate cognitive evolution and may ultimately help us to infer important stages in the emergence of human cognition. The notion that attention mediated processes of organisation are a key factor in understanding human cognitive evolution finds support also in the mounting evidence for interspecies and developmental differences in the propensity to attend to the global structural aspects of visual patterns rather than their local elements in isolation (Fagot and Deruelle, 1997; Spinozzi, De Lillo and Truppa, 2003; De Lillo et al., 2005). Adult humans do so spontaneously and with ease (Navon, 1977). By contrast, monkeys preferentially attend to disconnected local elements of visual patterns (Fagot and Deruelle, 1997; Spinozzi, De Lillo and Castelli, 2004). Young children and chimpanzees show an intermediate pattern between these two extremes (De Lillo et al., 2005; Fagot and Tomonaga, 1999; Hopkins and Washburn, 2002). The use of organisational principles seems therefore to be of considerable comparative interest in both working memory and perception.

The full characterisation of the cognitive processes and brain functions which mediate the ability to use structure in humans and the extent to which they pertain to different species and different cognitive domains is an exciting challenge for comparative research carried out within the approach discussed in this chapter.

Acknowledgements

Part of the work on serial recall in humans described in this chapter was supported by Biotechnology and Biological Sciences Research Council (BBSRC) grant BB/C007840/1 to the author who is also very grateful to Tony Andrews for developing the software used in all the experiments with human participants described here.

References

Aadland, J., Beaty, W.W. and Maki, R.H. (1985). Spatial memory of children and adults assessed in the radial maze. *Developmental Psychobiology*, 18, 163–172.

Astur, S., St. Germain, S. A., Baker, E. K., Calhoun, V., Pearlson, G. D. and Constable, R. T. (2005). fMRI hippocampal activity during a virtual radial arm maze. *Applied Psychophysiology and Biofeedback*, 30(3): 307–17.

Baddeley, A. D. (1996). Exploring the central executive. *Quarterly Journal of Experimental Psychology*, 49A(1): 5–28.

Baddeley, A. D. (2003). Working memory: looking back and looking forward. *Nature Reviews Neuroscience*, 4: 829–39.

Bartolomucci, A., de Biurrun, G. and Fuchs, E. (2001). How tree shrews (*Tupaia belangeri*) perform in a searching task: evidence for strategy use. *Journal of Comparative Psychology*, 115(4): 344–50.

Berch, D. B., Krikorian, R. and Huha, E. M. (1998). The Corsi block-tapping task: methodological and theoretical considerations. *Brain and Cognition*, 38: 317–38.

Bor, D., Duncan, J., Wiseman, R. J. and Owen, A. M. (2003). Encoding strategies dissociate prefrontal activity from working memory demand. *Neuron*, 37: 361–7.

Bousfield, A. K. and Bousfield, W. A. (1966). Measurement of clustering and sequential constancies in repeated free recall. *Psychological Report*, 19: 935–42.

Chater, N. (1996). Reconciling simplicity and likelihood principles in perceptual organization. *Psychological Review*, 103(3): 566–81.

Chater, N. (1997). Simplicity and the mind. *The Psychologist*, November: 495–8

Corsi, P. M. (1972). Human memory and the medial temporal region of the brain. *Dissertation Abstracts International*, 34(02): 891B (University microfilms No. AA105-77717).

Cowan, N. (2005). *Working memory capacity*. New York: Psychology Press.

Cramer, A. E. and Gallistel, C. R. (1997). Vervet monkeys as travelling salesmen. *Nature*, 387: 464.

Cronbach, L. J. (1951). Coefficient alpha and the internal structure of tests. *Psychometrika*, 16(3): 297–334.

De Lillo, C. (1994). The logic of memory search in non-human primates *(Cebus apella)*. PhD thesis. The University of Edinburgh.

De Lillo, C. (1995). Strategic Search Behaviour in Monkeys. In E. Alleva, A. Fasolo, H. P. Lipp, L. Nadel and L. Ricceri (eds). *Behavioural brain research in naturalistic and semi-naturalistic settings: possibilities and perspectives*. NATO ASI Series, Series D: Behavioural and Social Sciences, p. 442.

De Lillo, C. (1996). The serial organisation of behaviour by non-human primates: an evaluation of experimental paradigms. *Behavioural and Brain Research*, 81: 1–17.

De Lillo, C. (2004a). Search strategies and spatial memory for clustered sites: a comparison of young children (*Homo sapiens*), capuchin monkeys (*Cebus apella*) and rats (*Rattus norvegicus*). *Folia primatologica*, 75(suppl 1): 231–353.

De Lillo, C. (2004b). Imposing structure on a Corsi-type task: evidence for hierarchical organisation based on spatial proximity in serial spatial memory. *Brain and Cognition*, 55(3): 415–26.

De Lillo, C., Aversano, M., Tuci, E. and Visalberghi, E. (1998). Spatial constraints and regulatory functions in monkey's (*Cebus apella*) search. *Journal of Comparative Psychology*, 112: 353–62.

De Lillo, C. and Lesk, V. (2009). Spatial clustering and hierarchical coding in immediate serial recall. *European Journal of Cognitive Psychology*, in press. DOI: 10.1080/09541440902757918.

De Lillo, C. and McGonigle, B. O. (1997). The logic of searches in young children (*Homo sapiens*) and tufted capuchin monkeys (*Cebus apella*). *International Journal of Comparative Psychology*, 10: 1–24.

De Lillo, C., Spinozzi, G., Truppa, V. and Naylor, D. M. (2005). A comparative analysis of global and local processing of hierarchical visual stimuli in young children and monkeys (*Cebus apella*). *Journal of Comparative Psychology*, 119(2): 155–65.

De Lillo, C., Visalberghi, E. and Aversano, M. (1997). The organisation of exhaustive searches in a 'patchy' space by capuchin monkeys (*Cebus apella*). *Journal of Comparative Psychology*, 111: 82–90.

Dubreuil, D., Tixier, C., Dutrieux, G. and Edeline, J.M. (2003). Does the radial arm maze necessarily test spatial memory? *Neurobiology of Learning and Memory*, 79(1): 109–117.

Fagot, J. and Cook, R. (2006). Evidence for large long-term memory capacities in baboons and pigeons and its implications for learning and the evolution of cognition. *PNAS*, 103(46): 17564–7.

Fagot, J. and Deruelle, C. (1997). Processing of global and local visual information and hemispheric specialization in humans (*Homo sapiens*) and baboons (*Papio papio*). *Journal of Experimental Psychology: Animal Behaviour Processes*, 23(2): 429–42.

Fagot, J. and Tomonaga, M. (1999). Global and local processing in humans (Homo sapiens) and chimpanzees (Pan troglodytes): use of a visual search task with compound stimuli. *Journal of Comparative Psychology*, 113(1): 3–12.

Foreman, N., Arber, M. and Savage, J. (1984). Spatial memory in preschool infants. *Developmental Psychobiology*, 17(2): 129–37.

Foti, F. Mandolesi, L., Aversano, M. and Petrosini, L. (2007). Effects of Spatial Food Distribution on Search Behavior in Rats (*Rattus norvegicus*) *Journal of Comparative Psychology*, 121(3): 290–299.

Fuster, J. M. (1989). *The pre-frontal cortex*. 2nd edn. New York: Raven Press.

Glanzer, M. and Clark, W. (1964). The verbal-loop hypothesis: conventional figures. *American Journal of Psychology*, 77: 621–6.

Helmoltz, H. Von (1924). *Helmoltz's treatise on physiological optics* (translated from the 3rd German ed.; edited by J.C.P. Southhall).Rochester NY, Optical Society of America.

Hopkins, W. D., and Washburn, D. A. (2002). Matching visual stimuli on the basis of global and local features by chimpanzees (*Pan troglodytes*) and rhesus monkeys (*Macaca mulatta*). *Animal Cognition*, 5: 27–31.

Kirsh, D. (1995). The intelligent use of space. *Artificial Intelligence*, 73: 31–68.

Krebs, J. R. and Davies, N. B. (1993). *An introduction to behavioural ecology*. 3rd edn. Oxford:, Blackwell.

Leeuwenberg, E. and Boselie, F. (1989). How good a bet is the likelihood principle. In A.G. Elsedoorn and H. Bouma (eds), *Working models of human perception*. London: Academic Press, pp. 363–79.

Navon, D. (1977). Forest before the trees: the precedence of global features in visual perception. *Cognitive Psychology*, 9: 353–83.

MacArthur, R. H. and Pianka, E. R. (1966). On optimal use of a patchy environment. *The American Naturalist*, 100: 603–9.

MacDonald, S. E. and Wilkie, D. (1990). Yellow-nosed monkey's (Cercopithecus ascanius whitesidei) spatial memory in a simulated foraging environment. *Journal of Comparative Psychology*, 104(4): 382–7.

MacDonald, S.E., Pang, J. and Gibeault, S. (1994). Marmoset (*Callithrix jacchus jacchus*) spatial memory in a foraging task: win-stay versus win-shift strategies. *Journal of Comparative Psychology*, 108(4): 328–44.

MacPhail, E. M. and Bolhuis, J. J. (2001). The evolution of intelligence: adaptive specializations vs. general processes. *Biological Review*, 76: 341–64.

Martin, R. D. (1990). *Primate origins and evolution: a phylogenetic reconstruction*. Princeton, NJ: Princeton University Press.

McGonigle, B. O. (1984). Intelligence from the stand-point of comparative psychology. *Paper presented at the Thyssen Philosophy Group*, Matlock, September.

McGonigle, B. and Chalmers, M. (1998). Rationality as optimised cognitive self-regulation. In M. Oaksford and N. Chater (eds), *Rational Models of Cognition*. Oxford: Oxford University Press.

McGonigle, B., De Lillo, C. and Dickinson, A. (1992). A comparative and developmental analysis of serially motivated organisation in young children and *Cebus apella*. Paper presented at V European Conference on Developmental Psychology. Seville, 6–9 September.

Menzel, E. W. (1973). Chimpanzee spatial memory organization. *Science*, 182: 943–5.

Milner, B. (1971). Interhemispheric differences in the localisation of psychological processes in man. *Cortex*, 27: 272–7.

Milton, K. (1981). Distribution of patterns of tropical plant food as an evolutionary stimulus to primate mental development. *American Anthropologist*, 83: 534–48.

Milton, K. (1993). Diet and primate evolution. *Scientific American*, 269(2): 70–7.

Olton, D. S. and Samuelson, R. J. (1976). Remembrance of places passed: spatial memory in rats. *Journal of Experimental Psychology: Animal Behavior Processes*, 2: 97–116.

Pomencrantz, J. R. and Kubovy, M. (1986). Theoretical approaches to perceptual organization: simplicity and likelihood principles. In K. R. Boff, L. Kaufman, and J. P. Thomas (eds), *Handbook of perception and human performance. Volume II: Cognitive processes and performance*. New York: Wiley, pp. 1–45.

Pothos, E. M., and Chater, N. (2002). A simplicity principle in unsupervised human categorization. *Cognitive Science*, 26: 303–43.

Schenk, F., Grobety, M. C., Lavenex, P. and Lipp, H. P. (1995). Dissociation between basic components of spatial memory in rats. In E. Alleva, A. Fasolo, H. P. Lipp, L. Nadel and L. Ricceri (eds), *Behavioural brain research in naturalistic and semi-naturalistic settings: possibilities and perspectives*. NATO ASI Series, Series D: Behavioural and Social Sciences, pp. 277–300.

Spetch, M.L. and Edwards, C.A. (1986). Spatial memory in pigeons in an open-field feeding environment. *Journal of Comparative Psychology*, 100: 266–278.

Spinozzi, G., De Lillo, C. and Castelli, S. (2004). Detection of 'grouped' and 'ungrouped' parts in visual patterns by tufted capuchin monkeys (*Cebus apella*). *Journal of Comparative Psychology*, 118(3): 297–308.

Spinozzi, G., De Lillo, C. and Truppa, V. (2003). Global and local processing of hierarchical visual stimuli in tufted capuchin monkeys (*Cebus apella*). *Journal of Comparative Psychology*, 117, 15–23.

Sutherland, S. (1989). Simplicity is not enough. In A. G. Elsedoorn and H. Bouma (eds), *Working models of human perception*. London: Academic Press, pp. 382–40.

Valsecchi, P., Bartolomucci, A., Aversano, M. and Visalberghi, E. (2000). Learning to cope with two different food distributions: the performance of house mice (*Mus musculus*). *Journal of Comparative Psychology*, 114: 272–80.

Visalberghi, E. and De Lillo, C. (1995). Understanding primate behaviour: a cooperative effort of field and laboratory research. In E. Alleva, A. Fasolo, H. P. Lipp, L. Nadel and L. Ricceri (eds), *Behavioural brain research in naturalistic and semi-naturalistic settings: possibilities and perspectives*. NATO ASI Series, Series D: Behavioural and Social Sciences. Dordrecht: Kluwer Press, pp. 413–24.

4
The Emergence of Linear Sequencing in Children: A Continuity Account and a Formal Model

Maggie McGonigle-Chalmers and Iain Kusel

4.1 Introduction

Complex biological systems display behaviour that is situated in a space of variation between brittle finite state automata (e.g. a spider building a web) and executive skills based on the fine-tuning of learned voluntary behaviours, such as piano playing. In the case of the latter, changes occur in evolution and development that relate not only to the acquisition of well-behaved serial ordering behaviours, but also to concomitant changes in the way that the perceptual world is represented as knowledge. A chess master, for example, perceives and represents the position of pieces in terms of long-term moves and gambits, with consequent superiority for spatial memory of the pieces as compared with a less experienced player (de Groot and Gobet, 1996).

Even the apparently untutored skills of searching and sorting are indicative of increasing levels of organisational development in memory. Sorting into 'like' items by children has long been observed to be related to changes in the way that the object relations of similarity and difference defining classes and subclasses are understood (Inhelder and Piaget, 1964; Vygotsky, 1962). Similarly, the ability to represent linear order as captured by Piaget's size seriation task, where a disordered set of differently sized rods have to be serially ordered into an ascending or descending series, also undergoes prolonged changes during human development (Kingma, 1983; Piaget, Inhelder and Szeminska, 1960). Stages in its development reveal a gradual change from incomplete and/or incorrect series, followed by a correct series achieved by trial and error, and, finally, expert seriation at around the age of seven, at which point rods are selected systematically from one end-point of the size range and placed in sequence without error. An essential and adaptive feature of human memory organisation, without which much of logic and most of mathematics would have no foundation, the ability to order monotonically along a direction of dimensional change, is both an achievement of executive control as well as an indication of how the child comes to perceive and represent the size relations among the items. It is, in short, a prime example of how serial ordering expertise at the level of overt behaviour can be an important mirror on the emergence of the complex mind.

Although attempts have been made to explain how complex systems grow from weak to strong in their inferential capacity as development proceeds (Fischer, 1984; Halford, 1993; Piaget, 1970; Zelazo et al., 2003), it is still an open question as to how to disentangle the products of learning within a situated activity and the informing culture within which that learning occurs (McGonigle and Chalmers, 2008). Research with highly evolved non-humans can point us, however, towards those competences in humans that owe at least as much to the natural knowledge-gaining architecture of the primate brain as they do to specific aspects of our human culture. Seriation is one such competence. In our lab, we have demonstrated the capability of non-human primates to order up to 12 sizes monotonically without error. The monkeys (*Cebus apella*) were trained over a succession of increasingly complex problems using three different classes of size stimuli presented in random alternation on a touch screen, and showed a 'learning to learn' effect in doing so that would certainly contraindicate any explanation based on learning by simple associative 'chaining' (McGonigle, Chalmers and Dickinson, 2003). The emergence of seriation was apparent in monkeys as an economic solution to a problem of information management rather than as an uphill struggle against a combinatorially explosive set of possible orders. For, while set size increases were causing an exponential rise in the number of possible orders, learning, by contrast, started to plateau.

The emergence of this solution was prompted in the monkeys by a long period of exposure to tasks whose structure and increased complexity had of necessity to be imposed according to the requirements of experimental design, informed at each stage by the monkeys' success thus far. In human development, seriation competence emerges with no such obvious supervision and we are left with analogous behaviours but no access in the child to the stages of self-discovery or the task factors that might promote such emergence. To help understand the growth and dynamics of linear ordering behaviour within human development as well as within a comparative context, we presented school-age children with an explicitly trained version of the size seriation task similar to that used with monkeys. By using a simple version of the task, however, we could relax the level of supervised training. The children were allowed free rein as to how to start the task, and the stages of learning were not parsed into sub-stages as they were with the monkeys. This exercise provided insights into the mechanisms of change during development that have eluded previous accounts, and – through its transparency of process – was amenable to formal modelling. This has resulted in a behavioural microanalysis as well as a formal model of how seriation competence might emerge in children in terms of the same economy-preserving features that we argued for in monkeys (McGonigle and Chalmers, 1996, 2001). We describe this model in this chapter within the context of giving both empirical and formal warrant for the view that the adoption of a linear monotonic search through a space of multiple size relations is an emergent and deceptively 'simple' property of the complex mind.

4.2 The task

Classically, seriation was measured by requiring children to make an ordered series by selecting the rods and arranging them into 'a staircase' either by copying a model or through verbal instruction (Piaget and Szeminska, 1941). Piaget's task was deliberately beyond the scope of serendipitous success; the numbers of elements to be ordered were eight at least, and the differences between them were barely discriminable.

As the developmental profile obtained using this method is largely defined in terms of pass/fail criteria, the terms under which younger children might succeed were not revealed. The select-and-place procedure of the classic task, furthermore, obscures the micro-behaviours of comparison and contrast among the items; when the child takes an item from a test pool, the set to be evaluated is constantly altered and diminished, confounding errors of selection with difficulties in repair (Neapolitan, 1991).

The classic paradigm thus fails to provide any way of systematically exploring the cycle of successful behaviour and representational change that might offer a clue as to how and why expert seriation might emerge during development. In an attempt to capture the immediate precursors to the end-state behaviour, we used tasks that have a fair chance of being

Figure 4.1 The touchscreen seriation task and learning results from five- and seven-year-olds on five and seven items, showing the relatively lengthy training required by five-year-olds in contrast to virtually error-free performance by older children

solved within the context of a short learning episode. Five- and seven-year-old children were presented with a computer touch screen based version of the task, using (at first) only five easily discriminable items (squares). The stimuli were placed in a random linear arrangement on a touch screen; hereafter [12345 where 1 = largest] (see Figure 4.1). The subject must touch them in the correct sequence learning to always proceed from either biggest to smallest or smallest to biggest, as reinforced on a touch-by-touch basis by auditory feedback from the computer. Training continued until the child could execute an errorless sequence as evinced by a learning criterion of 4/5 completely correct trials across two sets of different absolute sizes (i.e. across at least ten trials in total).

4.3 The task ecology

The presumption is that the performance obtained is a reflection of the acquisition of a particular type of serial control, that is, the management of size relations using an iterative principle and a constant direction of change. To ensure that this was the case we presented seriation within a cohort of other tasks in order to:

a. distinguish size-sequencing from the acquisition of an arbitrary serial list (see Treichler in this section) based on colour rather than size differences;
b. distinguish the learning of a monotonic sequence from the acquisition of an 'uneconomic' non-monotonic size sequence;
c. distinguish the ability to sequence sizes from the ability to identify each and every ordinal size within the set (e.g. second biggest, middle-sized, etc.), an ability seen by Piaget to be a corollary of seriation skills only by the age of seven;
d. discover the limits on any learning evinced by measuring transfer to monotonic sequences of greater length.

Using children across the age range at which seriation becomes spontaneous and principled, we sought age-related changes within each task as well as changing task–task relationships as seriation improves. The resulting analyses across these conditions are described elsewhere (McGonigle and Chalmers, 1996, 2001, 2006), but collectively they indicated:

Age-related changes in monotonic size sequencing and as a function of set size. All seven-year-olds started their criterion run of correct trials within a maximum of one trial (where they had simply started with the wrong end-point); five-year-olds were extremely variable – some succeeding quickly, but others taking tens of trials.

This shows that there is a development stage in which the relational sequence, biggest to smallest or vice versa, has to be actively and laboriously discovered through corrective feedback. This does not seem to be a matter of simply being slow to 'latch on' to the relevant rule. Even when trained on five items and then, with the addition of two sizes on the end, transferred to seven items, seven out of a group of twelve five-year-olds required a mean of an extra 38 trials over and above their original learning.

Monotonicity as a 'privileged' form of serial search. The expertise of the seven-year-olds on size seriation yielded an unsurprising superiority over their learning of a completely arbitrary string. While five-year-olds showed no such superiority in terms of trials and errors to criterion, other measures suggested that they were not simply performing the monotonic task *as if* it were an arbitrary list. This advantage for monotonic size orders was particularly clear from the comparison with non-monotonic tasks which were all but impossible, even for seven-year-olds, when seven items were used (in marked comparison with adults). This suggests a further stage in the development of size-sequencing and we shall return to the implications of this at the end of this chapter.

Size-sequencing as prior to ordinal-size comprehension. This was assessed using both relational matching to sample tasks as well as colour-conditional tasks (e.g. if all the items are red, always choose middle-sized; if green, take smallest, etc.). Monotonic sequencing was always in advance of ordinal-size matching for five-year-olds (McGonigle and Chalmers, 2002); indeed most five-year-olds failed outright after 200 trials. Seven-year-olds, by contrast, were virtually error-free and as adept on ordinal-size matching as they were on the monotonic size-sequencing, even for seven item sets.

4.4. Overall characterisation

In the five-year-old subject, we appear to have captured a non-arbitrary form of serial control where the size information is utilised in the learning but where the learning has not yet been fossilised into a routine. In any model of this learning, we must be aware that it is *not* supported by the semantics of ordinal-size relations such as second smallest, third biggest, and so forth. otherwise our ordinal identification tasks would have been considerably easier. Similarly, it is constrained in its extendability to longer sets. It therefore seems to occupy the space of learning that might inform us as to the mechanisms of change that will ultimately develop into the seven-year-old profile.

As for what produces these changes, we could only speculate in past work (McGonigle and Chalmers, 1996, 2001; McGonigle and Chalmers, 2002). It is of necessity an inferential exercise of filling in the gaps in both evolution and development that detail how simple behaviours based on elementary perception become expert behaviours based on efficient information processing and high information gain. This chapter aims to convert our speculations into a formal model.

4.5 Model-building: objectives and previous attempts

If we are to establish where to draw the line between the emergent products of situated learning and skills acquired (or honed) through some ancillary route such as language or counting (McGonigle and Chalmers, 2008), it is crucial to know how far situated learning alone can take us. In this context, formal models of seriation have been devised to explain seriation developments as a consequence of seriation behaviour per se. However, as yet there is no model we know of that does not either simply redescribe stages of success on the classic task in formal terms or already presume a crucial property of the end-state representation, that is, that there already exists an ordinal 'place' for every item in the set.

An early Production Systems model (Young, 1971) followed Piaget's basic findings through three main stages of seriation performance (incomplete to trial and error and then principled seriation), accounting for performance at each stage by a set of condition-action production rules, such as:

Condition: Goal = add first block and holding block in hand
Action: put block at far left and pop goal stack.

Young's model simulates stage progression by adding rules that reflect the stage changes that Piaget describes, but in handcrafting them to fit stage behaviour, it does not address how such behaviour emerges as a product of the agent's own activity and its internal state. Moreover, Young's model has no learning dynamics. To represent development, it assumes rule stacks which can be changed at any point. In short, the problem is that although Young's model is useful in classifying behaviour at time t, it cannot help us understand the source of change in behaviour from time $t - n$ to t.

Mareschal and Shultz (1999) developed a model to replicate stage changes by training a modular neural network architecture with examples of which item to place where within a set. The network responded with a signal for 'which' (identified by the ordinal position in the set [123456]) and a signal for 'where' to place an item (similarly identified) corresponding to operational seriation behaviour, that is, placing each item in the correct position. Using cascade correlation as an error minimisation technique, the network gradually responds correctly to a sufficient number of test examples. This model was successful in showing how a computational model can reproduce seriation behaviour via small incremental changes, in this case weight changes within a network.

However, the Mareschal and Schulz connectionist architecture includes strong assumptions that the developmental profile does not warrant. Among these is the presumption that the seriator already has access to a determinate set structure with n items where n = the place for each item in the set. For example, at each training episode the fitness of the 'which' array and the 'where' array correspond to a real number calculated from a known order, that is, 1 > 2 > 3 > 4 > 5 > 6. The system's ability to benefit from the ordinal position in the set is thus assumed, yet it is the emergence of seriation in the absence of such prior knowledge that we should be trying to explain in our models.

Also, Mareschal and Schulz include a further presumption of the effects of learning from a model, stating that 'if the items are largely ordered, the child is more likely to view the items as a completed series' (Mareschal and Schultz, 1999: p. 679). 'Good figure' in the form of a completed model to copy has not in fact been found to be prescriptive of ordering ability in the child (McGonigle and Chalmers, 2001); the child must be able to integrate a series before spatial structure can benefit their ordering. Before they reach this stage, children are robustly indifferent to their 'bad' spatial productions (Neapolitan, 1991).

It is crucial therefore to make as few assumptions as possible to bring to a formalism. Accordingly, we assert that seriation is the emergent property of a perceptual data stream gathered in real time, using minimalist codes at each point: the binary relations of 'bigger than' and 'smaller than'. These assumptions, however, already bring certain constraints and biases based on what we know empirically about binary relational codification and how they operate in multiple stimulus situations.

4.6 Initial constraints on proposed model

Binary relations. The visual system of the primate seeks connectivity between objects in the world in terms of rules of relation (see Jones, this section). The simplest application of this to a set of multiple items would suggest an initially crude separation of [12345] into [12] and [4 5], that is, 'big' and 'small' sets respectively, with [3] having indeterminate status. Sorting into big and small categories is indeed characteristic of the earliest stages of seriation development (Piaget et al., 1960).

Embodied 'pointing'. The visual system of the primate also transforms a simultaneous presentation of a task scene into a sequential one, the eye acting as a 'pointer' to environmental objects, during successive saccading sweeping of the visual field. Thus our minimalist assumption is that the system does not start with a representation of the full set of objects, just enough to retrieve size information on demand (Ballard et al., 1997).

Inhibition. Once a stimulus has been selected from a task scene, it is inhibited from subsequent selection by the primate motor system (Jeannerod, 1999). This is shown by a lack of backwards errors (e.g. 121 in a 12345 set) in both human and non-human subjects during sequencing tasks (Terrace and McGonigle, 1994).

Perceptual bias enabling end-point identification. Further differentiation of big and small items in order to discover the 'start' item is enabled by the special properties attaching to end-points. There are several psychological accounts of how this can work in the case of 'biggest', including a 'skyline' effect, in which the one item is detected as rising higher above the ground plane than anything else (Clark, 1970), and also the proximity of the contours of the largest item to an external frame of reference such as a computer screen (Bryant, 1974). Similar arguments based on greatest distance from the frame surround can apply to the smallest, though as Clark (1970) and others have suggested 'small' may have a derivative 'not big' status based on scanning to a 'vanishing point' (McGonigle and Chalmers, 2002), producing a further bias towards 'biggest' over 'smallest'. In children's seriation, as well as in size-rule learning by children and monkeys (McGonigle and Chalmers, 2002), there is certainly faster learning of end-points, and a bias towards starting with the biggest item has been observed in seriation (Leiser and Gillierion, 1990).

4.7 From binary codification to ranking: a micro-behavioural analysis

The first rule that the child has to discover when learning to seriate is that a binary rule subdividing the set into big ones and small ones is insufficient to identify the start item in the set, and that these binary categories must therefore be further subdivided, first of all to identify the start item. An asymmetry might apply in the relative salience of the biggest versus the smallest start item as described above, but, this asymmetry notwithstanding, isolating the end-points within the big and small categories would be deemed to be the fastest element in acquisition.

Thereafter, however, the child has to separately identify the next in the sequence from an (as yet) undifferentiated set; the internal elements do not yet have a separate identity until they are discovered through the application of an iterative rule. The next task for the child is to establish the minimal difference between the end-point and the 'next one'. This suggests that confusion errors during learning will peak around the item adjacent to the one being learned but only in the direction already specified by a big/not big asymmetric relation, that is, as forwards rather than backwards errors. As learning progresses further, the set of confusable items diminishes and learning should accelerate towards the other end-point. A similar profile would occur when searching from smallest.

Figure 4.2 shows that the pattern of acquisition of a five-item series by five-year-olds (with seven-year-olds in comparison) is indeed accelerated as measured by the point at which each rule was at criterion level.

As for error distributions during the course of learning, Figure 4.3a shows the confusion errors during learning as illustrated with the five-item set. Both directions of learning in Figure 4.3a show the errors proceeding in a

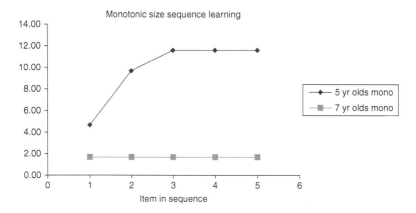

Figure 4.2 The point at which each item within a five-item monotonic size sequence reached criterion for five- versus seven-year-olds

forward direction (first to last item) across the set. The larger spread of errors for 'smallest' in the smallest to biggest direction suggests an asymmetry of natural choice favouring starting from the 'big' end consistent with the prediction derived from Clark (1970).

4.8 Comparisons with arbitrary sequence learning

When the error profiles are computed for the colour sequence (see Figure 4.3b), two differences emerge. The first is that the error is not now expressed as a distribution that peaks around the 'next' item in the set; the second is that the absolute accumulation of error is greater in the arbitrary sequence as if there is an internal rule-based constraint in the case of size over and above the effects of direct reinforcement (from response-produced feedback).

The slow but accelerated forward drive towards learning the monotonic rule appears to be characteristic of an early stage of development, but we have no way of assessing how such learning then moves as *a consequence of its own success* towards rapid solution and a change in the representation of the size relations as apparent in the seven-year-old. The modelling described below aims to specify how such change might occur by first of all simulating the learning of the five-year-old. We restrict our formalism to the learning profile of the five-item monotonic size sequence learning and the five-item colour sequence by way of comparison.

4.9 Modelling monotonic seriation as a game

It is possible to formalise how the child moves from having a broad, binary classification of set elements of big and small, through to having a ranked

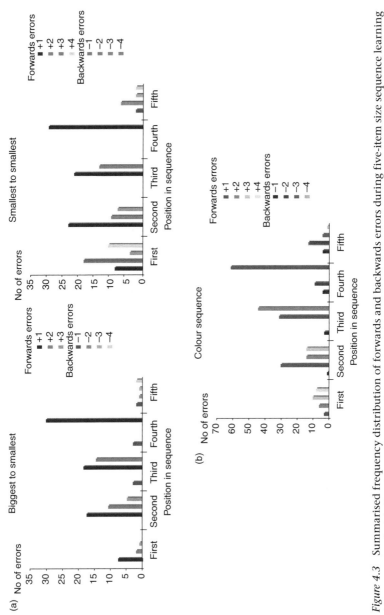

Figure 4.3 Summarised frequency distribution of forwards and backwards errors during five-item size sequence learning (a) as compared with five-item arbitrary colour sequence learning (b) which shows almost double the frequency of errors and, unlike the size tasks, no ranking of forward errors in terms of their proximity to the target

representation of these set elements. We do not make the assumption of an order relation being provided within the formalism, but only assume the repeated application of one binary rule which can detect an asymmetric size relation over time.

First, we represent the acquisition of monotonic seriation as a type of game with which an artificial agent engages throughout a simulation. We can define sets to which the task elements belong as this game proceeds:

- BIG
- SMALL
- SELECTED
- CANDIDATE
- FRAME

We can distinguish two distinct phases that the five-year-old goes through when selecting from a five-item monotonic set, which we can split into two types of game. Initially, the first game starts and all stimuli enter the set CANDIDATE. Then there is a search for the end-point of the sequence, supported by making a broad perceptual discrimination by size into BIG and SMALL categories and then refining the selection further towards the 'biggest' or 'smallest'. The decision to focus the end-point search within BIG or within SMALL is determined by a coin flip.

To refine the broad binary classification first of all into *biggest* and not biggest (or *smallest* and not smallest), we assume that all stimuli are compared against a referent external to the stimuli which we term the FRAME. This search is confined only to the selection of one distinct item from the set, and so we call this **Game A**, which ends when the participant selects the appropriate end-point to start the sequence. The selected stimulus has then entered SELECTED, and been removed from CANDIDATE. The element in SELECTED is inhibited from competing in **Game B**, which then starts. We shall describe the subsequent processes with regard to the biggest-to-smallest task.

Any of the remaining set [2345] (where 1 represents the largest stimulus) are now contenders for entry into SELECTED, and a series of pair-wise binary comparisons are made in a search for the 'next' in the direction specified by the end-point, but this time items are compared against the last element added to SELECTED. Thus when 1 is selected following Game A, it prompts a search for a 'smaller than' stimulus; when 2 is (ultimately) selected, it becomes the perceptual referent for the next search for a 'smaller than' stimulus, and so forth. This takes the participant beyond a single selection situation based on an external frame of reference and into a series of internal binary comparisons.

4.10 Game and learning dynamics

The essence of the model is representing seriation's emergence as the gradual, self-regulated inference of a **global rank** across the task stimuli based on multiple **localised binary** comparisons. As core model parameters we can distinguish 'in the world' variables, which represent the stimuli of unchanging size and colour as an invariant, Platonic reality, with 'in the head' variables we term stimulus urgencies. Urgency variables represent the 'plastic' tendency of a complex biological system to select a particular stimulus that changes over time (see Hofstadter, 2001, for a similar concept as applies to human conceptual slippage). The urgency variables all start at the same value, but are updated based on a scoring mechanism which includes size difference information derived from the stimuli *as well as the urgency variables themselves*, resulting in a cycle of causation (McGonigle and Chalmers, 2001) between action, representation and perception being incorporated into the formalism.

We represent the stimuli as a set of objects held in a task lattice which comprises the task environment, and hence the system's 'niche', as Figure 4.4a illustrates.

We represent the 'in the world' size property as Gaussian variables characterised by mean (M) values ranging from 50 (small) to 90 (big), and a variance (V) of 10, and hence is fixed for each stimulus. This 'noisy' way of representing variables in a model adheres to Marr's 'principle of least commitment', which states that one should represent information in a biological system as a probability distribution instead of an absolute value whenever possible (Marr, 1976). These variables allow us to represent the detection of a size difference (Δ) resulting from a binary comparison by calculating the difference between samples from any two of these variables (Figure 4.4). We represent the 'in the head' urgency values for each stimulus as Gaussian random variables, initialised to a prior mean (μ) of 25 and a variance (σ) of 25 / 3, which are chosen to keep mean values within three standard deviations of 25.

The urgency variables gradually acquire a rank as time proceeds based on multiple, binary comparisons between the 'in the world' variables, each

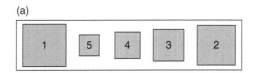

Figure 4.4a The model representation of stimuli as objects held within a task lattice displaying a permutation of monotonic set [12345]. For the monotonic case, the size varies but all other properties remain the same. For the arbitrary case, the size remains the same. The numbers are for illustration of target ordinal position

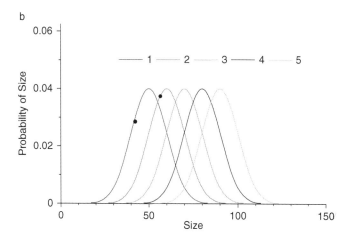

Figure 4.4b The representation of the size property of the stimuli for monotonic set [12345] as Gaussian probability distributions. This allows the representation of a size difference (Δ) between any two variables by sampling from the respective distributions and taking the difference between these values, samples illustrated by two black dots for stimulus 1 and 2 respectively

comparison computing a size difference. These urgency variables thus represent the system's acquired knowledge about the set, and an adaptation to the physical properties of the task. An inference about the global order of the stimuli is made gradually as the agent interacts with the task objects and acquires 'in the world' information about how they relate to one another in terms of a size difference. This information is combined with the urgency value of each stimulus to give a value, SCORE:

SCORE = $(\mu / \sigma) / \Delta$

The value SCORE thus incorporates the urgency variables in addition to the size difference computed between the task objects, and so combines the influence of ongoing learning with the object–object relationships computed 'online'. The FRAME variable is modelled as a dummy stimulus–urgency variable pair. SCORE is larger for a smaller urgency variance and smaller referent-stimulus size difference and so the less uncertainty combined with the smallest objective difference results in a higher score (see Table 1 for an example).

4.11 Modelling the action selection

The exhaustive scoring of a set of candidates generates a list of scores, one for each of the candidate stimuli as they are being scanned by the visual system. This list is scaled to the range 0 – 1 to allow a ranked and weighted representation of the available scores. The stimulus being selected can then be simulated by generating a random number (r) in the range 0 – 1, and using it to select from this list of scores. For example, on comparing referent–stimulus relations in Game A, we get data as illustrated in Table 1a.

The weighting of each SCORE value ensures that the stimulus with the highest score will be selected much of the time, but those with lower scores will be selected some of the time also. The scores for set [123] are split across the unit interval with weights 0.38, 0.35 and 0.28 respectively. If r <= 0.37, we choose stimulus 1; if r > 0.37 and <= (0.37 + 0.35), we choose stimulus 2; if r > (0.37 + 0.35) and r <= 1 we choose stimulus 3. This is known as 'softmax' action selection and it is a way to maintain an exploration and exploitation balance within a set of possible actions (Sutton and Barto, 1998). The ranked weightings do not have to be represented by the agent, but they will affect the probability with which an item is selected at every binary comparison during the sequence. When a stimulus is selected, we term this action a 'bet'.

After each bet, feedback is given to the system in terms of an environmental signal to say if the stimulus selected is of the correct ordinal position. If the feedback is positive, the urgency value of that stimulus is increased and it then becomes the visual reference point against which the remaining items are compared. A new list is calculated with respect to the remaining items, and another bet is made. If the feedback is negative, the system continues making bets until the correct stimulus is selected. We see the bet outcome data as analogous to the stream of perceptual data that informs the complex biological system executing a seriation, an example of which can be seen in Table 1b.

A serialised betting procedure changes the urgency values with regard to the correct bets that are made during a simulation. Utilising the principle of multiple serial bets within a trial based on making comparisons and actions on objects, we can also represent the complex biological system self-regulating its tendency over successive trials to select one stimulus over another. This is described in the next section.

Table 4.1a. Example sequence of binary comparisons for Game A, in which stimuli 1, 2 and 3 are compared against the 'frame' reference point. The size difference (Δ) is combined with the mean (m) and variance (s) of the urgency variables for each of 1, 2 and 3 to generate the SCORE. In the model, the random number generator is compared against the weighted scores to determine the selection.

Stimulus	Referent	Δ	μ	σ	SCORE	Weight
1	Frame	10.8	25.98	5.22	0.46	0.37
2	Frame	18.2	23.74	7.25	0.39	0.35
3	Frame	31.8	24.15	5.69	0.13	0.28

Table 4.1b. Each selection is represented as a bet outcome. An example is given below of a stream of bet outcome data in terms of stimulus and referent pairs within one monotonic trial. When a correct selection is made the referent changes and the games progress until the trial completes on the selection of stimulus 5.

Stimulus	Game A						Game B				
Referent	Frame	Frame	Frame	Frame	1	2	2	2	3	3	4
Stimulus	2	2	2	1	2	4	4	3	5	4	5
Outcome	Error	Error	Error	Correct	Correct	Error	Error	Correct	Error	Correct	Correct

4.12 Bayesian inference of order

To represent the gradual differentiation of urgency values over time, that is, over trial repetition, we take the referent–stimulus pairs generated on positive feedback as observations that inform a Bayesian inference procedure. Bayesian inference allows the prediction of probable causes, which are unknown and hidden from view, from observations, which are data that we can see (see Luger, Chapter 6 of this volume). By creating a joint probability distribution consisting of urgency variables (unknown, and set to an equal prior) and observation variables (known, based on bet data) on which Bayesian inferences can be executed, we can thus predict an urgency 'league table' from a sequence of bet outcomes. As we know the bet outcomes, for example, '1 beats 2', we make them part of the urgency joint probability distribution and make the Bayesian assumption that they tell us something useful about the urgency values. The result is that, over time, the urgency variable distributions evolve into a rank, computed using the standard Bayesian equation for generating posterior values (left-hand side of the equation) from known values (right-hand side of the equation). In the equation, u stands for urgency and b for bet outcome:

$$P(u \mid b) = p(b \mid u) \cdot p(u) / p(b)$$

To model the above, we infer the values of these urgency variables by representing the bet outcome data as variables and linking them together in a factor graph. A factor graph is a mathematical structure that allows complex, joint probability distributions to be represented, linked and simplified via functions termed factors (see Figure 4.5a). Messages are generated from these factors that update the urgency and bet outcome variables via the links, which is possible as the messages themselves are Gaussian variables (see Luger, Chapter 6, this volume, on message passing and loopy belief propagation).

In the model, the urgency variables are linked to the bet outcome variables via a noise factor such that a change in the bet outcome variables result in a similar change in the urgency variables. This noise factor is needed to separate initially unknown urgency variable values of stimuli from perceptual

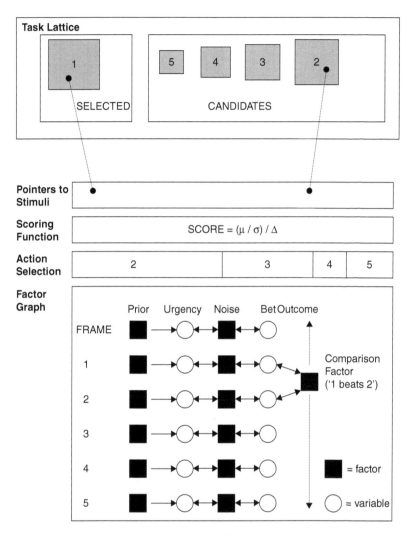

Figure 4.5 Model architecture and information flow (perception at top, inference at base) within a single binary comparison between stimuli 1 and 2. We see that stimulus 1 is the referent and so Game B is in progress. Stimulus 2 is currently selected from CANDIDATE, and as an exhaustive search is in progress, we assume stimuli 3, 4 and 5 have been scored. SCORE is computed for each and the most probable stimulus is selected, in this case 2. The bet outcome information, in this case '2 is next best to 1', is used to adjust the bet outcome and urgency variable distributions within the factor graph

judgements made about these stimuli that are known. The source of the change in the urgency variables thus derives from the bet outcome variables, and ultimately from a comparison factor which truncates and readjusts the shape of the bet outcome variables, resulting in a positive shift across trials

The Emergence of Linear Sequencing in Children 71

in the mean value for the bet winner, which are 1, 2, 3, 4 and 5 respectively within each trial (for biggest to smallest). The linked urgency variables are then updated via a message which contains the adjusted Gaussian distribution (see Figure 4.5).

Linking the bet outcome and urgency variables in this way thus allows us to capture noisy scenarios in which the most urgent stimulus is not always

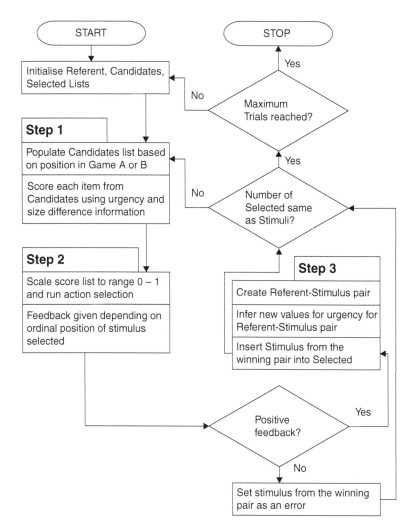

Figure 4.6 Flow of information (Steps 1 to 3) within a simulation trial through the components illustrated in 4.5. We can see the cycle in which urgency variables computed in one trial affect those in the next trial in a 'cycle of causation' (McGonigle and Chalmers, 2001)

chosen when it should be, and thus the model will generate errors as well as correct selections. The bets are initially won with equal probability, due to the urgency variables having the same initial values. As the urgency values change, however, they influence the SCORE values positively, and so the more urgent stimuli are selected more often. However, this selection is noisy as there is still the possibility of stimuli with lower urgency values being selected in any one bet.

In addition to representing the noisy choice we see in seriation, the acceleration of learning over time is represented. The forward propulsion in learning to seriate comes from the initial biases and the tendency to inhibit a choice, and is accelerated by these changing urgency values. The SCORE values are thus affected by the start-up weights and biases, but also become altered every time a stimulus wins a bet. From multiple serial bets, a global ranking, or stimulus 'league table', can be inferred by the artificial agent, the bet outcomes effectively being converted over time into a rank order.

The model components that underlie the simulation dynamics are illustrated in Figure 4.6 and the information flow through each of these components is shown as a simulation cycle in Figure 4.7. The information flow can be seen as taking place in three steps, which are labelled on these figures.

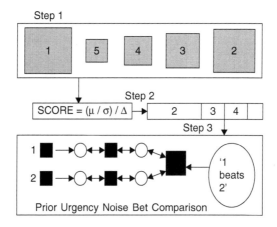

Figure 4.7 Model architecture and information flow (Steps 1 to 3) within a single binary comparison between stimuli 1 and 2. Stimulus 1 is the referent and Game B is in progress. SCORE is computed for each of CANDIDATES and the most probable stimulus is selected, in this case 2. The bet outcome information, in this case '1 beats 2', is used to adjust the bet outcome and urgency variable distributions within the factor graph. Open circles represent the variables, and filled squares the factors

4.13 Simulation conditions

A simulation has 50 trials. Simulations were run for monotonic stimulus sets, wherein the size is different, and so permutations of [12345], and arbitrary stimulus sets, where there is no size difference, and so permutations of [ABCDE]. Each trial begins with a permutation of the stimuli being generated, for example [24351] for the monotonic condition, and [BDECA] for the arbitrary condition. Twenty-five simulations were run for each of the two conditions.

The evaluation function was the same for both conditions, however the SCORE values will not be influenced by a size difference in the arbitrary condition with respect to the [ABCDE] set, as all stimuli are the same size. Thus, any of set [ABCDE] is an equal contender for entry into SELECTED from the outset, as there is no basis for selecting an end-point and a subsequent 'direction of difference'. Both conditions allow a small probability (0.1) for selection from the selected items to represent the almost complete inhibition of these items from further consideration. Bet outcome winners cause their corresponding urgency variables to increase in mean value over time. As variance decreases with evidence accrual in Bayesian inference, we posit a threshold of certainty of variance as a signal for the currently targeted stimulus to be reliably selected. This threshold maps onto the child's criterion condition of correct selection on 4 / 5 of the trials.

4.14 Simulation results

Monotonic. We can see by inspection that the errors for the monotonic case are principled, having a 'wave' effect centring on the next target stimulus from the current one, and decrease in number as the set is executed, analogous to the child data described above. This growing individuation of set items in the monotonic case is possible as the requisite size relations are available to gradually infer a rank (Figure 4.8a). This 'wave' effect is a robust one and consistent over many simulations, with a low variance for each ordinal position.

Arbitrary. The arbitrary case does not allow the system to benefit from the relational structure inherent in the task and so relies on 'brute force' updating based on environmental feedback alone. Again, by inspection we can see that the error patterns are less principled (i.e. a larger variance) and considerably more numerous (Figure 4.8b). In the arbitrary case, the benefit of relational computation and hence a ranking of the set according to size based on material relations in addition to urgency values is not possible. Specifically, the errors for ordinal position 1 have no 'wave' effect as we see in the monotonic case. Akin to the monotonic case, errors for positions 2, 3

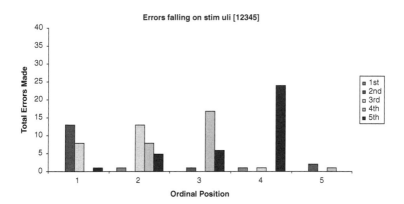

Figure 4.8a The distribution of incorrect selections across set [12345] for the simulations

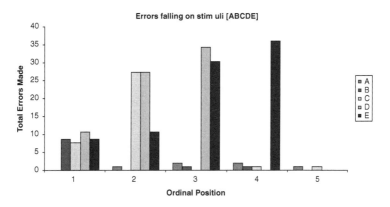

Figure 4.8b The distribution of incorrect selections across set [ABCDE] for the simulations

and 4 do show a ranking. However, the variance is much more pronounced than in the monotonic case, and on many individual trials we do not see this ranking.

4.15 Overview of urgency variable dynamics

We can visualise the urgency variable distribution which emerges in both monotonic and arbitrary conditions as trials proceed (Figure 4.9).

The set [12345] elements start with urgency distributions of the same mean and high variance, with a slight bias for bigger stimuli. Confidence

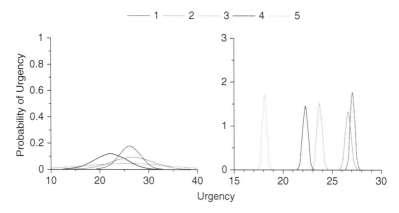

Figure 4.9 Model representation of the urgency (u) variables for monotonic set [12345] as Gaussian probability distributions for a typical simulation. We see the distributions within a monotonic trial at the start (left; trial 1) and at the end (right; trial 50) of a single simulation

in the correct order of the target stimuli increases due to the accumulation of evidence as to 1 scoring best overall, 2 scoring next best to 1, 3 scoring next best to 4 and 4 scoring next best to 5. We can see from inspection that the urgency values start to peak and shift, thus increasing mean according to the target ordinal position and decreasing variance across the board. This means that the system is not only confident that stimulus 1 is the first one to select, but crucially that 2, 3, 4 and 5 are ranked runners-up. We can thus understand better the gradual shift of the set items from having no differential status to each item having a specific rank. The arbitrary case shows the same differentiation of urgency variables, but without the initial bias to the 'bigger' items. This is justified as the behavioural result is the same in both conditions, that is, the discovery of a serial route through the task space. However, the route the system takes to get there is more error-prone in the arbitrary case.

4.16 General discussion

Success on the seriation task is hard-won by non-human primates and young children, despite the illusion of simplicity in acquiring the knowledge of how to arrange elements in a particular order. We know that in the young child, ordering must proceed without any advance knowledge of ordinal position within a set. Somehow the primitives for carving up the world into, for example, big and small must evolve into a pattern of selection that seems able to place any number of elements into its precise, unique slot. At some point prior comes the ability to acquire a modest version of this pattern, as evinced by our five-year-old participants.

More constrained, and thus more principled, than arbitrary list learning, the acquisition of five-item size seriation by five-year-olds opens a window on how seriation may evolve from the act of attempted seriation itself. Informed by an empirical microanalysis of such learning and also by known psychological constraints applying to the perception of size relations, we have seen that it is possible to replicate the monotonic seriation error profile of the five-year-old child. In doing so, we have given formal warrant to the idea that binary rules of relation generate a direction of search and a constraint on selection further down the set which is more economic than arbitrary list learning and which has the capability of becoming applicable to any number of elements.

Central to our approach has been the incorporation of the situated and embodied dimension of a complex biological system. The very act of manually selecting an interrogated stimulus biases the system to a forwards direction of processing by inhibiting parts of the representation that have been selected, and simultaneously reduces the number of items to be selected. Embodied, situated learning transforms into expertise, such that the internal representation of solutions to complex ordering tasks, as in games like chess (de Groot and Gobet, 1996), undergo transformations that literally alter how a problem is perceived. We can speculate that the knowledge structures acquired during monotonic seriation acquisition serve as a filter, allowing the best move as ranked by the system's 'life history' to be applied. Akin to a chess master seeing 'chunks' of moves in a rank, a child perhaps develops the ability to see analogous ranked routes through information space.

While Piaget strove to identify this perceptual/representational shift as the consequence of action-based learning, he never truly elaborated an embodied stance based on visual interrogation and manual selection. To achieve seriation expertise, according to Piaget, children undergo a general internal shift in the nature of their representations of object relations. This shift enables them to grasp the reciprocity of the relations leading to an ordered set, 'bearing in mind that a given element, say E, is both longer than those already in the series (E>D, C) and shorter than the ones yet to follow (E<F, G)' (Inhelder and Piaget, 1964: 257). The internal grasp of logical relations were seen as the inseparable accompaniment to the actionable consequences of that insight. Our model makes no such demands on the need for logical insight in explaining how a child learns to seriate.

One of the outcomes of our behavioural fractionation, therefore, is that we have been able to distinguish between expert behaviour and the type of multi-relational representation that Piaget envisaged. Children are clearly on a path to a simplifying linear solution to dealing with multiple size relations long before they a) apply this rule spontaneously to any set of elements, and b) grasp the ordinal and cardinal properties of all the items in the set. If we are to maintain our search for continuity, we still have some way to travel even within human development before we can place the achievements of the operational child on the same learning path as the five-year-old.

A missing link, therefore, in all accounts, including our own, is the possible importance of converting principle serial ordering behaviours into static structures with 'good form'. We as yet have no way of knowing to what extent the externalisation of principled serial control into a neat linear assembly is a crucial element in the advancement of the next stage of seriation, and one that would perhaps always elude a non-human without the necessary manipulative skills. This would certainly qualify as a case of 'augmented reality' or a simple inspectable structure (see Clark, Chapter 11) with new properties that are not available in the jumbled set of items prior to ordering. One of these properties is that it would (additionally) afford the ready application of cultural devices such as counting. It is certainly the case that spatial 'good form' greatly enables the use of the count alphabet in children up to school age (Fuson, 1988; Nunes and Bryant, 1996).

We are thus aware that our modelling of seriation has yet to simulate the appearance of 'spontaneous' seriation and its corollary of ordinal and cardinal understanding and may only do so when 'new' properties such as the acquisition of spatial ordinal codes and/or count words map onto these. And completely outside the scope of our behavioural analysis as yet is the fact that adult human subjects (but not seven-year-olds) can cope with some limited forms of non-monotonic sequence which would appear to be based on assigning number-based values to the exemplars and recoding the sequence to be ordered as, for example, 24315 (McGonigle and Chalmers, 2006).

Somewhere along the line we will have to identify how much of the developmental achievements are not actually generated by the individual but are fostered and probably greatly accelerated by enculturation. And as Clark (Chapter 11) points out, it is not just 'devices' but the actual behaviours that can be the by-products of a culture – especially true in the case of a Goldilocks and Three Bears nursery world where children are constantly invited to count, rank and assemble according to physical properties.

But what remains of central importance to the issue of evolutionary continuity, we would argue, is that any and all symbolic, linguistic, or perceptual devices proposed as runaway mechanisms generated through cultural rather than biological evolution are seen in the context of the cognitive foundations *without which they could not possibly be grounded*. The count alphabet would have no generativity whatever without presupposing both a monotonic number line and, in the case of the base ten system, a hierarchical structure affording the possibility of recursion.

It is therefore crucial to avoid ascribing the products of a human culture as in some sense the apotheosis of cognitive evolution that 'explains' why we are different from animals. For, in the case of seriation, it seems that much of the hard work, for the individual at least, is over by the stage at which the behaviour has become principled. The divergence from the trial-and-error learning (and non-continuity theorists are always gleeful to count learning trials in the large numbers that often apply to animals) that then occurs in human development may look as if it is something that is more important, more specifically 'human'. More specifically human possibly, but not more important. Unless we have a clear and detailed idea of how simple solutions arise in the individual as a hard-won situated response to the demands of a complex perceptual world, we will never have an insight into the cognitive architectures that support our language and our logic.

References

Ballard, D., Hayhoe, M., Pook, P. and Rao, R. (1997). Deictic codes for the embodiment of cognition. *Behavioral and Brain Sciences*, 20: 723–767.

Bryant, P.E. (1974) *Perception and Understanding in Young Children*, London Methuen

Clark, H. H. (1970). The primitive nature of children's relational concepts. In J. R. Hayes (ed.), *Cognition and the development of language*. New York: Wiley.

de Groot, A. D. and Gobet, F. (1996). *Perception and memory in chess. Studies in the heuristics of the professional eye*. Assen, NL: Van Gorcum.

Fischer, K. W. (1984). *Human development: From conception through adolescence*. New York: Freeman.

Fuson, K. C. (1988). *Children's counting and concepts of number*. New York: Springer.

Hofstadter, D. (2001). Analogy as the Core of Cognition. In D. Gentner, K. J. Holyoak and B. N. Kokinov (eds), *The Analogical Mind: Perspectives from Cognitive Science*. Cambridge, MA: The MIT Press/Bradford Book, pp. 499–538.

Halford, G. (1993). *Children's understanding: The development of mental models*. Hillsdale, NJ: Erlbaum.

Inhelder, B. and Piaget, J. (1964). *The early growth of logical thinking in the child*. Trans. E. A. Lunzer and D. Papert. London: Routledge and Kegan Paul.

Jeannerod, M. (1999). To act or not to act: perspectives on the representation of actions. *Quarterly Journal of Experimental Psychology*, 52A: 1–29.

Kingma, J. (1983). Seriation, correspondence and transitivity. *Journal of Educational Psychology*, 75(5): 763–71.

Leiser, D. and Gillierion, C. (1990). *Cognitive science and genetic epistemology*. New York: Plenum.

Mareschal, D. and Shultz, T. R. (1999). Children's seriation: A connectionist approach. *Connection Science*, 11, 153–188.

McGonigle, B. and Chalmers, M. (1996). The ontology of order. In L. Smith (ed.), *Critical Readings on Piaget*. London: Routledge.

McGonigle, B. and Chalmers, M. (2001). Spatial representation as cause and effect: Circular causality comes to cognition. In M. Gattis (ed.), *Spatial schemas and abstract thought*. London: MIT Press.

McGonigle, B. and Chalmers, M. (2002). The growth of cognitive structure in monkeys and men. In S. B. Fountain, M. D. Bunsey, J. H. Danks and M. K. McBeath (eds), *Animal cognition and sequential behaviour: Behavioral, biological and computational perspectives*. Boston: Kluwer Academic, pp. 269–314.

McGonigle, B. and Chalmers, M. (2006). Ordering and executive functioning as a window on the evolution and development of cognitive systems. *International Journal of Comparative Psychology, 19*: 241–67.

McGonigle, B. and Chalmers, M. (2008). Putting Descartes before the horse (again!). *Behavioral and Brain Sciences,* 31(2): 142–3.

McGonigle, B., Chalmers, M. and Dickinson, A. (2003). Concurrent disjoint and reciprocal classification by *Cebus apella* in serial ordering tasks: Evidence for hierarchical organization. *Animal Cognition, 6,* 185–97.

Neapolitan, D. M. (1991). *A micro-analysis of seriation skills.* University of Edinburgh, Edinburgh.

Nunes, T. and Bryant, P. (1996). *Children doing mathematics.* Oxford: Blackwell.

Piaget, J. (1970). *Genetic epistemology.* New York: Columbia University Press.

Piaget, J., Inhelder, B. and Szeminska, A. (1960). *The child's conception of geometry.* London: Routledge and Kegan Paul.

Piaget, J. and Szeminska, A. (1941). *La genese du nombre chez l'enfant.* Neuchatel: Delachaux and Niestle.

Sutton, R. S. and Barto, A. G. (1998). *Reinforcement learning: an introduction.* The MIT Press, Cambridge, MA.

Terrace, H. S. and McGonigle, B. O. (1994). Memory and representation of serial order by children, monkeys and pigeons. *Current Directions in Psychological Science,* 3(6): 180–5.

Vygotsky, L. S. (1962). *Thought and language.* Cambridge, MA: Harvard University Press.

Young, R. M. (1971). *Children's seriation behaviour: A production systems analysis.* Pittsburgh: Carnegie-Mellon.

Zelazo, P. D., Muller, U., Frye, D. and Marcovitch, S. (2003). The development of executive function: Cognitive complexity and control-revised. *Monographs of the Society for Research in Child Development,* 68(3): 93–119.

5
Sensitivity to Quantity: What Counts across Species?
Sarah T. Boysen and Anna M. Yocom

5.1 Introduction

The burgeoning literature on numerical skills, counting and quantity judgements by animals over the past two decades attests that the long shadow of Clever Hans has finally dissipated. Not only have the types of tasks expanded, but the number of species, particularly those outside the mammalian order, is beginning to become more diverse. The primate order has been well represented, including studies with capuchin monkeys (Judge, Evans and Vyas, 2005), cotton-top tamarins (Uller, Hauser and Carey, 2001), rhesus monkeys (Beran, 2001, 2007a, b; Brannon and Terrace, 1998; Cantlon and Brannon, 2006; Hauser, Carey and Hauser, 2000); squirrel monkeys (Thomas and Chase, 1980; Terrell and Thomas, 1990), chimpanzees (Beran and Beran, 2004; Beran, Evans and Harris, 2008; Biro and Matsuzawa, 2001; Boysen and Berntson, 1989; Matsuzawa, 1984; Tomanaga, 2008), gorillas (Hanus and Call, 2007) and orang-utans (Call, 2000). However, some of the most interesting new studies have explored numerical questions with such disparate species as salamanders (Uller et al., 2003), chickens (Abeyesinghe et al., 2005), parrots (Pepperberg, 1994; Vick and Bouvet, in press), horses (Uller and Lewis, 2009), pigeons (Emmerton, Lohmann and Neimann, 1997; Roberts, 2005), mangabey monkeys (Albaich-Serrano, Guillen-Salazar and Call, 2007), cotton-top tamarins (Kralik, 2005), dolphins (Kilian et al., 2003), dogs (West and Young, 2002), sea lions (Gentry and Roeder, 2006; Gentry, Palmier and Roeder, 2004), cleaner wrasse fish (Danisman, Bshary and Bergmüller, 2010) and mosquitofish (Agrillo, Dadda and Bisazza, 2007).

Significant inroads have also been made theoretically. One recent proposal is that animals may share an ancient system for detecting and comparing numerosities, or a 'number sense' (Brannon, 2005; Dehaene, Dehaene-Lambertz and Cohen, 1998; Hauser et al., 2003; Nieder and Miller, 2003), including recent evidence for the neurobiological underpinnings of numerosity (Nieder, 2005; Nieder and Miller, 2004a, b). For example, Nieder and Miller (2004a) demonstrated that monkeys configured for single cell recordings showed the capacity of individual neurons to encode numerical quantity. During a delayed matching-to-sample task using numerosities, the greatest proportion of neurons showing numerical selectivity were located in the lateral prefrontal cortex of rhesus monkeys. Remarkably, the neurons tested represented some 31 per cent of randomly selected cells (Nieder and Miller, 2004a). Another area where numerosity-tuned neurons were located was the fundus of the intraparietal sulcus, with some 18 per cent found to encode numerosity. Neurons in these two cortical areas were specifically tuned for selectivity to numerosity, and showed maximum responsivity to one of five numerosities shown in a visual display. Moreover, they formed a bank of overlapping numerical filters in a sequential arrangement that also encoded the ordinal features of cardinality. Essentially what was revealed was a number line of cortical neurons that fired selectivity with increasing or decreasing numerosity, expressly mapping what is typically regarded metaphorically as a mental number line. Thus recent quantitative assessments of the neuroanatomical and electrophysiological parameters of numerical selectivity in the brain is emerging in support of current theoretical hypotheses for the functional properties of counting, ordinality and other aspects of numerical capacities (e.g., see Dehaene, Dehaene-Lambert and Cohen, 1998; Brannon, 2005).

Extensive research on non-human animals' understanding of numerosity has demonstrated that a variety of animals, ranging from salamanders to chimpanzees, are also capable of responding to numerosity (Beran, 2001; Boysen and Berntson, 1995; Matsuzawa, 1984; McComb, Packer and Pusey, 1994; Smith, Piel and Candland, 2003; Uller et al., 2003). Evidence for numerically related skills has been documented in naturalistic settings, including evidence that female lions can assess the size of potential intruding groups based on the number of perceived lions they hear roaring (McComb et al., 1994), and chimpanzees have been shown to assess their potential for success during intergroup conflict based upon the number of individual vocalisations of potential individuals from a neighbouring community (Wilson, Hauser and Wrangham, 2001). Under captive conditions, typical tasks have included numerically related judgements, including a variety of discrimination tasks by apes (e.g., Beran, 2001; Boysen and Hallberg, 2000; Matsuzawa, 1984; Tomanaga, 2008), the judgement of relative weight by squirrel monkeys (McGonigle and Chalmers, 1977), and spontaneous representation of number by tamarins (Uller, Hauser and Carey, 2001), among many others. While most studies have explored animals' abilities to judge small numerosities, it has been shown more recently that discrimination abilities in monkeys remain accurate even for larger numerosities up to 30 items (Cantlon and Brannon, 2006). While not all experimental approaches to numerosity have controlled for the various potential confounding variables, such as equating surface area across stimulus choices, brightness of visual arrays, the simultaneous versus sequential presentation of items, or overall rate of stimulus presentation, more recent studies have implemented appropriate experimental controls of the range of continuous variables that also change with the number of items presented. Even under such controlled conditions, however, both rhesus macaques and tamarin monkeys maintained the same level of accuracy in their performance (e.g., Beran, Evans and Harris, 2008; Cantlon and Brannon, 2006; Hauser et al., 2003).

A closer look at the performance of non-human primates (rhesus monkeys) reveals that the animals exhibit size and distance effects in their responses (Brannon, 2005), as seen in human infants (Starkey and Cooper, 1980; Xu and Spelke, 2000; Xu, Spelke and Goddard, 2005) and adults (Barth, Kanwisher and Spelke, 2003), for example, numerosity judgements by rhesus monkeys on a delayed match-to-numerosity task which consisted of successive visual displays that contained the same small number of pseudo-randomly placed items (Nieder, Freedman and Miller, 2002). The animals were required to determine the quantity of displayed items, despite the fact that the items were highly varied in appearance, and retain this information over a brief delay. Results indicated that the monkeys relied on quantity information that was abstract, rather than other lower-level perceptual characteristics of the display such as area, geometric configuration, or density. Additional experiments looked at non-match numerosities, and revealed that the monkeys made more errors when numerosities were adjacent, and performed better when the numerical distance between the two displays was increased, known as the numerical distance effect. These findings suggest that, for rhesus monkeys, differences among small numerosities were easier to determine than comparable differences among large numerosities.

While several theories have been proposed to account for the size and distance effects (e.g. Brannon et al., 2001; Dehaene, 2001), recent neurophysiological work has provided evidence that animals possess a logarithmically scaled representation of number (Nieder and Miller, 2003). According to the logarithmic hypothesis, small numbers are more easily discriminated because they have less cognitive 'overlap', while larger numbers are compressed, resulting in more cognitive overlap, making them more difficult to discriminate (Dehaene, 1997). Behavioural studies have supported these findings and theoretical position by demonstrating a logarithmically scaled number representation for pigeons (Roberts, 2005), rhesus macaques (Nieder and Miller, 2003) and young children (Siegler and Opfer, 2003). It has been suggested that there may be an 'upper limit' to the number of stimuli that could be differentiated without a verbal numerical system (Hauser et al., 2000). However, Nieder and Miller (2003) found that the ratio between two sets, as opposed to their absolute numerical value, controlled the level of discrimination by rhesus monkeys. The difficulty of the discriminations became more challenging, with errors more common when the ratio between items decreased. That is, when the larger numeral of a set is divided by the smaller numeral, and the resulting ratio decreases, the more difficulty the monkeys had discriminating between the two numbers (Nieder and Miller, 2003). These findings suggest that there is not an upper limit on the number of items that can be discriminated, but instead representations become more approximate (e.g. less accurate) for larger numbers (Nieder and Miller, 2003). While it is possible that, while all species may share a non-verbal 'number sense,' different species may each have a different threshold discriminability ratio for number. If this is the case, one would expect to see slight changes in numerical capacities in accordance with phylogenetic differences, including possible differences among a family of species (e.g. great apes). For example, humans may have the lowest or most fine-tuned discrimination threshold ratio, followed by other apes. Similar differences may exist between the various non-human monkeys, with comparable similarities and differences across the phylogenetic continuum. Indeed, adult humans have a very low discrimination threshold of approximately 1.14 (Barth et al., 2003). Human infants show a slightly higher discriminability ratio threshold, between 1.5 and 2.0 at six months of age, although such skills improve quickly with age (approximately 1.25 to 1.5 at nine months old (Lipton and Spelke, 2003). Tamarins, a small New World (Central and South America) primate species, and rhesus monkeys, representing the Old World (Africa and Asia) both have a threshold ratio that is similar to that of a nine-month-old human infant (Beran, 2007b; Hauser et al., 2003). Alternatively, it is possible that specific ecological pressures, such as availability of food, may have caused some species to have a lower discriminability ratio than others, irrespective of phylogeny.

5.2 The quantity judgement task

An intriguing phenomenon has emerged from the experimental work on numerical comparisons with chimpanzees (*Pan troglodytes*). A quantity judgement task required the ape subjects to choose between two arrays of candy, differing in quantity, and included a reversed-contingency reward structure, as first reported by Boysen and Berntson (1995). A priori, the paradigm seemed straightforward and was originally designed to be incorporated into a social cognition task, specifically strategic deception. However, unanticipated difficulties with the acquisition of the task by the animals emerged. In the original version of the quantity judgement task, two chimpanzees were tested together, although only one animal was an active respondent. The chimpanzee pair was presented with two separate dishes containing different amounts of candy arrays, positioned in clear view, but out of reach. The active subject, the Selector, was given the opportunity to choose a dish. However, a new twist was added to what was an otherwise straightforward numerical discrimination. Instead of giving the Selector chimpanzee the contents of the dish she chose, the selected candy array was immediately given to the *other* chimpanzee (the Receiver), who simply had to wait as patiently as she could (often a great challenge for a chimpanzee) for the Selector to make a choice in each trial. The Selector was then rewarded with the contents of the *remaining* dish, the very dish that she had passed up at the start of the trial. These were the only task demands, and there had been no anticipation that the animals, particularly the older chimpanzee, Sarah, would have any difficulty whatsoever. Sarah was perhaps the most test-sophisticated chimpanzee in the world, having been taught an artificial language at an early age, with continuous teaching and testing until her mentor, David Premack, retired in 1987. The language system she used was based on a system of coloured plastic shapes that stood for words, including names of her teachers, foods, objects, concepts like same or different, as well as operator words that allowed Sarah complete complex analogies (e.g. Premack and Woodruff, 1978). Following her language training, she had participated in a range of new cognitive tasks since that time in the Boysen laboratory at Ohio State. The second chimpanzee, Sheba, had participated in the same cognitive tasks (e.g. Boysen and Berntson, 1989), and she, too, was exceptionally test-sophisticated, including the use of a computer-interfaced touch-frame system that displayed graphics, photographs, numerals and words that were used across a wide variety of cognition experiments.

During the initial training session, the ratio between the two candy arrays was small (1 versus 2), to help establish the basic rule structure of the task. Sarah was chosen to be the first Selector, and promptly failed miserably across the entire block of eight sessions by continuing to choose the larger array on nearly every trial. Increasing the ratio between the candy arrays from 1:4, and then 1:6, did not enhance her performance, which remained at chance. Because the chimps' understanding of the task demands of the quantity judgement task (with reversed contingency [RCT]) was critical to the design of the strategic deception study to follow, it was necessary to step back and reappraise the approach. The plan was to teach the animals the RCT task, and incorporate it into a social/cognitive framework by imposing a physical barrier between the two chimpanzees, such that only the Selector knew the two quantities of candy available. The hypothesis was that the Selector would more often insure that she received the larger reward, that is, use strategic deception to garner the most reward, in true Machiavellian style. However, the first subject failed to grasp the relationship between choosing the larger array and being rewarded with the smaller one. Sheba, the Receiver, had been a passive participant for three sessions, but despite the potential confound of her prior experience as the reward recipient, the animals switched roles in the task. Despite starting Sheba with the same choice of 1 versus 2 candies, it was immediately apparent that any previous participation had apparently contributed little to her understanding of the task either. She, just as Sarah had done, failed to receive the larger reward, but instead persisted at choosing the bowl containing more candy. Her performance mirrored Sarah's across the subsequent increases in the reward ratios of candy (1:4 and 1:6). Because both chimpanzees repeatedly failed to choose the smaller array first, indicating a complete lack of understanding as to how they could reap the greater number of candies, the social cognition study was scrapped. Instead, the animals' inability to acquire the RCT task, in and of itself, became the focus of study.

Additional efforts to investigate the animals' limitations with the task were undertaken, although it is important to note that subsequent tasks were *not* designed to find some reinforcement procedures that would impose conditions on the chimpanzees to respond optimally on the original task. Indeed, a number of 'replications' of the RCT task included such training regimens (e.g. Silberberg and Fujita, 1996). Rather, it was the subjects' inability to acquire facility with the task demands that was the most interesting outcome, and became the impetus for all further studies. In addition, individual chimpanzees were tested by themselves once it was determined that inter-animal competition for the rewards was not a variable, and that the same interference mechanism that disrupted the animals' ability to respond optimally under the conditions of the RCT still emerged. The persistence with which both subjects continued to choose the array that resulted in a smaller reward was remarkable to observe. Within milliseconds of responding, the subjects often responded behaviourally with great dismay, banging on the apparatus and vocalising in distress. Yet, within seconds when the next trial occurred, they persisted in choosing the larger array. The apparent automaticity of their choices was quite significant. At the time of testing, concurrent training with the Arabic numeral symbols had been ongoing for several years, and most of the chimpanzees had functional use (e.g. for counting, summation, addition and other numerical concepts) of representations from 0 to 7 or 8. Their understanding of numerals suggested an obvious choice to incorporate into the quantity judgement task, as numbers would abstract several variables inherent in the physical candy arrays, and allow for testing alternative hypotheses that might help explain the animals' poor performance.

Immediately, replacing the candy arrays with numerals resulted in a dramatic shift in performance. The animals' use of symbols instead of the actual candy arrays allowed them to override whatever interference process was inherent when actual food arrays were employed (see Figure 5.1). The shift in their behaviour was instantaneous.

The experiment using numerals was set up as an ABBA design, so some numeral sessions were followed by trials during which candy arrays were used. Performance plummeted each time the candy was used. Reinstatement of the numerical symbols resulted in significant accuracy, such that the chimpanzees would readily select the smaller array, and thus earn the contents of the second dish with the larger reward. An immediate question arose whether the animals were responding to the obvious hedonic value of

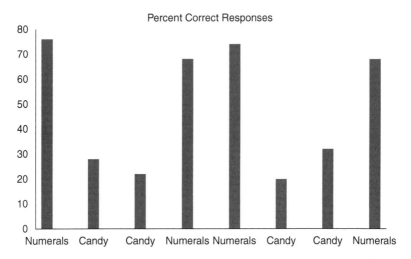

Figure 5.1 Introduction of Arabic numerals as choice stimuli in the Quantity Judgment Reverse Contingency Task (RCT) compared with performance when candy arrays were presented. Note that successful performance with candy stimuli did not generalize following effective responses with numerals

the candy right in front of them or if some other perceptual factors might be operating. To test this, collections of non-edibles (small rocks) were used in a follow-up experiment, again in an ABBA design with candy and rock arrays. Using these stimuli, performance was not significantly different when either candy or rocks were presented, suggesting that the perception of greater mass, and not the incentive value of the candy, was driving the chimps' choices (see Figure 5.2).

However, despite the animals' success using numerical representations, accurate performance with numbers did not generalise to subsequent sessions (or additional experiments) when edible arrays were reinstated. Instead, any change between the use of symbols to represent the arrays and the actual candy arrays themselves resulted in comparable levels of performance shown when each type of stimulus array was first introduced (e.g. Boysen et al., , 1996). Such disparate levels of performance persisted for several years following the original study. However, it is again important to reiterate that subsequent experiments after the initial findings of this powerful interference effect were not focused on finding a training regimen that would essentially force the animals to learn to choose the smaller array. Rather, studies that followed were aimed at examining the potential contributing variables that might provide a clearer understanding of the factors that drove the animals' suboptimal choices, compared with those that facilitated receipt of the larger reward (see Boysen, Berntson and Mukobi, 2001;

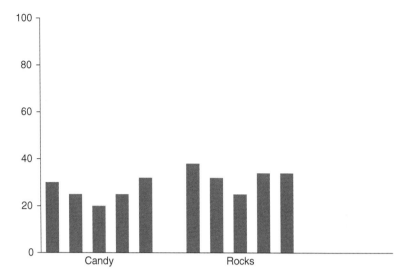

Figure 5.2 Quantity judgment reverse contingency task (RCT) by chimpanzees using candy and rock arrays. Comparison of performance between two non-representational stimulus arrays revealed no significant differences

Boysen, Mukobi and Berntson, 1999). For example, one follow-up experiment introduced arrays that were composed of the same type of candy, but in larger- and smaller-size combinations (Boysen, Berntson and Mukobi, 2001). When these mixed-size arrays were used, the chimpanzees revealed a significant bias towards arrays that contained even a single larger element, even when the mass of the combined smaller elements was greater than the single larger candy. Overall, element size had greater impact on performance than numerosity, although both factors would be relevant to a foraging animal, especially one whose diet was comprised significantly of fruit.

5.3 Studies using a modified reverse contingency task (RCT)

Since publication of original quantity judgement task (RCT), a number of experiments have been undertaken by other investigators using similar or related methodologies, as well as an ever-widening array of species, and the first theoretical paper discussing RCT (see excellent review and discussion by Shifferman, 2009, although the terminology used is not identical). It is not surprising that some of the first replications were undertaken with non-human primates, specifically Japanese snow monkeys (Silberberg and Fujita, 1996). In this study the animals were presented with only trials that had a 1:4 ratio for each pair of quantities. However, the subjects were unable to inhibit selecting the larger array despite 200 training trials.

At this point, the investigators sought to devise a training and reinforcement schedule whereby the monkeys would only receive a reward of any kind if they selected the smaller array. Thus, incorrect trials (ones in which the larger array was selected first) were unreinforced, and resulted in extinguishing the animals' responses to the larger array, as would be expected under conditions of non-reward. Responses to the single food item resulted in a reward of 4 raisins, referred to later as a large-or-none contingency procedure (Anderson, Awazu and Fujita, 2000). Another methodological difference in Silberberg and Fujita (1996) was the use of correction trials during which the spatial position of the stimulus pairs (left/right) were maintained (in the same position) until a correct trial was completed. Subsequent replications with the same procedures, including the large-or-none reward structure and correction trials, were reported by Anderson et al. (2000, 2004) with squirrel monkeys, a small New World species, and also by Genty, Palmier and Roeder (2004) using two species of lemurs (*Eulemur fulvus* and *E. macaco*). A significant point is that monkeys tested to date essentially fail to perform optimally when tested under the conditions of the original RCT (Boysen and Berntson, 1995), and none of the squirrel monkeys or lemurs from the above studied were trained or tested on the symbolic phase of the original paradigm.

5.4 Recent quantity judgement RCT studies

Subsequent studies, two with cotton-top tamarins (Kralik, 2005; Kralik, Hauser and Zimlicki, 2002) and another with all four species of great apes (Vlamings, Uher and Call, 2006) completed the RCT and incorporated abstract representations in the form of colour dishes associated with specific quantities. Because an effort was made to maintain more similar procedures to the original RCT task with the ape subjects, a more detailed discussion of their respective results is warranted.

Vlamings et al. (2006) investigated an RCT with all four species of great apes (chimpanzees, bonobos, gorillas and orang-utans) using more limited array pairs (0 versus 4; 1 versus 4) and in another condition, food visibility or covered pairs. In addition to the previous reports of the RCT used with several monkey species and chimpanzees, these authors were intrigued by results from two captive orang-utans tested by Shumaker et al. (2001), who reported their subjects readily solved the RCT using array pairs in the same array pairs as the original Boysen and Berntson (1995) study. Indeed, the orang-utans' performance was reported to be superior to that of the chimpanzees, even when the chimps used symbolic representations (numerals). Shumaker et al. (2001) suggested that species differences between the two great apes could be attributable to significant differences in their social structure in the wild, including more solitary foraging that would result in reduced competition for resources. Behaviourally, chimpanzees are demonstrably more hyperactive than the typical orang-utan, and, consequently, greater impulsivity under the testing conditions of the RCT would not be altogether surprising. However, a careful reading of the methodology described by Shumaker et al. (2001) reveals that a considerable number of preliminary 'training sessions', from which no data or procedures are reported, thus renders their findings inconclusive. In addition, the small number of animals tested and the potential differential contribution of previous cognitive training may also have been factors (Shumaker et al., 2001; Vlamings et al., 2006).

Nonetheless, the archival reports for other non-human primates' performance with modified versions of the RCT and the more recent findings reported by Shumaker et al. (2001), some of which were at odds with previous chimpanzee performance (Boysen and Berntson, 1995; Boysen, Mukobi and Berntson, 2001), served as a jumping-off point for the Vlamings et al. experiments. The aim of their replication was to follow the same methods of Boysen and Berntson (1995), with the hypothesis that the orang-utans they tested would optimise their performance, based on the Shumaker et al. (2001) study. That is, the investigators hypothesised that the orang-utans would be more likely to inhibit selection of the larger array, and would be able to choose the smaller of the two arrays, in order to be rewarded with the remaining array that consisted of the larger reward. A second goal was to compare performance across the other two great ape species, bonobos and lowland gorillas, and, finally, a third goal was to compare the relative impact of food quantity and the visibility of the food in affecting the animals' responses. In their experiments, Vlamings et al. (2006) presented their ape subjects with open dishes of raisin arrays that were paired with covered, coloured food dishes that had been previously associated with different quantities. Food quantity was assessed with two food comparisons, between arrays composed of 0 and 4 raisins, and 1 versus 4 raisins. The original procedures of the RCT were maintained such that any choice of array by the subjects resulted in them receiving the amount of reward that made up the second array. Thus, the apes always received some amount of reward on every trial, but could only maximise their reward if their choice was the smaller of the two available raisin arrays. All animals had previously been shown reliably to select the larger of two different quantities (Hanus, Call and Tomasello, 2003), and underwent two phases of colour training during which three coloured dishes (orange, purple and black) and the three food quantities (0, 1 and 4 raisins) were associated together. This allowed the use of the covered coloured dishes as test stimuli and incorporated some of the perceptual features of the raisin arrays, and thus served as abstract representations.

Once criterion performance for each of the colour-training phases was reached, subjects moved on to formal testing. During these comparisons, the animals were tested with four conditions that included two open dishes containing 0 and 4 raisins; two open dishes with 1 and 4 raisins; two covered dishes with coloured lids with 0 and 4 raisins; and two covered dishes with coloured lids filled with 1 and 4 raisins. Overall, all ape subjects, with the exception of two orang-utans that were unsuccessful with the first phase of colour training and thus were dropped from further testing, learned to associate the relationship between the colours and their respective quantities, and were 100 per cent correct in their choice of the larger array across the three training sessions of the second colour phase. During formal testing, there were no differences in performance levels among all four species, and performance improved across session blocks. Subjects performed better with 1 versus 4 pairings than with 0 versus 4 pairs, and were better on trials with the dishes covered than those where the raisins were visible. In addition, there were interactions between the variables of Visibility and Quantity, with the effect of visibility a function of the quantity combinations, and also an interaction between the visibility of the raisins and session block, indicating that animals' performance improved at different rates, depending on the perceptual features of the food arrays (covered versus visible). There were also large individual differences in performance among the subjects, as well. In summary, using similar procedures as the original RCT, but using a more limited number of paired comparisons, some of the four species of great apes tested by Vlamings et al. (2006) were able, to some extent, to choose the smaller quantity in order to get the larger reward. However, performance was affected by food quantity, with better performance on comparisons of 1 and 4 raisins, and incentive value such that the subjects performed better when the dishes were covered, rather than open, with the food visible, particularly for the 0 to 4 combinations. In addition, there were significant individual differences in performance.

Vlamings et al. (2006) proposed that their findings suggested that the results of the visible and covered trials corresponded to three separate levels of quantity processing, including perceptual, representational and conceptual levels. In the conclusions drawn from their findings, Boysen and Berntson (1995) had proposed two opposing processes, one perceptual and the other cognitive, with the powerful interference effect resulting from a behavioural predisposition for the chimpanzees to choose the larger quantity. Given their performance when Arabic numerals were substituted for the candy arrays, this suggested that the perceptual process interfered with the cognitive learning process that allowed the animals to choose the smaller array when perceptual features were subsumed by numerical symbols. In both studies, quantity and representational information had an impact on the animals' performance. The results of the Vlamings et al. (2006) study, given that some of the apes tested were successful in optimising their responses on the RCT, suggest that the wider range of comparisons presented to the chimpanzees in the original Boysen and Berntson (1995) study represented a greater challenge to the animals' performance. In addition, further testing of additional subjects, including new species, with a range of variables such as quantity, visibility and varied levels of abstraction, may be fruitful towards understanding the intriguing features of the quantity judgement task, particularly when reverse contingencies (RCT) are introduced (Boysen et al., 1996).

Perhaps the most interesting comparisons of the reported results with chimpanzees and the RCT task were revealed when the results from a highly similar (and concurrent, though unknown to each group of investigators) set of experiments was reported by Russell and colleagues (Russell et al., 1991). Working with four groups of children, including a group of normal three-year-olds, a group of normal four-year-olds, a group of Down's children who were age-matched for four years, and a group of autistic children, the authors devised a game for two children to play that would allow them to test for their understanding of strategic deception. Recall that it was also strategic deception that was the original impetus for the Boysen and Berntson (1995) study, although the chimpanzee subjects were never able to demonstrate an understanding of the task demands. In the Russell et al. (1991) study, two boxes were presented on a table to a pair of children, with one child on one side of the table and the other child sitting on the opposite side. The children were not aware that one of the boxes contained a Smarties chocolate candy, and, like the task designed for the chimps, one child was permitted to choose a box. When the experimenter opened the box, if the Smartie was there, it was given to the other child. All the children eventually understood the rules for sharing the candy, and the experimenters then moved on to the next testing phase. Now the boxes had a clear window facing the side of the child who was allowed to choose, thus insuring that the selecting subject always knew where the candy was located. Russell was interested in comparing the four experimental groups to see which children would use strategic deception to insure that they always got the Smartie for themselves.

The results were remarkable, and broke down along some of the same dimensions that allow some children, but not others, to pass tests for Theory of Mind. That is, normal four-year-olds and the Down's children who were age-matched with the four-year-olds all consistently used strategic deception to insure they got the rewards, while the normal three-year-olds and the autistic group always pointed to the box containing the Smartie. This insured that their partner always received the candy, and, despite watching as the experimenter gave the chocolate away, these two groups of children failed to grasp the idea that their partner did not share the same perspective as they did. Without this understanding, the younger children and those on the autism spectrum were not able to invoke the use of strategic deception to their advantage. What is intriguing about these findings is the comparison with our normal chimpanzees' performance. By simply manipulating the nature of the test stimuli and presenting real candy arrays, it was possible to represent the cognitive domain of normal three-year-old children and those along the autism spectrum, both groups that were unable to employ strategic deception. In contrast, the older children and the Down's population that was age-matched with the four-year-olds who had the opportunity to choose had no difficulty whatsoever recognising that they alone had knowledge of the candy's location on each trial, and thus could respond optimally in order to always get the reward for themselves. Thus, when abstract representations of the candy arrays were used with the chimpanzees, in the form of Arabic numerals, the animals' performance was comparable that of the four-year-old children and the Down's group. That is, the animals were freed from the biological imperative that had such a powerful interference effect on their choices when actual candy arrays were used, thus affecting their ability to make the optimal choice of the smaller array. In the latter case, their poorer performance looked much like the younger children and autistic group's inability to inhibit pointing to the windowed box that clearly showed the location of the Smartie, thus insuring that their partner would reap the benefits of their response.

Consequently, the use of representational symbols in place of the actual arrays and their perceptual characteristics allowed the chimpanzees to bootstrap their choices, and when compared to Russell et al.'s (1991) findings with different experimental groups of children, suggests that comparable cognitive mechanisms may support both kinds of processing in the ape subjects, depending on the stimuli that are available as alternatives. The different children's groups, however, responded according to their cognitive/developmental level, indicating that abstract reasoning capacities, such as strategic deception, are supported by a maturational bootstrapping system that allows for representational schemas to facilitate the use of social cognition. It is fascinating that both systems were available to the chimpanzees, depending on the nature of the choice stimuli. However, for the children, brain maturation over time and the accompanying cognitive strides in information processing allowed for two different mechanisms, one that would not allow the younger or autistic group to inhibit choosing the visible reward, and the other that permitted the older groups of children (normal and Down's) to withhold their responses so as to maximise the reward via strategic deception, were simply a function of age. For the apes, within the same test subjects, perceptual versus abstract depiction of the arrays resulted in response characteristics that looked like three-year-old or four-year-old children. And, unlike with the children, maturation alone did not facilitate the animals' performance, as subsequent studies over a period of several years revealed comparably poor performance with actual candy arrays compared with abstract symbols (Boysen et al., 2001). It was as though the two neural-processing mechanisms were unable to communicate with one another, and existed in parallel formats, while changes in children's cognitive processing through maturation provided a hierarchical set of mechanisms that unfolded developmentally through a series of integrated mechanisms. The evolutionary underpinnings as evidence for continuity between the two species is clear, but the environmental requirements for expression of similar cognitive mechanisms appear to differ in some important, yet still elusive domain.

References

Abeyesinghe, S. M., Nicol, C. J., Hartnell, S. J. and Wathes, C. M. (2005). Can domestic fowl, *Gallus gallus domesticus*, show self-control? *Animal Behaviour*, 70: 1–11.

Agrillo, C., Dadda, M. and Bisazza, A. (2007). Quantity discrimination in female mosquitofish. *Animal Cognition*, 10: 63–70.

Albaich-Serrano, A., Guillen-Salazar, F. and Call, J. (2007). Mangabeys (*Cercrocebus torquatus lunulatus*) solve the reverse contingency task without a modified procedure. *Animal Cognition*, 10: 387–96.

Anderson, J. R., Awazu, S. and Fujita, K. (2000). Can squirrel monkeys (*Saimiri sciureus*) learn self-control? A study using food array selection tests and reversed reward contingency. *Journal of Experimental Psychology: Animal Behavioral Processes*, 26: 87–97.

Anderson, J.R., Awazu, S. and Fujita, K. (2004). Squirrel Monkeys (Saimiri sciureus) Choose Smaller Food Arryas: Long-Term Retention, Choice with Nonpreferred Food, and Transposition. *Journal of Comparative Psychology*, 118: 58–64.

Barth, H., Kanwisher, N. and Spelke, E. (2003). The construction of large number representation in adults. *Cognition*, 86: 201–21.

Beran, M. J. (2001). Summation and numerousness judgments of sequentially presented sets of items by chimpanzees (*Pan troglodytes*). *Journal of Comparative Psychology*, 115: 181–91.

Beran, M. J. (2007a). Rhesus monkeys (*Macaca mulatta*) succeed on a computerized test designed to assess conservation of discrete quantity. *Animal Cognition* 10: 37–45.

Beran, M. J. (2007b). Rhesus monkeys (*Macaca mulatta*) enumerate large and small sequentially presented sets of items using analog numerical representations. *Journal of Experimental Psychology*, 33: 42–54.

Beran, M. J., and Beran, M. M. (2004). Chimpanzees remember the results of one-by-one additions of food items to sets over extended time periods. *Psychological Science*, 15, 94–9.

Beran, M. J., Evans, T. A., and Harris, E. H. (2008). Perception of food amounts by chimpanzees based on the number, size, contour length and visibility of items. *Animal Behavior*, 75: 1793–1802.

Biro, D., and Matsuzawa, T. (2001). Use of numerical symbols by the chimpanzee (*Pan troglodytes*): cardinals, ordinals, and the introduction of zero. *Animal Cognition*, 4, 193–9.

Boysen, S. T. (1991). Counting as the chimpanzee views it. In: *Cognitive aspects of stimulus control. The Fourth Dalhousie Conference*. W.K. Honig & J.G. Fetterman (Eds.). Hillsdale: NJ: Lawrence Erlbaum Associates pp. 367–383.

Boysen, S. T., and Berntson, G. G. (1989). Numerical competence in a chimpanzee (*Pan troglodytes*). *Journal of Comparative Psychology*, 103: 23–31.

Boysen, S. T., and Berntson, G. G. (1995). Responses to quantity: perceptual versus cognitive mechanisms in chimpanzees (*Pan troglodytes*). *Journal of Experimental Psychology: Animal Behavior Processes*, 21: 82–6.

Boysen, S. T., Berntson, G. G., Hannan, M., and Cacioppo, J. T. (1996). Quantity-based interference and symbolic representations in chimpanzees. *Journal of Experimental Psychology: Animal Behavior Processes*, 22: 76–86.

Boysen, S. T., Berntson, G. G., and Mukobi, K. L. (2001). Size matters: impact of item size and quantity on array choice by chimpanzees *(Pan troglodytes)*. *Journal of Comparative Psychology,* 115: 106–10.

Boysen, S. T., Mukobi, K. L., and Berntson, G. G. (1999). Overcoming response bias using symbolic representations of number by chimpanzees (*Pan troglodytes*). *Animal Learning and Behavior,* 27: 229–35.

Boysen, S. T., and Hallberg, K. I. (2000). Primate numerical competence: contributions toward understanding nonhuman cognition. *Cognitive Science,* 24: 423–43.

Brannon, E. M. (2005). What animals know about numbers. In J. I. D. Campbell, *Handbook of Mathematical Cognition.* New York: Taylor and Francis, pp. 85–107.

Brannon, E. M., and Terrace, H. S. (1998). Ordering of the numerosities 1 to 9 by monkeys. *Science,* 282: 746–9.

Brannon, E. M., Wusthoff, C. J., Gallistel, C. R., and Gibbon, J. (2001). Numerical subtraction in the pigeon: evidence for a linear subjective number scale. *Psychological Science,* 12: 238–43.

Call, J. (2000). Estimating and operating on discrete quantities in orangutans (*Pongo pygmaeus*). *Journal of Comparative Psychology,* 114: 136–47.

Cantlon, J. F., and Brannon, E. M. (2006). Shared system for ordering small and large numbers in monkeys and humans. *Psychological Science,* 17: 401–6.

Danisman, E., Bshary, R., and Bergmüller, R. Do cleaners learn to feed against their preference in a reverse reward contingency task? Animal Cognition. In press.

Dehaene, S. (2001). Subtracting pigeons: logarithmic or linear? *Psychological Science,* 12: 244–6.

Dehaene, S. (1997). *The number sense: how the mind creates mathematics.* New York: Oxford University Press.

Dehaene, S., Dehaene-Lambert, G., and Cohen, L. (1998). Abstract representations of numbers in the animal and human brain. *Trends in Cognitive Neuroscience,* 21: 355–61.

Emmerton, J., Lohmann, A., and Niemann, J. (1997). Pigeons' serial ordering of numerosity with visual arrays. *Animal Learning and Behavior,* 25, 234–44.

Genty, E., Palmier, C., and Roeder, J. J. (2004). Learning to suppress responses to the larger of two rewards by two species of lemurs, *Eulemur fulvus* and *E. macaco.* *Animal Behavior,* 67, 925–32.

Gentry, E., and Roeder, J. J. (2006). Self-control: why should sea lions, *Zalophus californianus*, perform better than primates? *Animal Behaviour,* 72: 1241–7.

Hanus, D. and Call, J. (2007). Discrete quantity judgments in the great apes (*Pan paniscus, Pan troglodytes, Gorilla gorilla, Pongo pygmaeus*): the effect of presenting whole sets versus item-by-item. *Journal of Comparative Psychology,* 121: 241–9.

Hanus, D., Call, J., and Tomasello, M. (2003). Quantity-based judgments by orangutans (*Pongo pygmaeus*), gorillas (*Gorilla gorilla*), and bonobos (*Pan paniscus*). *Folia Primatologica,* 74: 196–7.

Hauser, M. D., Carey, S., and Hauser, L. (2000). Spontaneous number presentation in semi-free-ranging rhesus monkeys. *Proceedings of the Royal Society of London B,* 267: 829–33.

Hauser, M., Tsao, F., Garcia, P., and Spelke, E. S. (2003). Evolutionary foundations of number: spontaneous representation of numerical magnitudes by cotton-top tamarins. *Proceedings of the Royal Society of London B,* 270: 1441–6.

Judge, P. G., Evans, T. A., and Vyas, D. K. (2005). Ordinal representation of numeric quantities by brown capuchin monkeys (*Cebus apella*). *Journal of Experimental Psychology: Animal Behavior Processes,* 31, 79–94.

Kilian, A., Yaman, S., von Fersen, L., and Güntürkün, O. (2003). A bottlenose dolphin discriminates visual stimuli differing in numerosity. *Learning and Behavior*, 31: 133–42.

Kralik, J. D. (2005). Inhibitory control and response selection in problem-solving: how cotton-top tamarins (*Saguinus oedipus*) overcome a bias for selecting the larger quantity of food. *Journal of Comparative Psychology*, 119: 78–89.

Kralik, J. D., Hauser, M. D., and Zimlicki, R. (2002). The relationship between problem-solving and inhibitory control: cotton-top tamarins (*Saguinus oedipus*) performance on a reversed contingency task. *Journal of Comparative Psychology*, 116: 39–50.

Lipton, J. S., and Spelke, E. S. (2003). Origins of number sense: large-number discrimination in human infants. *Psychological Science*, 14: 396–401.

Matsuzawa, T. (1984). Use of numbers by a chimpanzee. *Nature*, 315: 57–9.

McComb, K., Packer, C., and Pusey, A. (1994). Roaring and numerical assessment in contests between groups of female lions, *Panthera leo*. *Animal Behaviour*, 47: 379–87.

McGonigle, B., and Chalmers, M. (1977). Are monkeys logical? *Nature*, 267: 694–7.

Nieder, A. (2005). Counting on neurons: the neurobiology of numerical competence. *Nature Reviews Neuroscience*, 6: 177–90.

Nieder, A., Freedman, D. J., and Miller, E. K. (2002). Representation of the quantity of visual items in the primate prefrontal cortex, *Science*, 297: 1708–11.

Nieder, A., and Miller, E. K. (2003). Coding of cognitive magnitude: compressed scaling of numerical information in the primate prefrontal cortex. *Neuron*, 37, 149–157.

Nieder, A. and Miller E. K. (2004a). Analog numerical representations in rhesus monkeys: evidence for parallel processing. *Journal of Cognitive Neuroscience*, 16: 889–901

Nieder, A., and Miller, E. K. (2004b). A parieto-frontal network for visual numerical information in the monkey. *Proceedings of the National Academy of Science*, 101: 7457–62.

Pepperberg, I. M. (1994). Numerical competence in an African Grey parrot (*Psittacus erithacus*). *Journal of Comparative Psychology*, 108: 36–44.

Premack, D., and Woodruff, G. (1978). Does the chimpanzee have a theory of mind? *Behavioral and Brain Sciences*, 4: 515–26.

Roberts, W. A. (2005). How do pigeons represent numbers? Studies of number scale bisection. *Behavioural Processes*, 69: 33–43.

Russell, J., Mauthner, N., Sharpe, S., and Tidswell, T. (1991). The 'windows task' as a measure of strategic deception in preschoolers and autistic subjects. *British Journal of Developmental Psychology*, 9, 331–49.

Shifferman, E. (2009). Its own reward: lessons to be drawn from the reversed-reward contingency paradigm. *Animal Cognition*, 12: 547–58.

Shumaker, R., Palkovich, A., Beck, B. B., Guagnano, G. A., and Morowitz, H. (2001). Spontaneous use of magnitude discrimination and ordination by the orangutan (*Pongo pygmaeus*). *Journal of Comparative Psychology*, 115: 385–91.

Siegler, R. S. and Opfer, J. E. (2003). The development of numerical estimation: evidence for multiple representations of numerical quantity. *Psychological Science*, 14: 237–43.

Silberberg, A. and Fujita, K. (1996). Pointing at smaller food amounts in an analogue of Boysen and Berntson's (1995) procedure. *Journal of the Experimental Analysis of Behavior*, 66: 143–7.

Smith, B. R., Piel, A. K., and Candland, D. K. (2003). Numerity of a socially housed hamadryas baboon (*Papio hamadryas*) and a socially housed squirrel monkey (*Saimiri sciureus*). *Journal of Comparative Psychology*, 117: 217–25.

Starkey, P., and Cooper, R. G. (1980). Perception of numbers by human infants. *Science*, 210: 1033–5.

Thomas, R. K., and Chase, L. (1980). Relative numerousness judgments by squirrel monkeys. *Bulletin of the Psychonomic Society*, 16: 79–82.

Terrell, D. F., and Thomas, R. K. (1990). Number-related discrimination and summation by squirrel monkeys (*Saimiri sciureus sciureus* and *S. boliviensus boliviensus*) on the basis of the number of sides of polygons. *Journal of Comparative Psychology*, 104: 238–47.

Tomanaga, M. (2008). Relative numerosity discrimination by chimpanzees (*Pan troglodytes*): evidence for approximate numerical representation. *Animal Cognition*, 11: 43–7.

Uller, C., Hauser, M. D. and Carey, S. (2001). Spontaneous representation of number in cotton-top tamarins (*Saguinus oedipus*). *Journal of Comparative Psychology*, 115: 248–57.

Uller, C., Jaeger, R., Guidry, G., and Martin, C. (2003). Salamanders (*Plethodon cinereus*) go for more: rudiments of number in an amphibian. *Animal Cognition*, 6: 105–12.

Uller, C., and Lewis, J. (2009). Horses (*Equus caballus*) select the greater of two quantities in small numerical contrasts. *Animal Cognition*, 12: 733–8.

Vick, S-J., and Bovet, D. African grey parrot (*Psittacus erithacus*) self-control is task-dependent: performance on a reverse contingency-contingency task. *Animal Cognition*. In press.

Vlamings, P. H. J. M., Uher, J. and Call, J. (2006). How the great apes (*Pan troglodytes*, *Pongo pygmaeus*, *Pan paniscus*, and *Gorilla gorilla*) perform on the reversed contingency task: the effects of food quantity and food visibility. *Journal of Experimental Psychology: Animal Behavior Processes*, 32: 60–70.

Wilson, M. L., Hauser, M. D., and Wrangham, R. W. (2001). Does participation in intergroup conflict depend on numerical assessment, range location, or rank for wild chimpanzees? *Animal Behaviour*, 61: 1203–16.

West, R., and Young, R. J. (2002). Do domestic dogs show any evidence of being able to count? *Animal Cognition*, 5: 183–6.

Xu, F., and Spelke, E. S. (2000). Large number discrimination in 6-month-old infants. *Cognition*, 74: B1–B11.

Xu, F., Spelke, E. S., and Goddard, S. (2005). Number sense in human infants. *Developmental Science*, 8, 88–101.

Part II
Complexity in Robots

Introduction
David McFarland

In 1990 there was a meeting in Paris of biologists, ethologists, roboticists and AI people; strange bedfellows at that time. This meeting led to the formation of the International Society for Adaptive Behaviour, and its journal, *Adaptive Behaviour*. In the past twenty years, behaviour-based artificial intelligence has evolved into embodied intelligence, a term widely used by roboticists today. The collaboration between biologists and roboticists has come a long way, albeit with a very long way to go.

Twenty years ago few people had the idea that animals and robots should have things in common. Since then there has been considerable progress in incorporating biological 'inventions' into robots, a field which is sometimes called 'animal robotics'. Some of these developments are discussed in this book. But animal robotics has also had a major impact on biology. At the turn of the century publications began to appear describing the use of robots as tools in biological research, and some long-standing biological theories received a jolt as a result. For example, the snapshot theory was initially developed to explain how bees relocate a food source using a 'snapshot' of the surrounding scenery taken during a previous visit (Cartwright and Collett, 1987). This type of theory has been applied widely in animal behaviour studies, and techniques based on it have also been used by roboticists. In some cases, however, roboticists have been able to mimic the homing behaviour of animals without implementing the snapshot (e.g. Lambrinos, 2003). In other words, building a robot can show that the snapshot idea is not necessary in certain cases.

Non-necessity is not non-existence. The demonstration that a robot can behave, in a particular situation, in the same way as an animal, without the robot having a certain knowledge structure, does not prove that the robot and the animal control their behaviour in the same way, or that, in our example, the bee does not have a snapshot (McFarland, 2008). It is perfectly possible that the robot does it one way and the animal does it another way. What the engineering debunks are claims that animals *must be* using a particular mechanism. Roboticists know what mechanisms their robots are deploying, because they designed and made them.

One of the difficulties with this area of research is that related disciplines have very different terminology. Still worse, they have different interpretations of the same terminology. For example, in the field of cognitive ethology, the word 'cognition' means 'information processing'. When one thinks of the information processing that goes on in a flying bird (detailed information sent to each wing and tail feather), the word 'cognition' does not seem appropriate. Another usage is that cognition involves 'manipulation of explicit representations'. The problem here is that one has to know what it is for representations to be explicit. Unfortunately some animal cognitive psychologists use the term 'declarative representation' (or declarative knowledge). This is a term borrowed from philosophy, meaning 'that which people can declare'. This usage might be clarified, but often isn't, by appeal to computational principles where the notion of declarative representations (ones not dedicated and limited to specific narrow procedural uses) are well understood. To use the term 'declarative' for animals that can declare nothing, without further elaboration, does seem bizarre. Unfortunately terminology does matter. If one imagines an evolutionary sequence, such as: associative (procedural) processes only > associative + cognitive (explicit but not declarative) > associative + cognitive + declarative (humans), one can see that poor terminology might lead one to overlook certain important stages of evolutionary development.

In Chapter 6 of this book, Mark Lee, Ulrich Nemzhow and Marcos Rodrigues point to the gulf between the behaviour produced by our best robotic efforts and the richness of behaviour, learning and adaptability manifest in living systems. They, like most roboticists these days, are well aware of the ground yet to be covered in designing intelligent robots. Their approach is to characterise and model cognitive processes whose validity for engineering applications can be assessed through the building of autonomous robots capable of working in the real world. They take the view that behaviour is the consequence of tracking reference goals by reducing perceptual mismatch. This approach leads on to an architecture that offers a framework in which low-level reactive behaviour can be integrated with higher-level schemas in a control hierarchy. The relevance for real applications is illustrated by robot grasping experiments where a natural food product is to be grasped by a commercial assembly robot. No prior knowledge of the product shape is available to the robot but a grasping algorithm is described that solves this problem by adapting the robot gripper orientation to perceived features of the product without operator intervention. Building on these ideas, further developments are described involving 3D vision and pattern recognition, 3D registration, fast 3D reconstruction, real-time tracking of faces and eyes, and a fully automatic real-time 3D face-recognition system. The implications of the research are considered for a range of application areas such as 3D CCTV systems, 3D animation and entertainment.

By adopting a variety of principles concerning the construction of cognitively minimalist systems, the New AI made progress where traditional AI had failed. Ironically, some of the most successful strategies of New AI, such as modularity and subsumption, stood in direct contradiction to some of the principles held by many New AI researchers, such as an opposition to hierarchical representation. In Chapter 7, Joanna Bryson explores the role of hierarchical and modular structures in generating intelligent behaviour in both natural and artificial intelligence. Her contribution gives a contemporary perspective on the debates of the last 25 years on these subjects. She adds the perspective of another discipline to the New AI mix – evolutionary developmental theoretical biology, commonly known as evo-devo. Evo-devo has shown how genetic control is also, to a considerable extent, modular and hierarchical, so that even single mutations can control complex developmental processes, a theme taken up again in section 3, with regard to human evolution.

Whether an artificial agent could understand its place in the world is an epistemological issue, involving a blending of empiricist and rationalist traditions. In Chapter 8, George Luger discusses ways in which such epistemological sophistication might be modelled, and eventually become embodied within robots. He takes issue with both the traditional planning-oriented approach to AI and robotics and the more modern approach that includes subsumption architecture. He addresses the general question of how an agent, whether human or robotic, interacts with its environment, and he uses a form of Bayesian mathematics to describe this agent/environment synergism. Finally, he considers the agent's ability to understand its interactions with the non-self, the problem of epistemological access.

In Chapter 9, Alan Bundy points out that appropriate representation is the key to successful problem solving. He argues that representations should be fluid; their choice, construction and evolution should be under the control of the autonomous agent, rather than being predetermined and fixed. Quite a good summary of human problem solving (outside technical spheres) is that humans spend a lot of effort finding a good representation, or sometimes several, and when they are done, that usually makes the problem easy. If they can't find one, then they give up. Long chains of reasoning within a single problem representation are very unusual. But automating this search for good representation (with all the contributions that that would make towards understanding how people do it) is still a very challenging computational problem. This chapter describes Lakatos and Pólya's historical contributions and their current relevance to automation.

References

Cartwright, B. A. and Collett, T. S. (1987). Landmark maps for honeybees. *Biological Cybernetics*, 57: 86–93.

Lambrinos, D. (2003). Navigation in desert ants: the robotic solution. *Robotica*, 21: 407–26.

McFarland, D. (2008). *Guilty robots, happy dogs: the question of alien minds*. Oxford: Oxford University Press.

6
Towards Cognitive Robotics: Robotics, Biology and Developmental Psychology

Mark Lee, Ulrich Nehmzow and Marcos Rodriguez

6.1 Introduction – early robotics research

The question of how 'intelligent' controllers can be designed for machines has always attracted much interest and research activity. The idea of reproducing *some* facets of human cognition in a designed artefact has fascinated scientists, philosophers and charlatans throughout history. But despite the enormous efforts that were directed at this issue in the twentieth century, only very recently has any significant progress been made. The robots of the last century typically were brittle; they failed in even simple tasks as soon as these tasks deviated even slightly from the original specifications; they were slow; and they needed constant attention from software engineers.

Up to the late 1980s and early 1990s, robotics research was dominated by work originating from the control theory and cybernetics communities, which meant that the fundamental assumptions of how intelligent behaviour could be achieved were very similar if not identical. The Sense-Think-Act cycle (Nehmzow, 2003) was one such assumption: sensor signals ('sense') would be perceived by the robot, processed through various processing stages ('think') and result in motor action ('act'). This cycle would then be repeated. Another common assumption was the Physical Symbol System Hypothesis (Newell and Simon, 1976), which claimed that general intelligent action could be achieved by manipulating symbols (which represented states of the real world) through a set of 'the right kind' of rules. Assumptions like these influenced much of early robotics research towards a 'programming approach' to intelligent behaviour, involving rules, knowledge-bases, agents and simulated environments.

This changed drastically when Rodney Brooks introduced the concept of 'Behaviour-based Robotics', and many laboratories started to pursue research in the 'New AI', or 'Embodied Intelligence' as it became known. This approach considered work with physically embedded machines as essential, used little or no symbols (using neural or 'sub-symbolic representations'), and saw the behaviour of a robot as emergent from the robot's interaction with the environment, the robot's morphology, and the many unknown factors that influence robot behaviour. In the last decade this approach has blossomed into the 'Embodiment' movement which argues that truly autonomous intelligent agents must be situated, embedded, and embodied, and, currently, the only exemplars are to be found in the natural world. This has spurred much biologically inspired robotics research that has taken ideas and models from brain science (neurology, anatomy, physiology), psychology (behaviour, perception and psychophysics), cognitive science, ethology, and even evolutionary theory.

This chapter has its origins in an early collaborative robotics research project conducted jointly by the universities of Edinburgh and Aberystwyth that led not only to new insights in robot control, but also stimulated novel research in related areas.

The project was unusual for its time in that it integrated robotics expertise from a Computer Science Department at Aberystwyth University in Wales with the interests of the Laboratory for Cognitive Neuroscience in the Psychology Department at Edinburgh University. Details of the project are given in the Acknowledgements.

Our main motivation was, and still is, to gain an understanding of how robot systems could achieve some of the rich, flexible behaviour seen everywhere in the autonomous agents of the animal kingdom. There still remains a large gulf between the behaviour produced by our best robotic efforts and the richness of behaviour, learning, and adaptability so obviously manifest in living systems. In the 1990s we were unsatisfied with the current methods for designing and engineering intelligent systems, and found a lack of general principles for embodied intelligence research. In particular we saw psychology as the potential missing link, with its higher-level models, emphasis on behaviour, and relative openness to the problems of complexity. Since then we have developed our approach further and have followed three principled lines of research, each quite different but all relating in their own way to the central problem of understanding and designing complex autonomous systems.

In the next section, we briefly expand on the important issue of autonomy in robotics. Then follows a summary of our early founding work, before the three following sections each describe in turn our three lines of attack: steps towards a science of mobile robotics; an approach for developmental learning; and the potential role of full 3D geometric knowledge of the world.

6.2 Autonomy and embodiment

The concept of Autonomous Systems can have many realisations but a central characteristic is the ability to sense, understand, and act upon the environment in which the system operates. Thus, any internally processed information must be grounded in meanings ultimately derived by sensing and acting in the environment. This is why robotics is an excellent framework for autonomous systems research, as it forces issues like sensing, perception, action, error-recovery, and survivability to be faced in an integrated and challenging format.

The important paradigm shift in robotics brought about by the Embodied movement has been the rejection of simplified 'toy worlds', or artificial simulated environments, and the emphasis on the 'real world'. Furthermore, unstructured environments are the required proving ground for modern experiments, bringing not only realistic noise, disturbance and uncertainty to the fore, but also opening up the enormous complexity that autonomous systems must ultimately face and manage.

For example, consider the following scenario: A robot is to patrol the sea-facing pedestrian area at a popular coastal resort. The robot might be required to perform coastal surveillance, environmental audit, assist the public, monitor local conditions, and search for missing persons (e.g. via heat sources). Such assistive robots will need a wide variety of sensors (to monitor local conditions: weather, waves, tide etc.), some form of interface for interactions with the public, a link to remote services (control room or coastguard), and sophisticated perceptual processes (tracking and awareness functions to raise security warnings; novelty analysis for unusual events or objects).

It is clear that such a system must survive unattended for long periods thus requiring genuinely autonomous operation. But autonomy is not dependent on any group of specific functions or capabilities. Rather, autonomy is the ability to cope with changing situations and circumstances, and this in turn depends upon gaining a grounded understanding of those very situations from experience. For example, a seafront robot that is temporarily unable to deliver a warning message directly may communicate with other robots or agents to recruit their help and achieve the task by different means. Similarly, actions that fail in one context (e.g. sandstorm) might be reconfigured through experience with other actions in related conditions (e.g. fog).

Embodiment is a vital property of autonomous systems because any understanding of the environment must be built up from sensory-motor activity and the morphology and physical hardware of a robot is an essential factor in determining both its behaviour and the extent of its cognitive abilities. We take the view of others (e.g. Lakoff, 1987) that all cognitive competencies are grounded in sensory-motor acts and even higher functions such as language are intimately related to basic sensory and motor experience (Rizzolatti and Arbib, 1998).

The achievement of autonomy is seen by the robotics community as one of the most pressing research challenges and is essential for the successful deployment of robots in service, domestic and health-care scenarios as well as in hostile and remote environments.

6.3 Control and perception in mobile robots

In the early 1990s we began exploring new approaches for building autonomous robots working in everyday unstructured environments. Our joint task concerned the development of a courier robot capable of locating and identifying targets in order to carry out a given cognitive task, namely the delivery of mail or messages via mapping, route finding, and navigation of its local environment.

We used mobile robots of various manufacture. Figure 6.1 shows a typical modern wheeled robot with a cylindrical body and a range of sensors (ultrasonic, infrared, laser) mounted around its circumference. There are two main wheels, set on a diagonal, so that by driving these in contra or similar directions the robot can either rotate about its centre or move forward.

The difference between normal and unexpected events is of vital importance for autonomous systems and so one of our first investigations concerned the notion of novelty. We defined 'novel' events as those not conforming to the current model of usually experienced events, and we built a variety of data-driven novelty filters that adapted through real-world robot-environment interaction. Using sonar as the sensing modality, a self-organising feature map performed clustering on the preprocessed sonar perceptions of the robot. This was demonstrated in a corridor-exploring task where the robot initially only sees closed doors in a corridor and so treats open doors as 'novel', but later treats both open and closed as 'normal'.

This work was extended for dynamic environments where dynamical objects (e.g. other robots) are present and need to be learned as 'normal'. Using visual input from cameras on the robot we adopted a model-based approach that combines expectations from a model with current sensory perceptions. If either an abnormal sensory perception is perceived, or a behaviour is detected that is unusual in the robot's current context, the situation is classified as novel (Neto and Nehmzow, 2004). A robot behavioural model was obtained using the Narmax system of non-linear polynomial identification (described in section 4.3) and then the robot was tested in a complex environment that had been encountered before. Novelty is signalled by a large error difference between the input perceptions and the model predictions. When any significant change was made to the environment, such as removing a small pillar, adding a barrier between two pillars, or displacing a pillar, all were successfully detected as novel events (Ozbilge et al., 2009).

Another emphasis was on the growth of competence necessary if an autonomous robot is to build on its experience. An illustration of this can be seen in our early implementation of a visual layer of competence. A camera and

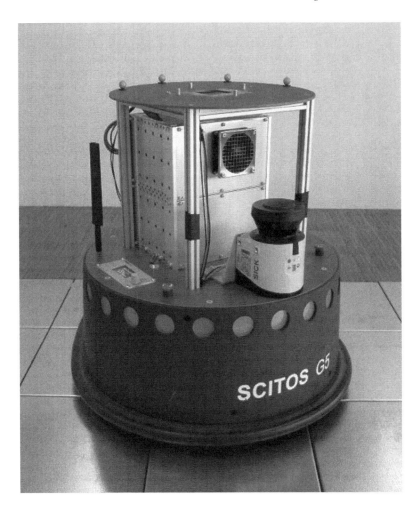

Figure 6.1 A typical modern mobile robot

visual processing system on a mobile robot were used to develop adaptive and learned responses to obstacles, recognise objects, and recognise target locations (Nehmzow et al., 1993). The cognitive levels in the final system consisted of three major parts: a reactive layer, a locative competence, and a visual ability. The interaction, growth, and development of these stages were our focus of interest. The three stages of competence can be summarised as in Table 6.1.

The table shows the behaviours growing in complexity from a simple propensity to move forward to a many-layered ability to find a specific object

Table 6.1. Development of competencies

BEHAVIOUR	MAPPING	LEVEL OF VISION
Reactive Stage		
go_forwards	None	threshold sonar
Locative Stage		
go_forwards	position of defaults	threshold sonar
go_forwards	sonar values + defaults	distance-value sonar
goto_point (using go_forwards)	(uses map data)	distance-value sonar
Vision Stage		
find_any_object	object position + sonar values + defaults	object/non-object sensing + distance-value sonar
label_specific_object	object name and position + sonar values + defaults	distant object sensing + feature identification + distance-value sonar
find_specific_object (uses map data) (using goto_point without avoid)	as above	as above

while avoiding intermediate objects and obstacles. Sonar and vision competence also develops gradually from a simple threshold sonar device that can only ask the question 'Am I too close to a surface?' to advanced vision that can discriminate between two different objects of interest to the agent and ask, 'Is this the object I want?'

The mapping process largely reflects the maturation of the other two areas of development. It starts by plotting where defaults occur during the 'go_forward' behaviour and then adds the information that it gets from the visual/sonar system as it becomes more complex. At the same time, the robot is using the lowest-level, and computationally simplest, visual information that it can to perform a task, and thus costly high-level visual information is only used when necessary to open up new possibilities.

6.4 Control in organically unstructured environments

After taking account of relevant considerations from ethology we developed for robotics the servo-based model of behavioural control proposed by William Powers (Powers, 1973). In classical control theory any feedback signals (usually negative) are fed back from the output of the controlled plant. Thus a motor speed controller will monitor the speed of the output shaft and use the error value to adjust the input power accordingly. The basic conceptual difference in Powers' model was to take the feedback signal from the environment, that is, after the effects of the plant had been felt on the environment. Thus the feedback loops are closed by environmental interaction. The difference from the behaviour-based methodology is that the

Another feature of this work was the application requirement for absolutely no programming to be allowed, not only for the grasping of variable items but also for setting up the equipment for different batch runs. This was because of the hygiene regime and the management obsession to completely minimise human involvement. We achieved this by a system of 'teaching by showing'. A single exemplar sandwich would be shown to the camera system in each of its partial assembly stages, and then the robot could deduce the action required by noting the difference between the current stage S_i and the next successive stage S_{i+1}. The robot action was automatically generated by the goal of reducing the difference to zero. In this way any assembly sequence could be defined and the quality of the result would reflect the quality of the examples initially shown to the system.

6.5 Scientific methods in robotics: towards a theory of robot–environment interaction

The literature in robotics research contains many impressive feats of engineering. A typical paper will describe the problem analysis, design and performance results of a system that tackles some challenging task. The modus operandi has often been one of prototype building with successive refinement and the cumulative growth of expertise and knowledge (on the part of the experimenter). This process might characterise the very early days of Victorian engineering but is unsatisfactory for modern science. In particular, the most glaring omission is the lack of reproducibility of the published results. It is almost impossible to test claimed results by reproducing experiments because (a) the full details of the equipment, software, and conditions are not given (usually because they would take far too much space); (b) it is impossible to duplicate the exact same laboratory apparatus as the equipment is often part original or modified; and (c) the initial conditions for experiments (including ambient conditions like lighting) are not fully recorded or are otherwise unavailable.

We believe that this situation must change if we are to attain a more scientific approach to building robot systems. We must build up an organised body of scientific knowledge that facilitates a much better understanding of such systems and allows for proper evaluation of our understanding and progress.

Taking mobile robotics as an example, assume that a small mobile robot is to be used in experiments to trace around the walls of an arena; this is known as a wall-following task. A program will cause the robot to act but the actual path taken will depend upon the combined influences of the control program running on the robot, the robot's physical properties, and the environment itself. Hence robot behaviour cannot be reduced simply to the output of a program but is the result of interactions of three entities: Robot/

feedback is now within the perceptual process, thus behaviour becomes the consequence of acting to reduce perceptual mismatch. We produced a variation on Powers' theory, and implemented several models to investigate its properties (Rodrigues and Lee, 1994). This approach offers a framework in which low-level reactive behaviour can be integrated with higher-level schemas in a control hierarchy – a challenging issue in Embodied Intelligence.

We applied these ideas in various real applications with cluttered environments for both mobile robots and robotic manipulator arms and hands. One example was the development of robot grasping strategies for unprocessed natural food products, including packing fish portions into boxes and sandwich assembly. A typical task is to grasp and place natural food items at target locations with an industrial quality assembly robot, but with no prior knowledge of the product size or shape. Natural food items are extremely variable in their shape, consistency and quality and these features preclude any preprogrammed grasping pattern. Using the Powers approach we designed a grasping algorithm that solves this problem by adapting the robot gripper orientation to perceived features of the product without operator intervention. Figure 6.2 shows a layer of cucumber and tomato slices being located during the assembly of a sandwich.

Figure 6.2 Sandwich assembly. A laser stripe (visible at extreme lower left) produces a profile of the product on an imaging camera when the belt is moving. This is then used to deduce the items that are missing and direct the robot to fetch and place such components of the sandwich

Environment/Task. This is why robotics research is hard, as can be appreciated by the following case study based on real trials. Our wall-following robot is programmed to move parallel to any wall, while maintaining a constant distance from the wall. Unfortunately we soon find that the robot's path – its trajectory around the arena – is not repeatable. Each time we try to start it off in exactly the same location with exactly the same speed and heading we always find that after a while its path deviates from the previous one. Figure 6.3 shows an example of the deviation between two trajectories in four increasing time snapshots.

At first the paths are near identical but they gradually diverge and eventually may become completely different trajectories. This effect is caused because of very slight variations in friction, material properties, or ambient conditions.

Such microscopic effects are unavoidable and uncontrollable, and consequently present a problem for our desire for scientifically repeatable results.

This problem is one of chaos. However careful we are, we have no way of controlling the conditions so that identical behaviour is produced. This

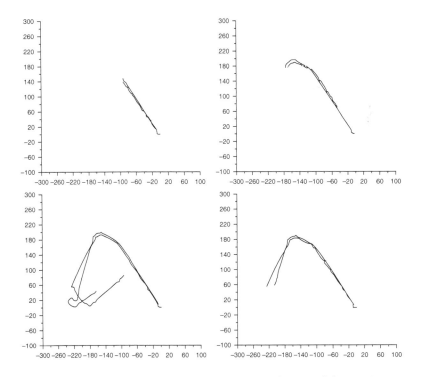

Figure 6.3 Four snapshots of two trajectories. Starting from top left, time increases clockwise

is the sign of a chaotic system. Note that this is the case even though all the components of the experiment may be deterministic in nature. A solution might involve a way of measuring behaviour that can take account of these chaotic effects. We need a suitable 'Behaviour Meter' that can be used to compare different behaviours. If we consider just the dynamics of our mobile robot behaviour then various possible quantitative methods can be considered. We have experimented with quantitative descriptions of phase space and found three particular measures very helpful (Nehmzow, 2009).

6.5.1 Phase space

Phase space is a concept from dynamical systems theory and is used to describe all possible states of a dynamical system. Not only spatial position but velocities will typically be involved, thus phase spaces may have high dimensionality. It turns out that if a log is taken of robot variables at regular time intervals during a behavioural trajectory then the phase space for the system can be reconstructed from the observations in these time series alone. Figure 6.4 shows an example of part of a reconstructed phase space (this is a five-dimensional phase space but only three can be shown). Note that this is not a robot spatial trajectory but a plot in phase space. The variation in the traces indicates the degree of unpredictability; a strictly periodic behaviour would produce a single curve.

6.5.2 A behaviour meter

Given a phase space we can then apply our three measures from our 'Behaviour Meter'; these are: the Lyapunov exponent; the Prediction Horizon; and the Correlation Dimension. The Lyapunov exponent is a measure of the phase space that describes the rate of divergence of two trajectories that

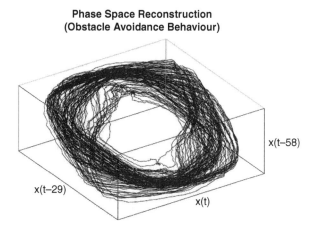

Figure 6.4 Phase space example (showing 3 of 5 dimensions)

started infinitesimally close to each other. Figure 6.5 shows the trajectories of a mobile robot executing two different behaviours: wall-following and obstacle-avoidance. It is clear that wall-following is the more predictable behaviour as the trace shows. Obstacle-avoidance causes the wall to be avoided as soon as it is detected and therefore looks more unpredictable and chaotic. The Lyapunov exponent for the wall-following case was calculated as between 0.02 and 0.03 bits/sec, while in the obstacle-avoidance trial it was between 0.11 and 0.13 bits/second. This shows that obstacle-avoidance is indeed more chaotic than wall-following.

The Lyapunov exponent is expressed as information loss per unit time; in other words it indicates the loss of information in the system (or degree of chaos) as one predicts the system state for longer and longer times ahead. We may be able to predict our robot's state in the next second fairly accurately but we would expect large errors if we try to predict several hours ahead. The Lyapunov exponent tells us how bad the situation is; if the Lyapunov exponent is zero then we have a noise-free, perfectly deterministic system whose behaviour we can accurately predict for any length of time. But as the Lyapunov exponent increases so our predictions get worse and eventually become no better than a guess. We refer to the point in time where complete loss of predictability occurs as the Prediction Horizon. Using information theory analysis we find that the Prediction Horizon for the wall-following behaviour was greater than 25 minutes (6,000 steps at sampling rate of 4Hz) and for the obstacle-avoidance was only 80 seconds. This shows that wall-following is much more predictable, and for any model of obstacle-avoidance, however good the model, it will not be able to predict the exact path of the robot for more than about 80 seconds.

The final measure of chaos is the Correlation Dimension, which gives a measure of aperiodicity or how close the system variables return to previous values (zero indicates strictly periodic data). For the wall-following behaviour the Correlation Dimension was calculated as around 1.5 while for obstacle-avoidance it was 2.5, again clearly indicating the increased chaotic nature of the obstacle-avoidance behaviour (Nehmzow, 2009).

6.6 Faithful and transparent modelling of robot–environment interaction

The issues we have discussed regarding real robot behaviour have some important implications for modelling and simulation. For a model to have any value it must be accurate and faithful, that is, it must predict or generate identical behaviour to the originally perceived behaviour. Models should also be reasonably transparent and analysable so that we can understand their meaning and what they represent. As an example of a non-transparent model, consider the case of a neural network that has been well trained on some specific task. The only information we have available is the weight

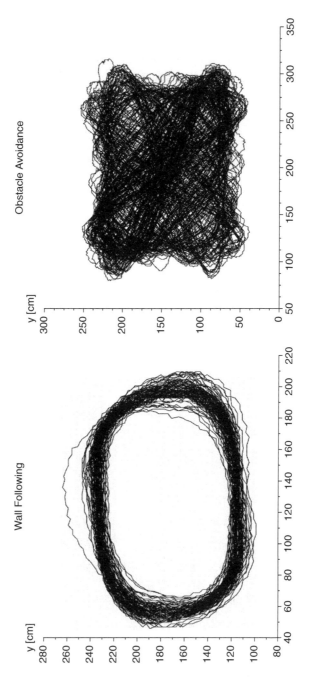

Figure 6.5 Trajectories of wall-following and obstacle-avoidance as recorded from an overhead camera

matrix that has captured the essence of the input–output relationship. But unfortunately the weight matrix is just an array of numbers that obscurely encode whatever the model has learned; they are quite unintelligible to humans. By comparison we are interested in transparent, parsimonious models that aid understanding of these complex situations.

Clearly, if there are indications of chaotic behaviour then the predictability of any models may be severely limited. However, it is possible to discern structure in even noisy data in many cases. The field involved in producing models of unknown systems is called System Identification, and a powerful and relevant method for robotics is the Narmax technique (Chen and Billings, 1989). This is a mathematical method that produces a non-linear polynomial of the system variables that expresses the relationship between a set of inputs and outputs. As a simple illustration consider a system with two inputs, $u1$ and $u2$, and a single output, y. We chose the system equation as:

$y(t) = u1(t)^2 + 0.5u1(t)u2(t)$, where t is the time-series index.

Thus this gives the ideal or theoretical output for any input values. But in the real-life situation, (a) the inputs contain a lot of noise, and (b) we have no idea of any equation that might represent the system. This is where system identification methods are useful and to demonstrate we use the Narmax method (Chen and Billings, 1989). First we collect some data by generating a set of values for $u1$ and $u2$, and add Gaussian distributed noise giving $u1$ and $\hat{u}2$, then we pass this data through our system (the above equation) to produce the output data $\hat{u}2$. We now have a set of input–output data for a noisy system. This data is then processed through the Narmax method, which delivers the following result:

$\hat{y}(t) = -0.01 + 0.05\hat{u}1(t) + 0.01\hat{u}2(t) + 0.91\hat{u}1(t)^2 + 0.52u1(t)u2(t)$

This is very close to the true underlying relationship: $y(t) = u1(t)^2 + 0.5u1(t)u2(t)$, and we now have a model of the system for further analysis, simulation, or prediction (Nehmzow, 2009).

6.7 Robot control

We have applied these techniques to a number of standard robotic problems. An interesting approach is to build the system model from data collected while human operators drive a robot to achieve a particular task. The resulting model then contains the expertise of the operators and can be used as a controller to drive the robot autonomously. Figure 6.6 shows a doorway navigation task viewed from above. On the left an operator has driven the robot from many starting points (in the region at the top) and navigated through the gap in the wall (the gap is two robot diameters wide).

Figure 6.6 Robot under manual control...and under model control

Narmax models produce polynomials with many terms for the various combinations of inputs and their products. In this case there were many laser and sonar inputs and the model polynomial came out with 38 terms. The robot was then run under control of the model, taking its actions from the model-processed sensory input. The traces in Figure 6.6 show 39 runs, and those of the human (on the left) are noticeably less smooth than those produced by the model (on the right). Doorway traversal is a delicate task, requiring a careful balance between several signals, but the Narmax model was successful every time. Post-analysis of the model showed that just a few of the sensors were playing a major role and these were all monitoring one side of the doorway. This strategy of following one side closely is clearly seen in Figure 6.6 and was an unexpected effect. The model-driven robot was much smoother and more accurate than a human operator!

6.8 The importance of development for cognitive robots

We have argued that true autonomy involves dealing with the new or unknown without external aid. This means that systems must not only adapt in accordance with current experience but must also be capable of adapting their learning processes themselves. Thus new competencies must emerge as conditions change and new demands are made. The way this problem has been solved in humans and other mammals is through processes of structured growth generally known as 'development'. In this section we consider the role of development in relation to cognitive aspects of robotics.

Developmental psychology has long studied human cognitive growth and produced many theories that could explain such growth. It is very surprising that, despite the vast body of psychological knowledge on learning and adaptation built up over the last century, very little work has considered implementing developmental processes in artificial systems.

This situation has finally changed and the topic of Developmental Robotics has recently become established as a new research area (Lungarella et al., 2003). This approach emphasises the role of environmental and internal factors in shaping adaptation and behaviour, and posits a developmental framework that allows the gradual consolidation of control, coordination, and competence (Prince et al., 2005).

6.8.1 Developmental stages

A key characteristic of human development is the centrality of behavioural sequences: no matter how individuals vary, all infants pass through sequences of development where some competencies always precede others. This is seen most strongly in early infancy as one pattern of behaviour merges into another. These regularities are the basis of the concept of behavioural stages – identifiable periods of growth and consolidation. Perhaps the most influential theories of staged growth have been those of Jean Piaget, who emphasised the importance of sensory-motor interaction, staged competence learning, and a constructivist approach (Piaget, 1973). It is recognised that stages tend to have vague boundaries and also vary greatly with individuals. Nevertheless, the existence of stages in development and their role in the growth of cognition appears to be very significant.

We believe that research into developmental algorithms for robotics should begin with and be firmly rooted in the early sensory-motor period. This is for several reasons: (1) it is logical and methodologically sound to begin at the earliest stages because early experiences and structures are highly likely to determine the path and form of subsequent growth in ways that may be crucial; (2) according to Piaget, the sensory-motor period consists of six stages that include concepts such as motor effects, object permanence, causality, imitation, and play; these are all issues of much relevance to robotics; (3) sensory-motor adaptation and learning is vital for autonomous robots; (4) it seems likely that sensory-motor coordination is a significant general principle of cognition (Pfeifer and Scheier, 1997).

Hence, we are investigating the earliest level of sensory-motor development: the emerging control of the limbs and eyes during the first three months of life. To the casual observer the newborn human infant may seem helpless and slow to change, but, in fact, this is a period of the most rapid and profound growth and adaptation. From spontaneous, uncoordinated, apparently random movements of the limbs, the infant gradually gains control of the parameters, and learns to coordinate sensory and motor signals to produce purposive acts in egocentric space (Gallahue, 1982). We believe there is much to learn for Embodied Intelligence and robotics from this scenario.

6.8.2 The key role of constraints

Any constraint on sensing, action, or cognition effectively reduces the complexity of the inputs and/or possible action. This reduces the task space and provides a frame or scaffold which shapes learning (Bruner, 1990; Rutkowska, 1994). When a high level of competence at some task has been reached, then a new level of task or difficulty may be exposed by the lifting of a constraint (Rutkowska, 1994). The next stage then discovers the properties of the newly scoped task and learns further competence by building on the accumulated experience of the levels before.

Various examples of internal sensory and motor constraints are seen in the newborn, for example the neonate has a very restricted visual system, with a kind of tunnel vision (Hainline, 1998) where the width of view grows from 30 degrees at two weeks of age to 60 degrees at ten weeks (Tronick, 1972). Although this may seem restricted, these initial constraints on focus and visual range are 'tuned' to just that region of space where the mother has the maximum chance of being seen by the newborn. When 'mother detection' has been established then the constraint can be lifted and attention allowed to find other visual stimuli.

Many forms of constraint have been observed or postulated (Hendriks-Jensen, 1996; Keil, 1990) and we have identified a range of different types in robotics. These include: anatomical or hardware constraints imposed by the system morphology; sensory-motor limitations (e.g. accuracy, resolution, bandwidth); cognitive/computational constraints; maturational constraints from internal and biological processes; and, not least, external or environmental constraints.

In our current research we are following these ideas of staged development and building robotic learning architectures in which sensory-motor competence grows cumulatively. Our method involves an overarching constraint network that restricts the ranges and number of parameters available to the robot during its early stages.

6.8.3 The LCAS approach

As mentioned before, we try to find explicit, abstract models and avoid preselected internal representations and methods. Accordingly, we have produced a general mechanism for staged development which takes 'constraint lifting' as a key process in allowing transitions between stages. We use novelty and expectation as the drivers and so any trigger for stage transitions is likely to be related to internal global states, not local events. Local stimuli (spatially and temporally) may cause local responses but global values can indicate levels of general experience or expectation.

Thus, global states such as global excitation can act as indicators that can detect qualitative aspects of behaviour such as when growth changes have effectively ceased or when a mapping between modalities has become saturated. They can then signal the need to enter a new level of learning by lifting a constraint or accessing a new sensory input. In this way, further exploration may begin for another skill level, thus approximating a form of Piagetian learning.

Our approach then consists of implementing the cycle: Lift-Constraint, Act, Saturate (LCAS), at a suitable level of behaviour. First, the possible or available constraints must be identified and a schedule or ordering for their removal decided. Next, a range of primitive actions must be determined together with their sensory associations. Also, any sensory-motor learning or adaptation mechanism is incorporated at this stage. Finally, a set of global measures needs to be established to monitor internal activity. When all this is implemented the initial behaviour may seem very primitive, but this is because all or nearly all constraints have been applied and there is little room for complex activity. During the Act process varying patterns of action effectively explore the scope for experience and any new experiences are learned and consolidated. Eventually there are no new experiences possible, or they are extremely rare, and this level becomes saturated. The global indicators then reach a critical level and the next constraint in the schedule is lifted and the cycle begins again.

The ideas reported here have all been explored in experiments on hand/eye robot systems. The details cover the learning of sensory-motor control for eye saccades (Chao et al., 2010), visual search (Hülse et al., 2009a), and hand-eye coordination (Hülse et al., 2009b; Hülse et al., 2010). The behaviours observed from our experiments display an increasing progression from initially spontaneous limb movements (known as 'motor babbling'), followed by more exploratory movements, and then directed action towards touching and grasping objects. Our research is continuing in a programme that aims to demonstrate autonomous cognitive growth on an iCub humanoid robot (Metta et al., 2008). We are exploring constraint networks as an overarching framework for orchestrating development and have build such networks by transposing a large sensory-motor constraint analysis of the human infant.

6.9 The role of geometric knowledge in recognition tasks

The application described in connection with robotic handling of natural food products, operated very successfully with only two-dimensional images of the environment. No prior information about the product shape was available and the only knowledge used was a set of control variables in a geometric relationship and the sensory input.

A follow-on research question addresses the opposite end of the knowledge spectrum: would full geometric knowledge of the shape allow us to explore in more detail the perception/anticipation/action model? In order to address this question, objects in the real world need to be directly perceived by artificial sensors and their geometries reconstructed from such perceptions.

Our research has focused on 3D data acquisition and exploitation from single 2D images using structured light methods (e.g. Robinson et al., 2004; Brink et al., 2008; Rodrigues and Robinson, 2009). We demonstrate that a robot can effectively track targets in 2D and reconstruct these into 3D as it wanders through the environment.

6.9.1 2D tracking and 3D reconstruction

The OpenCV Intel libraries (Bradski and Pisarevsky, 1999) provide built-in routines for real-time face detection based on Haar-like features. It is possible to train and use a cascade of boosted classifiers for rapid object detection for any arbitrary object. We have used the libraries to train classifiers for left and right eye. The general problem with such detection techniques is the number of false positives. For instance, in any image there could be various detected faces and some might not be real faces. Similarly for eyes, the routines normally detect more eyes than there are in the scene.

The solution is to run face and eye detection in separate threads and impose constraints: first, there should be only one face detected in the image and the face must be larger than a certain threshold; second, there should be only one left and only one right eye detected in the image, and these must be within the region of interest set by the face detection; third, the position of the face and eyes must not have moved more than a set distance since last detection so to avoid taking blurred shots due to undesirable motion. The routines are thus dedicated and continuously track multiple faces and eyes. Only when the above constraints are satisfied a shot is automatically taken; an example is depicted in Figure 6.7 (left).

We have developed a suite of routines for real-time 3D reconstruction using structured light patterns (Robinson et al., 2004). To avoid the need for accurate mechanisms and in order to speed up the acquisition process, a number of stripes can be projected at the same time and captured as a sequence of stripes in a single frame. However, it may be difficult to determine which captured stripe corresponds to which projected stripe when we attempt to index the captured sequence in the same order as the projected sequence. We call this the stripe indexing problem. For this reason methods have been devised to uniquely mark each stripe, by colour (Rocchini et al., 2001), stripe width (Daley and Hassebrook, 1998), and by a combination of both (Zhang et al., 2002).

Our research has shown the dependence between the stripe index and the measurement of a surface vertex defined by the common inclination constraint (Robinson et al., 2004). We also deal with occlusions (Wang and Oliveira, 2007) to improve the validity of the boundaries. Moreover, we have investigated how far the indexing problem can be solved with uncoded stripes, where the correct order of stripes in the captured image is

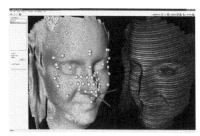

Figure 6.7 Left: when 2D constraints are satisfied for face and eye tracking, structured light is projected. Right: the captured 2D image is processed into 3D and texture mapping can be changed by 3D post-processing operations

determined by original algorithmic methods such as the maximum spanning tree algorithm (Brink et al., 2008). A number of 3D post-processing techniques can be applied, such as the ones discussed in Rodrigues and Robinson (2009); and Wang and Oliveira (2007) resulting in 3D models as depicted in Figure 6.7 (right).

6.9.2 Pattern recognition

Object recognition is based on feature extraction starting from three key feature points: the location of each eye and the tip of the nose. Our method is based on cutting oriented planes from the key features and detecting points on the mesh at the interception of those planes. We require pose alignment where the origin is placed at the tip of the nose, the eyes are aligned with the x-axis, and the y-axis is at a constant angle to a point at mid-distance between the eyes. This is achieved through an automatic iterative process, which has worked successfully even if the subject is not directly facing the camera; it has been tested on images facing up to 45 degrees to either side. A total of 43 points are located at the intersection of the various planes defined from the key features. An example of such points is depicted in Figure 6.8. Measurements are taken from such points as distances and ratios, in addition to area, volume, perimeter, and various types of diameters, such as breath and length, resulting in a set of 191 measurements per face model. The face models have been tested and recognised at 97 per cent accuracy for a database containing 276 models.

6.9.3 Discussion

Our research has demonstrated that defining specific robotic recognition tasks can be achieved in two distinct stages using appropriate constraints: 2D is well-suited for tracking objects in real time based on redundancy of information. Many objects (e.g. faces and eyes) can be tracked from a 2D scene and the actual selection of a particular scene for 3D reconstruction is made using constraints of size and number of objects detected.

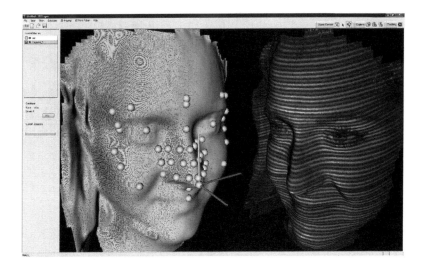

Figure 6.8 Automatic pose alignment and feature extraction

Once a scene is reconstructed in 3D, then more specific knowledge is required about the object of interest. We used the example of facial recognition that relies on geometric knowledge of a face object for pose normalisation and feature extraction. The need for further knowledge is likely to be true for the recognition of most 3D objects as their boundary conditions are only partially controlled through 2D tracking.

While the original question on deploying the perception/anticipation/action model remains largely unanswered, our research into 3D has opened new avenues of investigation. This includes games, animation, entertainment, security and engineering among others. The limitations of the method are related to projection issues, as objects over 4m from the camera cannot be reconstructed. We are currently working on new techniques to project sharp stripes over longer distances.

6.10 Summary

In our research journey we have learned much about the nature of autonomy and embodiment. It is interesting that our original research themes have continued, albeit in much advanced form, but still focused on certain key themes. We summarise these briefly.

In the quest for more fluid, flexible, and animate behaviour from our robots, we need to continue with the Embodied Intelligence approach to building artificial cognitive systems because the nature of the physical hardware of a robot is an essential factor in determining both its behaviour and the extent of its cognitive abilities. In addition, Developmental Robotics starts from the assumption that early experience provides a vital grounding for later competencies and thus offers an approach into the difficult problem of the growth of competence. Constraint networks could provide an environment in which behaviour develops without complex mechanisms for each stage; in other words: 'Gradual removal of constraint could account for qualitative change in behaviour without structural change' (Tronick, 1972).

Our interest in novelty as a learning spur has been justified by much current interest in Intrinsic Motivation (Oudeyer et al., 2007) which is examining the drivers for self-motivation, of which expectation and novelty are integral aspects. We have seen the value of insisting on natural, unstructured environments, and the high degree of complexity that this may entail. We have argued that complexity and chaos need to be faced, not avoided, and that tools for measuring behaviour in such situations are essential for good science. Our examples included tools to measure the degree of chaos and sensitivity to initial conditions. We hope that more robotic science and high-quality results will emerge from the use of new measurement techniques, proven authentic models, and take maximum advantage of psychological knowledge of cognitive growth.

Acknowledgements

The early joint work reported in the section on Control and Perception was carried out between 1993 and 1996 and was supported by the Medical Research Council through grant G9110835/G9125255 entitled 'Cognitive Modelling and the Design of Artificially Intelligent Systems'. The work was undertaken in the Centre for Intelligent Systems at the University of Wales, Aberystwyth (Principal investigator: Mark Lee) and in the Department of Psychology at the University of Edinburgh (Principal investigator: Brendan McGonigle). We also gratefully acknowledge the BBSRC, EPSRC and the European Commission for their support of many subsequent projects that developed the research further as described in the other sections. We also especially appreciated the help, advice, and encouragement of many colleagues, students, and friends.

References

Bradski, G. and Pisarevsky, V. (1999). Intel's computer vision library: applications in calibration, stereo, segmentation, tracking, gesture, face and object recognition. In *IEEE Computer Science Conference on Computer Vision and Pattern Recognition,* Volume 2. IEEE Computer Society; 1999.

Brink, W., Robinson, A. and Rodrigues, M. (2008). Indexing uncoded stripe patterns in structured light systems by maximum spanning trees. In *British Machine Vision Conference*, pp. 575–84.

Bruner, J. (1990). *Acts of meaning*. Cambridge, MA: Harvard University Press.
Chao, F., Lee, M. and Lee, J. (2010). A developmental algorithm for ocular-motor coordination. *Robotics and Autonomous Systems*, 58: 239–48.
Chen, S. and Billings, S. A. (1989). Representations of non-linear systems: the Narmax model. *Int. J. Control*, 49: 1013–32.
Daley, R. and Hassebrook, L. (1998). Channel capacity model of binary encoded structured light-stripe illumination. *Applied Optics*, 37: 17.
Gallahue, D. (1982). *Understanding motor development in children*. New York: John Wiley.
Hainline, L. (1998). How the visual system develops. In A. Slater (ed.), *Perceptual development: Visual, auditory, and speech perception in infancy*. Hove, UK: Psychology Press, pp. 5–50.
Hendriks-Jensen, H. (1996). *Catching ourselves in the act*. Cambridge, MA: MIT Press.
Hülse, M., McBride, S. and Lee, M. (2009a). Implementing inhibition of return: Embodied visual memory for robotic systems. In *Proc. 9th Int. Conf. on Epigenetic Robotics: Modeling Cognitive Development in Robotic Systems*. Lund University, Cognitive Studies, vol. 146, pages 213–14.
Hülse, M., McBride, S. and Lee, M. (2009b). Robotic hand-eye coordination without global reference: a biologically inspired learning scheme. In *Proc. Int. Conf. on Developmental Learning*, ICDL.
Hülse, M., McBride, S. and M., L. (2010). Fast learning mapping schemes for robotic hand-eye coordination. *Cognitive Computation*, 2(1): 1–16.
Keil, F. (1990). Constraints on constraints: surveying the epigenetic landscape. *Cognitive Science*, 14(4): 135–68.
Lakoff, G. (1987). *Women, fire, and dangerous things*. Chicago: University of Chicago Press.
Lungarella, M., Metta, G., Pfeifer, R. and Sandini, G. (2003). Developmental robotics: a survey. *Connection Science*, 15(4): 151–90.
Metta, G., Sandini, G., Vernon, D., Natale, L. and Nori, F. (2008). The iCub humanoid robot: an open platform for research in embodied cognition. In *PerMIS: Performance Metrics for Intelligent Systems Workshop*, August: 19–21.
Nehmzow, U. (2003). *Mobile robotics: a practical introduction*. London: Springer-Verlag.
Nehmzow, U. (2009). *Robot behaviour: design, description, analysis and modelling*. London: Springer-Verlag.
Nehmzow, U., Smithers, T. and McGonigle, B. (1993). Increasing behavioural repertoire in a mobile robot. In J. Meyer, H. Roitblat, and S. Wilson, (eds), *From animals to animats*, vol. 2. Cambridge, MA: MIT Press, pp. 291–7.
Neto, H. and Nehmzow, U. (2004). Visual novelty detection for inspection tasks using mobile robots. In *Proc. Brasilian Symposium on Neural Networks, IEEE Computer Society*, Los Alamitos.
Newell, A. and Simon, H. (1976). Computer science as empirical enquiry: symbols and search. *Communications of the ACM*, 19: 113–26.
Oudeyer, P., Kaplan, F. and Hafner, V. (2007). Intrinsic motivation systems for autonomous mental development. *IEEE Transactions on Evolutionary Computation*, 11(2): 265–86.
Ozbilge, E., Nehmzow, U. and Condell, J. (2009). Expectation-based novelty detection in mobile robotics. In *Proc. 'Towards Autonomous Robotic Systems' (TAROS)*.
Pfeifer, R. and Scheier, C. (1997). Sensory-motor coordination: the metaphor and beyond. *Robotics and Autonomous Systems*, 20(2): 157–78.

Piaget, J. (1973). *The child's conception of the world*. London: Paladin.
Powers, W. (1973). *Behavior: the control of perception*. Chicago: Aldine.
Prince, C., Helder, N. and Hollich, G. (2005). Ongoing Emergence: a Core Concept in Epigenetic Robotics. In *Proceedings of the 5th Int. Workshop on Epigenetic Robotics*. Lund, SE: Lund University Cognitive Studies, pp. 63–70.
Rizzolatti, G. and Arbib, M. (1998). Language within our grasp. *Trends in Neurosciences*, 21(5): 188–94.
Robinson, A., Alboul, L. and Rodrigues, M. (2004). Methods for indexing stripes in uncoded structured light scanning systems. *Journal of WSCG*, 12(3): 371–8.
Rocchini, C., Cignoni, P., Montani, C., Pingi, P. and Scopigno, R. (2001). A low cost 3D scanner based on structured light. In *Computer Graphics Forum*, vol. 20, pp. 299–308. Geneva: Eurographics.
Rodrigues, M. and Robinson, A. (24–27 July 2009). Novel methods for real-time 3D facial recognition. In *5th Int. Conf on Comp Science and Info Systems*.
Rodrigues, M. A. and Lee, M. H. (eds) (1994). Perceptual Control Theory. *Proceedings of the 1st European Workshop on Perceptual Control Theory*. Aberystwyth, UK: The University of Wales.
Rutkowska, J. (1994). Scaling up sensorimotor systems: constraints from human infancy. *Adaptive Behaviour*, 2: 349–73.
Tronick, E. (1972). Stimulus control and the growth of the infant's effective visual field. *Perception and Psychophysics*, 11(5): 373–6.
Wang, J. and Oliveira, M. (2007). Filling holes on locally smooth surfaces reconstructed from point clouds. *Image and Vision Computing*, 25(1): 103–13.
Zhang, L., Curless, B. and Seitz, S. M. (2002). Rapid shape acquisition using color structured light and multi-pass dynamic programming. In *3D Data Processing Visualization and Transmission, 2002. Proceedings. First International Symposium* IEEE, pp. 24–36.

7
Structuring Intelligence: The Role of Hierarchy, Modularity and Learning in Generating Intelligent Behaviour

Joanna J. Bryson

7.1 Introduction

Scientific social trends ranging from dynamical systems theories to postmodernism have called into question whether the apparent hierarchical structure of naturally occurring intelligent behaviour actually derives from structured intelligence. These questions and perspectives overlook substantial evidence from both neuroscience and biology more broadly that behaviour really is organised utilising both modularity and hierarchy. In mammal brains, modular structure starts from cellular composition, and continues conspicuously through the existence of discrete regions with differing processing capacities (Badre, 2008). We can discriminate a brain region by its consistent and regular pattern of nerve-cell type and intercell connectivity, while these same features vary between regions. Hierarchy derives from the interaction between these modules. This is particularly apparent when we can produce complex movements such as the production of particular words (Mateer et al., 1990) or grasping and transfer gestures (Graziano et al., 2002) by directly stimulating single cells.

Computer science tells us that a wide variety of possible computational architectures are able to produce the same computational outcome (Turing, 1936). Why then would evolution select something so elaborate as a highly regionally differentiated vertebrate brain? As scientists we believe that the most parsimonious explanation is the most probable. Similarly, the most simple structure sufficient is the most likely to be discovered and maintained by evolution. However, recognising parsimony is not always easy (Myung et al., 2000; Dawkins, 1997). Many factors need to be taken into consideration, such as path dependency (the impact of historical accident) and metabolic efficiency.

Theoretical computer science can help us understand why evolution reliably finds structured solutions. Efficiency is determined by task, in learning and information just as in physics (Wolpert, 1996). A modular brain addresses the variety of tasks intelligence requires with an efficiency that no single representation could provide. And animal-like intelligence does require varied capabilities. The most obvious example of this is the variety of sensory inputs we utilise. Light and sound each provide us with information the other cannot; the same is true of somatic touch and internal proprioception. The sensory organs that first detect these forms of information can already be thought of as specialised modules of our overall intelligence, but so too could the various cortical regions that process this information further downstream. Here the specialist regions may not be specialised just to sensory modality, but rather to information structure. For example, the 'visual cortex' in fact processes spatial information more generally; it is also invoked in learning Braille (Kauffman et al., 2002).

Following from the above brief introduction to the evidence of and explanations for structure in natural intelligence, this chapter will next describe the history of theories of structured control, attempting to explain their periodic losses and resurgence. I then address a few of the criticisms that underlie current scepticism concerning structured control. Finally, I discuss the origins of intelligent behaviour in the large, showing that modularity and hierarchy are pervasive in explaining the varieties of intelligent behaviour found in nature.

7.2 The history of theories of structured control

We often tend to think of scientific understanding as a monotonically increasing library of information, but this is not the case. While we may accumulate data more or less like a collection, real understanding requires the construction of theories. This construction is a social process. Scientists are trained to know that science is never about certainty. Science is rather a dynamic process concerned with approaching certainty through evidence and reason. Thus, at any particular time, leading experts on a topic will know a variety of hypotheses, the evidence for and against each of them, and experts associated with each of these matters. *Scientific understanding* then is really an aggregate term, reflecting a variety of dominant theories and individual hypotheses held at a particular time.

In this section I review the last century of history concerning the scientific evidence for and against structure playing a role in intelligent control. Given, as the previous section summarised, the prevalence of evidence for structure in intelligence, understanding why there is any doubt in the matter requires understanding the importance of theory on understanding, and the importance of personality, philosophy, politics and history on theory.

7.2.1 Structure

One of the defining features of intelligent behaviour is the ordering of expressed individual actions into coherent, apparently rational patterns.

From approximately 1951 until the mid 1980s, the dominant theories in both psychology and artificial intelligence for explaining intelligent behaviour held that hierarchical and sequential structures internal to the agent or animal underlie this ordered expression (e.g. McGonigle and Chalmers, 1996; Tinbergen, 1951; Hull, 1943; Lashley, 1951; Piaget, 1954; Dawkins, 1976). However, the last two decades have seen an increase of support for a more dynamic theory of intelligence (e.g. Botvinick, 2007; Port and van Gelder, 1995). This new theory holds that intelligence, like the brain itself, is actually composed of enormous numbers of small processes operating in parallel. Several researchers in this new paradigm have claimed that behaviour controlled by hierarchy is necessarily rigid, brittle and incapable of reacting quickly and opportunistically to changes in the environment (Goldfield, 1995; Seth, 2007; Hendriks-Jansen, 1996; Maes, 1991). They suggest that the apparent hierarchical organisation of behaviour is not the result of internal structured control, but it is rather only an inadequate model imposed on a far more complex dynamic process.

Hendriks-Jansen traces the hierarchical theory of behaviour organisation to the ethologist McDougall (1923), who presented a theory of the hierarchy of instincts. Ethological theory during this period, however, was dominated by Lorenz, who 'denied the existence of superimposed mechanisms controlling the elements of groups', instead believing that 'the occurrence of a particular activity was only dependent on the external stimulation and on the threshold for release of that activity.' (Baerends, 1976: 726 cited in Hendriks-Jansen, 1996: 233–4). This theory had obvious correlates with the then-popular black-box theories of Skinner (1935) and the behaviourists, though these were actually agnostic as to the underlying organisation of behaviour. The behaviourists' agnosticism was itself a scientific-political position set in opposition to Freud.

Lashley (1951) revitalised the hierarchical theory of behaviour with both data and reason. He demonstrated that hierarchy is the only explanation for the speed of some action sequences, such as those involved in human speech or the motions of the fingers on a musical instrument. Neural processes are simply too slow to allow elements of such sequences to be independently triggered in response to one another. Lashley therefore proposes that all the elements of such a sequence must be simultaneously activated by a separate process, the definition of hierarchical organisation. The exact neurological explanations of sequence learning are still an active area of research. However, Lashley's argument is well established. Items of a sequence are initially activated hierarchically, as a set, and then released by a second mechanism (Henson et al., 1996; Salthouse, 1986; Davelaar, 2007).

From roughly the time of Lashley's analysis, hierarchical models have dominated attempts to model intelligence. Particularly notable are the models of Tinbergen (1951) and Hull (1943) in ethology, Chomsky (1957) in linguistics and Newell and Simon (1972) in artificial intelligence and human problem solving. Mainstream psychology has been less concerned with creating specific models of behaviour control, but generally assumes hierarchical organisation as either an implicit or explicit consequence of goal-directed or cognitive theories of behaviour (Bruner, 1982). Staged theories of development and learning are also hierarchical when they describe complex skills being composed of simpler, previously developed ones (Piaget, 1954; Greenfield, 1991; Karmiloff-Smith, 1992).

In a tribute to Tinbergen (his PhD supervisor), Dawkins (1976) extended Lashley's argument for hierarchical control. Dawkins used computational parsimony to argue for hierarchical theories. Dawkins argues that it is more likely that a complex action sequence useful in multiple situations should be evolved or learned a single time, and that it is also more efficient to store a single instance of such a skill. Dawkins' arguments and proposals anticipate many of the techniques developed later for robust control in artificial intelligence. In particular, Brooks (1986,1991b) advocates the decomposition of intelligence into task-specific modules. These modules, termed *behaviours* are in turn organised into fairly egalitarian networks of competence, termed *levels*. As the name implies, the levels are themselves organised in a stack, with the highest level subsuming the goals of the lower levels, though these are assumed to be operating in parallel. For each level, the levels below it provide a reliable behavioural context for its operation. Ironically, although Brooks' system, the Subsumption Architecture, exploits both modular organisation and a simple, linear, goal-based hierarchy of module coordination, his approach has been taken by journalists and philosophers as evidence for homogeneous dynamical models intelligence (Kelly, 1995; Hendriks-Jansen, 1996). Shanahan (2005) points out a similar inconsistency in some people's perception of the Global Workspace Theory of conscious action selection.

More recently, many authors treat the hierarchical and modular structure of intelligence as obvious and accepted facts (Byrne, 1999; Prescott et al., 2007). Nevertheless, the controversy persists (Botvinick, 2007; Seth, 2007; Barsalou et al., 2007).

7.2.2 Emergence

The competing theory, that responsive animal intelligence cannot possibly be governed by hierarchical control, has emerged from some of the practitioners of the dynamic hypothesis of cognition (van Gelder, 1998). These researchers tend to seek a single mathematical system of representation to be sufficient for describing and learning intelligent behaviour. Not all researchers that fit this description are in principle opposed to hierarchical structure. For example, since the mid 1990s much machine learning has exploited hierarchical hidden Markov models (Fine et al., 1998). Although these are provably equivalent to single-layer Markov models, learning systems are simply easier to engineer when hierarchy is added. The same is true in neural network architectures, though here there are theoretical reasons to have at least two layers (Minsky and Papert, 1969). Nevertheless, most researchers choose at least three, for purely practical engineering reasons (Hansen and Salamon, 1990). Such approaches though are still relatively homogeneous, overlooking the example of nature favouring modularity as well as hierarchy.

The theory of dynamic action expression suggests that complex dynamic or chaotic systems operate within the brain producing the next behaviour not by selecting an element of a plan, but rather as an emergent consequence of many parallel processes (e.g. Seth, 2007; van Gelder, 1998; Brooks, 1991a; Goldfield, 1995; Kelso, 1995; Barsalou et al., 2007; Hendriks-Jansen, 1996; Maes, 1991). Evidence supporting the older hypothesis of structured hierarchical behaviour is seen to have been biased by the hierarchical and sequential nature of human explicit thought and language. In particular, because much theoretical work in psychology is conducted using computer models, theories may be biased towards the workings and languages of the serial processors of the machines available to most psychologists (Brooks, 1991a).

A fundamental appeal of the dynamic hypothesis is that it is necessarily correct, at least to some level. Assuming a materialist stance, intelligence is known to be based in the parallel operation of the body's neural and endocrine systems. It is nearly as well accepted that human and animal behaviour can be described as hierarchically ordered (Byrne and Russon, 1998; Dawkins, 1976; Greenfield, 1991; Prescott et al., 2007). Though note even here that in response to the Byrne and Russon (1998) target article on imitation of the hierarchical structure of behaviour, nearly a fifth of the commentaries chosen to appear with the article questioned the existence of hierarchical control (Vereijken and Whiting, 1998; Gardner and Heyes, 1998; Jorion, 1998; Mac Aogáin, 1998) of 21 commentaries.

The question is, given that there is statistically discriminable sequential and hierarchical ordering of behaviour, how and when are these behaviours so organised? Some argue that the order is emergent, which is to say not encoded explicitly but rather a consequence of the interaction of a system of simpler processes. Viewed from an atomic or even cellular level, this is of course necessarily true of all behaviour – in fact, of all matter. Further, few would argue that animals have complete structured behaviour plans that are expressed regardless of their environment. However, the structure of actions we see emerges not from an amorphous homogeneous being, but from the modular and hierarchical neural control system found in vertebrates. It is natural and parsimonious for us to conclude that this structure at least partially determines the patterns of action it generates.

7.3 The confusion of theories of structured control

Given the strong evidence of neuroscience for structured intelligent control, why does controversy continue in this area? There are several possible reasons. First, science is a political process carried out by humans with deep personal concerns and values. In particular, the twentieth century witnessed a large swing in human thought towards the importance of egalitarianism. Such political influence has had documented effect on evolutionary theory (Sterelny, 2007); it may have influenced ethology as well. Second, as the mathematics of chaos and complexity became known, it was correct scientific procedure as well as natural that its practitioners should try to account for as much data as possible with this new approach. But, finally, there have been a few specific influential theories that ultimately proved to be simply wrong. These wrong theories have turned some scientists away from concepts associated with the theories. In this section I address a few of these misconceptions.

7.3.1 Modularity does not require language, strict encapsulation or innateness

One reason scientists such as Elman et al. (1996) or Barsalou et al. (2007) reject modularity as an explanation of human intelligence is because they are really rejecting one particular characterisation of it. In psychology, the best-known description of modularity is due to Fodor (1983). One of the primary criteria Fodor provides for recognising a module is that a module must be innate, not learned. The lifelong interplay between genes and environment makes many developmental psychologists and biologists uncomfortable with the entire category of innateness (e.g. Elman et al., 1996; Thelen and Smith, 1994; Griffiths, 2001; Donnai and Karmiloff-Smith, 2000). Of course, innateness has no bearing on the functional or computational characteristics of modularity. Many proponents of psychological modularity believe that modules develop over the ontogeny of an organism (e.g. Carruthers, 2005; Sperber and Hirschfeld, 2006; Bates, 1999). Here the idea is that regions or subnetworks of the brain specialise to particular tasks in a response to experience, in an ontogenetic recapitulation of the evolution of the neural specialisation we see in the brain's various regions.

Similarly, Barsalou et al. (2007) reject modular theories because modules must be fully encapsulated; that is, they cannot affect each other at the level of data or representation. This characterisation is also due to Fodor (1983), though in AI it was particularly popularised by Brooks (1991b). Not every modular system shares this characteristic either, the brain being an obvious exception. But in AI, too, memory can be usefully stored in modules, possibly with special-purpose representations, yet still made accessible to other modules via interfaces (Metta et al., 2006; Bryson, 2001,2000a; Thórisson et al., 2004).

Another confusion that Fodor (1983) introduced is the claim that modules only support perception and action, and must be combined into a true mind via a layer composed of general intelligence that translates between them. This somewhat bizarre claim has not only been accepted but modified and extended to imply that modules can only be integrated with language, and that this explains human exceptionality (Carruthers, 2003; Chomsky, 2000; Spelke, 2003). While there is no question that language substantially affects human intelligence and culture, as has already been explained, modular systems exist throughout natural and artificial intelligence, in systems well integrated without language. Human exceptionalism in accumulating culture probably substantially predates language, and in fact must account for language itself (Bryson, 2009; Buckley and Steele, 2002).

7.3.2 Hierarchical organisation does not imply slow, autocratic or fragile control

In the Introduction I mentioned that the presence of hierarchical control in the brain was demonstrated in neuroscience by the production of complex movements by directly stimulating single cells. Examples include speaking particular words in humans (Mateer et al., 1990) and making grasp and transfer gestures appropriate to scramble feeding in macaques (Graziano et al., 2002). Considering such cells to be near an apex of a pyramid of top-down control is the classic error of reasoning that has motivated much anti-hierarchical rhetoric. Rather, such cells should be thought of as a juncture, an intersection between two overlapping hierarchies (see Figure 7.1). The first hierarchy generates behaviour when the cell is highly stimulated. The second hierarchy is the combination of motivation and perceived opportunity that would ordinarily bring activation and other information to the cell when neuroscientists were not the ones stimulating it. In terms of flow of information through the brain, the two hierarchies can be seen as inverted from one another (see Figure 7.1). Neighbouring cells in the same layer of the hierarchy receive nearly the same set of inputs and distribute nearly the same sort of outputs. These neighbouring cells will generally be interconnected to each other through patterns of mutual inhibition and excitation, simplifying the coordination problem between cells at the same level (Riesenhuber and Poggio, 1999).

The firing of each such cell is sufficient to generate behaviour involving the further processing of millions of cells; not only is the action executed by thousands of muscle cells signalled by thousands of afferent nerves, but also the basic signal is further modulated and controlled by brain structures such as the cerebellum and basic feedback control loops in the spine. But what causes these cells to fire in the first place is at least as complex – a system of not only perception and sensor fusion, but also attention and goal arbitration. The normal stimulus for such an action requires recognition of external stimuli, itself a hierarchical process fusing information from the many cells that compose sensory apparatus with learned categories. Any normal environment contains many such recognisable stimuli, and which one an animal attends to must itself be governed in a way appropriate to the animal's current needs (Redgrave et al., 1999).

The fact that the brain is essentially concurrent also allows an explanation for how hierarchically structured behaviour can avoid the fragility normally associated with single-point control, and also avoid latencies associated with having to revisit every step of a hierarchy (a criticism of hierarchy due to Maes, 1991). Forebrain gating systems may focus attention directly on computation at one sub-task, even though a hierarchically organised path brought attention to that task in the first place (Rensink, 2000; Bryson and McGonigle, 1998). The sort of distributed system described above and shown in Figure 7.1 is also robust to damage, as such networks adapt dynamically

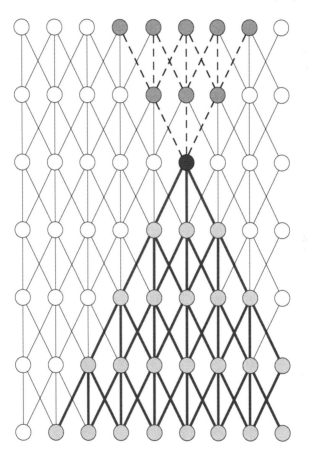

Figure 7.1 A gross abstraction of neural control. Hierarchical organisation can occur within columns, where peer nodes on the level of hierarchy each specialise as an apex for processing a certain set of sensory information, and for generating a certain form of behavioural control. Moderation between nodes at the same level of hierarchy happens both through mutual excitation and inhibition, and through executive control (cf. Redgrave, Prescott, and Gurney, 1999)

to account for data; they are perpetually learning in order to make more reliable predictions. Thus if damage removes part of the network, neighbouring cells adapt to fill in the gap in control.

My own work has shown in artificial intelligence that hierarchy working in interaction with concurrent modules can control mobile robots navigating ordinary office space (Bryson and McGonigle, 1998) and play complex real-time computer games like Capture the Flag against human opponents (Partington and Bryson, 2005). Further, I have shown that such hierarchical structure with switching attention can outperform a `fully aware' system at controlling an agent with multiple goals and threats in a fairly rich environment (Bryson, 2000b). In this last case I argue that my system worked better purely because it was easier to design and to adjust. Because the hierarchy reduced the complexity of the search space the designer has to parse, the program is sufficiently easier to optimise that this compensates for any loss of information through neglect of available options. The same combinatorics confront evolution and development in biology as confront human designers. Thus hierarchical structures provide advantages in all three cases.

7.4 Structural control extends beyond intelligence

So far I have described modularity as a decomposition of intelligent control for the purpose of reliability and/or efficiency. A sufficient though not necessary characteristic of modularity is for each module to have a distinct system of representation for storing acquired information. Once we begin to think about control as being decomposed modularly, we can also begin extending our understanding of intelligence into an understanding of the biological and evolutionary origins of cognition. This in turn can help us understand species-level differences in cognition and related traits, such as life history.

Modularity is already supported outside of psychology and neuroscience in evolutionary biology. Evolutionary developmental biology argues that physiological modularity, encoded by the genome, underlies and accelerates contemporary tetrapod evolutionary processes (Kirschner et al., 2006; Müller, 2008). Single mutations do not tend to make miniscule alterations to an animal, but rather to add, subtract or alter complexes of interrelated traits. For example, the shape of the beaks of Darwin's finches on the Galapagos is controlled by only two genes, and the same genes account for general head shape, skull thickness, and so forth, so that a bird will always be able to support its own beak (Campàs et al., 2010). My own work assigns exactly this role to modularity in the development of AI systems. Plasticity should only be made available where intelligence cannot practically be provided in advance by the developer, for example if a robot will be required to learn its own navigation routes in unknown buildings. Where such learning must take place, it should be performed in purpose-built modules that support the learning to the maximum extent possible (Bryson, 2001, 2003).

Can reasoning about modularity give insight into species-level differences in cognition? What needs to be explained? Cognition is often portrayed as being a universally useful ability. If this is true, we need to explain the diversity of cognitive capacities. We may be tempted to take an implicitly Lamarckian view, thinking that somehow humans are just particularly lucky or fittest, but in fact the discrepancies in fitness between extant species are not great (Kimura, 1985). Rather, extant species represent a position in a space of possible evolutionary trade-offs. The best-known example of such a trade-off is the quantity of offspring versus the amount of parental investment per offspring. We can think of species as ranged along an axis of variation between investing a great deal in a very small number of offspring (as with juvenile apes) or very little in a large number of offspring (as with fungus spores). Extant variation in cognitive abilities, ranging from apes through protozoa, indicates that there must be some compensating trade-off or trade-offs against cognition, creating axes of cognitive variation.

One of the axes of cognitive variation is the speed with which learning takes place. Even within species, the vertebrate brain consists of modules which learn at a variety of rates. Quick-learning systems can rapidly gather data from direct experience, but only because they can recognise and record that information with sparse representations (McClelland et al., 1995; Teyler and Discenna, 1986). The categories necessary for these sparse representations are formed through slower systems that extract generalisations from large quantities of data, possibly by considering repeated presentations (Louie and Wilson, 2001). This is one of the standard theories of the computational function of the hippocampal system: that it quickly gathers episodic memories which can then be used to train cortical systems on semantic knowledge. This episodic data itself relies on cortical/semantic category representation to provide a sufficiently sparse representation for storing large amounts of information in a confined brain region (Foster and Wilson, 2006; Rogers and McClelland, 2004; McClelland et al., 1995; Teyler and Discenna, 1986; Louie and Wilson, 2001).

Not every aspect of individual behaviour is learned. A great deal of behaviour is determined genetically, including (as the behaviourists eventually discovered) which sorts of stimuli can be associated with which types of action (Gallistel et al., 1991). This is true not only from the perspective of the architecture of the brain, but also the morphological characteristics of the body, which determine the possible range of an animal's actions, its capacities of perception, and its metabolic needs (Paul, 2004; Nehmzow et al., 1993; Pfeifer, 1999). Genetic evolution might also be viewed as a process of learning, a means by which species and therefore individuals within species acquire behaviour.

If we are willing to think of learning (or at least behaviour acquisition) in this way – as something that sometimes occurs outside of an individual – then we can postulate a continuum of adaptation processes running from individual development and learning through evolution and speciation. The computational costs and benefits of modularity and hierarchy are the same in all these cases. Thus we find modular architectures throughout nature: interacting but very different representations that generate behaviour. Genes in a genome, nerve cells in a brain, individuals in a society, species in an ecosystem – each of these levels of hierarchy has particular mechanisms for producing behaviour, and of acquiring and representing variation, and of performing optimisation across the possible behaviours. Each layer itself contains modules composed of relatively more homogeneous individual elements expressing relatively similar strategies across a relatively narrow spread of variation. For example, we can see in tetrapod embryos stable selection for two different types of development: a robust core spinal area where each cell is autonomous and controls its own expression, and the elaborated limb regions where hierarchical control of gene expression allows the relatively rapid evolution of novel forms (Winslow et al., 2007). Neither representation usurps the other, though one (the spinal version) is older. The hierarchical representation increases evolvability so is adaptive, but also depends on a more robust representation at the core. Within these modules, further sub-modules develop which become limbs, digits and organs (Müller, 2008).

We can similarly think of species as modules within an ecosystem. If all currently present life on earth has evolved from a single replicator, then the fact that the subsets of individuals that engage in recombination of their genotypes is not entirely arbitrary is interesting. Which couples co-parent offspring depends not only on physical proximity, but also on species. Thus a species might be thought of as a modularised search mechanism, which optimises a particular behavioural strategy which is encoded in that species' genotype.

One interesting trade-off helps bridge two of the levels I described earlier: the individual versus the society. This is the inclusive-fitness trade-off. This trade-off determines how much effort an individual gives to its own reproductive success versus its relatives' (see e.g. West et al., 2007). For example, one is twice as related to one's own children than to one's nieces or nephews, so inclusive fitness tells us that we should be twice as ready to invest effort in our own children as in a sibling's (Hamilton, 1964). Of course, breeding can incur significant costs. Normally, one is not considered to be altruistic if one invests effort in one's offspring, but is this simplistic definition really justified? Altruism is not a mystery of evolution, but rather a measure of where along a spectrum of potential inclusive-fitness investments a species happens to lie. Individuals from more altruistic species invest more heavily in the reproductive success of their extended kin, though still presumably the cost is in proportion to their kin's benefit and relatedness.

In humans and other cognitive social species, behaviour is acquired not only from genetics but also from culture, from social transmission. We might choose to define kin in terms of sharing any inherited traits that control the inclusive-fitness trade-off, whether biological or cultural. This leap allows us to generalise our explanation of altruism further. Levels of individual altruism may reflect not only species-level strategies, but also context- or personality-based variation in individual strategies. For example, Rutte and Taborsky (2007) have demonstrated that individuals of some species behave more altruistically when they experience or even witness more altruistic behaviour themselves. This presumably facilitates more rapid adaptation to social versus individual strategies, in response to the current environmental context.

Where levels of altruism are personality based, the distribution of such personalities in a population is still subject to selection. Where they respond to recent individual experience, this is an example of a trade-off being adjusted 'on the fly' based on individual experience. Of course, even where phenotypic plasticity is exploited this way, the range of possible individual variation and the mechanisms for assessing appropriate levels of investment are also subject to selection.

To return to our starting point, there must also be a set of fitness trade-offs that relate to the expression of cognition itself. Here too the trade-offs concern the intersection of two of my proposed levels for representing behaviour. Cognition is a special case of individual plasticity, and the trade-offs between individual plasticity and biological evolution are beginning to be well understood (Paenke et al., 2007; Hinton and Nowlan, 1987; Borenstein et al., 2006; Maynard Smith, 1987). Individual plasticity can accelerate biological evolution, but biological evolution provides greater reliability than individual learning. Thus, in a stable environment, selection generally favours the reduction of individual plasticity, but in periods of rapid environmental change (including where that change is caused by ecological dynamics, e.g. in the number or proportion of species), there will be more pressure for individual plasticity. Interestingly, even in stable environments, the presence of individual plasticity can sometimes decelerate evolution as the genome becomes very near the optimum. This is because learning at this point becomes sufficiently reliable that there is no selective pressure to eliminate the final amount of variation from the optimum. Thus phenotypic plasticity may be a key factor in maintaining sufficient genetic variation in the population to allow robustness in the case that a previously stable selective environment suddenly does change.

Trade-offs concerning cognition occur between species as well as within them (Lefebvre et al., 2004; Barrickman et al., 2008). Species that commit to highly cognitive strategies may compensate for the uncertainty of individual learning by providing epigenetic information transmission, such as social transmission of behaviour. Species that rely on social transmission, though, pay a penalty not only in terms of metabolic cognitive cost, but also in terms of extended life history which decelerates the rate of biological evolution. Culture could in theory allow species to take advantage of massive increases in both the rate of acquisition of new information and the reliability of learning due to concurrency, the ability to exploit the cognitive capacity of many individuals simultaneously (Bryson, 2009). However, in practice only humans seem to exploit this capacity fully and then only in very recent history from an evolutionary perspective (Ambrose, 2001).

The main constraints on using culture as a source of behaviour have to do with the probability of transmitting information into the next generation. This is determined by a number of factors, such as the rate of information transmission between individuals, the amount of time individuals spend in proximity, and the average and maximum lifespans of the individuals (Bryson et al., 2009). Of course, not every agent needs to learn all behaviours in a culture. There must only be a sufficient number of individuals in the population carrying a culture that a particular behaviour is reliably carried forward from generation to generation by at least some individuals. The probability of social learning can be even lower if the probability of individual discovery of the behaviour is sufficiently high to compensate.

In this section then I have argued that modularity can be seen as a useful organisational principle for developing adaptive behaviour across a wide range of representations. These representations – genes, nerve cells, individuals, societies, species – each support adaptation, and have a variety of mechanisms for operating within modules, between modules, and between the layers of hierarchy determined by the representational substrate. Neural or psychological modularity such as the majority of this chapter has discussed is then just a special case of generally useful organisational strategies.

7.5 Conclusion

My argument in this chapter has been that hierarchical structures, including modularity, are pervasive in natural intelligence, though so too are concurrent solutions. The reason for this is simple combinatorial complexity. Control and design are more easily addressed when broken into units, both as parallel modules and as layers of control. AI researchers often neglect the hierarchical nature of their own solutions, as they build homogeneous networks for control, yet keep these networks small and simple while having the items that they control be well-designed powerful operators. Similarly some natural scientists argue vociferously against modular models of mind while never doubting that the brain, the limbs and the sense organs operate in different ways and make different contributions into intelligent behaviour. In this chapter I have illustrated how modularity and hierarchy pervade intelligence, from genetic evolution through the neural structure of vertebrate brains, the accumulation of social learning by a society, and the differentiation of species' behaviour in an ecosystem.

References

Ambrose, S. H. (2001). Paleolithic technology and human evolution. *Science*, 291: 1748–53.

Badre, D. (2008). Cognitive control, hierarchy, and the rostro-caudal organization of the frontal lobes. *Trends in Cognitive Sciences*, 12(5): 193–200.

Baerends, G. P. (1976). The functional organisation of behaviour. *Animal Behaviour*, 24: 726–38.

Barrickman, N. L., Bastian, M. L., Isler, K. and van Schaik, C. P. (2008). Life history costs and benefits of encephalization: A comparative test using data from long-term studies of primates in the wild. *Journal of Human Evolution*, 54(5): 568–90.

Barsalou, L. W., Breazeal, C. and Smith, L. B. (2007). Cognition as coordinated non-cognition. *Cognitive Processing*, 8(2): 79–91.

Bates, E. (1999). Plasticity, localization and language development. In S. Broman and J. M. Fletcher (eds), *The changing nervous system: Neurobehavioral consequences of early brain disorders*. Oxford: Oxford University Press, pp. 214–53.

Borenstein, E., Meilijson, I. and Ruppin, E. (2006). The effect of phenotypic plasticity on evolution in multipeaked fitness landscapes. *Journal of Evolutionary Biology*, 19(5): 1555–70.

Botvinick, M. M. (2007). Multilevel structure in behaviour and in the brain: A model of Fuster's hierarchy. *Philosophical Transactions of the Royal Society, B – Biology*, 362(1485): 1615–26.

Brooks, R. A. (1986). A robust layered control system for a mobile robot. *IEEE Journal of Robotics and Automation*, RA-2: 14–23.

Brooks, R. A. (1991a). Intelligence without reason. In *Proceedings of the 1991 International Joint Conference on Artificial Intelligence*, Sydney, pp. 569–95.

Brooks, R. A. (1991b). Intelligence without representation. *Artificial Intelligence*, 47: 139–59.

Bruner, J. S. (1982). The organisation of action and the nature of adult-infant transaction. In M. von Cranach and R. Harré (eds), *The analysis of action*. Cambridge: Cambridge University Press, pp. 313–28.

Bryson, J. J. (2000a). Cross-paradigm analysis of autonomous agent architecture. *Journal of Experimental and Theoretical Artificial Intelligence*, 12(2): 165–90.

Bryson, J. J. (2000b). Hierarchy and sequence vs. full parallelism in reactive action selection architectures. In *From animals to animats 6 (SAB00)*. Cambridge, MA: MIT Press.

Bryson, J. J. (2001). Intelligence by Design: Principles of Modularity and Coordination for Engineering Complex Adaptive Agents. PhD thesis, MIT, Department of EECS, Cambridge, MA. AI Technical Report 2001-003.

Bryson, J. J. (2003). The behavior-oriented design of modular agent intelligence. In R. Kowalszyk, J. P. Müller, H. Tianfield and R. Unland (eds), *Agent Technologies, Infrastructures, Tools, and Applications for e-Services*. Berlin: Springer, pp. 61–76.

Bryson, J. J. (2009). Representations underlying social learning and cultural evolution. *Interaction Studies*, 10(1): 77–100.

Bryson, J. J., Lowe, W., Bilovich, A. and Čače, I. (2009). Factors limiting the biological evolution of cultural accumulation. Submitted.

Bryson, J. J. and McGonigle, B. (1998). Agent architecture as object oriented design. In M. P. Singh, A. S. Rao, M. J. and Wooldridge (eds), *The Fourth International Workshop on Agent Theories, Architectures, and Languages (ATAL97)*, Providence, RI: Springer, pp. 15–30.

Buckley, C. and Steele, J. (2002). Evolutionary ecology of spoken language: Co-evolutionary hypotheses are testable. *World Archaeology*, 34(1): 26–46.

Byrne, R. W. (1999). Imitation without intentionality. Using string parsing to copy the organization of behaviour. *Animal Cognition*, 2(2): 63–72.

Byrne, R. W. and Russon, A. E. (1998). Learning by imitation: A hierarchical approach. *Brain and Behavioral Sciences*, 21(5): 667–721.

Campàs, O., Mallarino, R., Herrel, A., Abzhanov, A. and Brenner, M. P. (2010).Scaling and shear transformations capture beak shape variation in Darwin's finches. *Proceedings of the National Academy of Sciences*, 107(8): 3356–60.

Carruthers, P. (2003). The cognitive functions of language. *Brain and Behavioral Sciences*, 25(6): 657–74.

Carruthers, P. (2005). The case for massively modular models of mind. In R. Stainton (ed.), *Contemporary debates in cognitive science*. Hoboken, NJ: Wiley-Blackwell.

Chomsky, N. (1957). *Syntactic structures*. The Hague: Mouton.

Chomsky, N. (2000). Linguistics and brain science. In A. Marantz, Y. Miyashita and W. O'Neil, (eds), *Image, language, brain: Papers from the first mind articulation project symposium*. Cambridge, MA: MIT Press, pp. 13-28.

Davelaar, E. J. (2007). Sequential retrieval and inhibition of parallel (re)activated representations: A neurocomputational comparison of competitive queuing and resampling models. *Adaptive Behavior*, 15(1): 51–71.

Dawkins, R. (1976). Hierarchical organisation: A candidate principle for ethology. In P. P. G. Bateson and R. A. Hinde (eds), *Growing points in ethology*. Cambridge, Cambridge University Press, pp. 7–54.

Dawkins, R. (1997). *Climbing mount improbable*. New York: W. W. Norton & Co.

Donnai, D. and Karmiloff-Smith, A. (2000). Williams Syndrome: From genotype through to the cognitive phenotype. *American Journal of Medical Genetics*, 97(2): 164–71.

Elman, J. L., Bates, E. A., Johnson, M. H., Karmiloff-Smith, A., Parisi, D. and Plunkett, K. (1996). *Rethinking innateness: A connectionist perspective on development*. Cambridge, MA: MIT Press.

Fine, S., Singer, Y. and Tishby, N. (1998). The hierarchical hidden Markov model: Analysis and applications. *Machine Learning*, 32(1): 41–62.

Fodor, J. A. (1983). *The modularity of mind*. Cambridge, MA: Bradford Books, MIT Press.

Foster, D. J. and Wilson, M. A. (2006). Reverse replay of behavioural sequences in hippocampal place cells during the awake state. *Nature*, 440(7084): 680–3.

Gallistel, C., Brown, A. L., Carey, S., Gelman, R. and Keil, F. C. (1991). Lessons from animal learning for the study of cognitive development. In S. Carey and R. Gelman (eds), *The epigenesis of mind*. Hillsdale, NJ: Lawrence Erlbaum, pp. 3–36.

Gardner, M. and Heyes, C. (1998). Splitting lumping and priming. *Brain and Behavioral Sciences*, 21(5): 695.

Van Gelder, T. (1998). The dynamical hypothesis in cognitive science. *Behavioral and Brain Sciences*, 21(5): 616–65.

Goldfield, E. C. (1995). *Emergent forms: Origins and early development of human action and perception*. Oxford: Oxford University Press.

Graziano, M. S. A., Taylor, C. S. R. and Moore, T. (2002). Complex movements evoked by microstimulation of precentral cortex. *Neuron*, 34: 841–51.

Greenfield, P. M. (1991). Language, tools and brain: The ontogeny and phylogeny of hierarchically organized sequential behavior. *Brain and Behavioral Sciences*, 14: 531–95.

Griffiths, P. E. (2001). Genetic information: A metaphor in search of a theory. *Philosophy of Science*, 68(3): 394–412.
Hamilton, W. D. (1964). The genetical evolution of social behaviour. *Journal of Theoretical Biology*, 7: 1–52.
Hansen, L. K. and Salamon, P. (1990). Neural network ensembles. *IEEE Transactions on Pattern Analysis and Machine Intelligence*, 12(10): 993–1001.
Hendriks-Jansen, H. (1996). *Catching ourselves in the act: Situated activity, interactive emergence, evolution, and human thought.* Cambridge, MA: MIT Press.
Henson, R. N. A., Norris, D. G., Page, M. P. A. and Baddeley, A. D. (1996). Unchained memory: Error patterns rule out chaining model of immediate serial recall. *Quarterly Journal of Experimental Psychology*, 49(1): 80–115.
Hinton, G. E. and Nowlan, S. J. (1987). How learning can guide evolution. *Complex Systems*, 1: 495–502.
Hull, C. (1943). *Principles of behaviour: An introduction to behaviour theory.* New York: D. Appleton-Century Company.
Jorion, P. J. M. (1998). A methodological behaviourist model for imitation. *Brain and Behavioral Sciences*, 21(5): 695.
Karmiloff-Smith, A. (1992). *Beyond modularity: A developmental perspective on cognitive change.* Cambridge, MA: MIT Press.
Kauffman, T., Theoret, H. and Pascual-Leone, A. (2002). Braille character discrimination in blindfolded human subjects. *Neuroreport*, 13(5): 571–4.
Kelly, K. (1995). *Out of control: The new biology of machines, social systems, and the economic world.* Boston: Addison-Wesley Longman.
Kelso, J. S. (1995). *Dynamic patterns: the self-organization of brain and behavior.* Cambridge, MA: MIT Press.
Kimura, M. (1985). *The neutral theory of molecular evolution.* Cambridge: Cambridge University Press.
Kirschner, M. W., Gerhart, J. C. and Norton, J. (2006). *The plausibility of life.* New Haven, CT: Yale University Press.
Lashley, K. S. (1951). The problem of serial order in behavior. In L. A. Jeffress (ed.), *Cerebral mechanisms in behavior.* New York: John Wiley & Sons.
Lefebvre, L., Reader, S. M. and Sol, D. (2004). Brains, innovations and evolution in birds and primates. *Brain, Behavior and Evolution*, 63(4): 233–46.
Louie, K. and Wilson, M. A. (2001). Temporally structured replay of awake hippocampal ensemble activity during rapid eye movement sleep. *Neuron*, 29(1): 145–56.
Mac Aogáin, E. (1998). Imitation without attitudes. *Brain and Behavioral Sciences*, 21(5): 696–7.
Maes, P. (1991). A bottom-up mechanism for behavior selection in an artificial creature. In J-A. Meyer and S. Wilson, (eds), *From animals to animats.* Cambridge, MA: MIT Press, pp. 478–85.
Mateer, C. A., Rapport II, R. L. and Polly, D. D. (1990). Electrical stimulation of the cerebral cortex in humans. In A. A. Boulton, G. B. Baker and M. Hiscock (eds), *Neuropsychology*, vol. 17 of *Neuromethods.* Clifton, NJ: Humana Press.
Maynard Smith, J. (1987). When learning guides evolution. *Nature*, 329: 761–2.
McClelland, J. L., McNaughton, B. L. and O'Reilly, R. C. (1995). Why there are complementary learning systems in the hippocampus and neocortex: Insights from the successes and failures of connectionist models of learning and memory. *Psychological Review*, 102(3): 419–57.
McDougall, W. (1923). *Outline of psychology.* London: Methuen.

McGonigle, B. O. and Chalmers, M. (1996). The ontology of order. In L. Smith (ed.), *Critical readings on Piaget*. London: Routledge, ch. 14.

Metta, G., Fitzpatrick, P. and Natale, L. (2006). YARP: Yet another robot platform. *International Journal on Advanced Robotics Systems*, 3(1): 43–8.

Minsky, M. and Papert, S. (1969). Perceptrons. Cambridge, MA: MIT Press.

Müller, G. B. (2008). Evo-devo as a discipline. In A. Minelli and G. Fusco (eds.) *Evolving Pathways: Key Themes in Evolutionary Developmental Biology* Cambridge: Cambridge University Press, pp 5–30.

Myung, J., Forster, M. R. and Browne, M. W. (2000). Special issue on model selection. *Journal of Mathematical Psychology*, 44(1).

Nehmzow, U., Smithers, T. and McGonigle, B. O. (1993). Increasing behavioural repertoire in a mobile robot. In *From animals to animats 2: (SAB '92)*. Cambridge, MA: MIT Press, pp. 291–7.

Newell, A. and Simon, H. A. (1972). *Human problem solving*. Upper Saddle River, NJ: Prentice-Hall.

Paenke, I., Sendhoff, B. and Kawecki, T. J. (2007). Influence of plasticity and learning on evolution under directional selection. *The American Naturalist*, 170(2): 47–58.

Partington, S. J. and Bryson, J. J. (2005). The behavior oriented design of an unreal tournament character. In T. Panayiotopoulos, J. Gratch, R. Aylett, D. Ballin, P. Olivier and T. Rist (eds), *The Fifth International Working Conference on Intelligent Virtual Agents, Kos, Greece*. New York: Springer, pp. 466-77.

Paul, C. (2004). Morphology and computation. In *From animals to animats 8 (SAB '04)*. Los Angeles. Cambridge, MA, pp. 33–8.

Pfeifer, R. (ed.) (1999). *Dynamics, morphology, and materials in the emergence of cognition*. New York: Springer.

Piaget, J. (1954). *The construction of reality in the child*. New York: Basic Books.

Port, R. F. and van Gelder, T. (eds) (1995). *Mind as motion: Explorations in the dynamics of cognition*. Cambridge, MA: MIT Press.

Prescott, T. J., Bryson, J. J. and Seth, A. K. (2007). Modelling natural action selection: An introduction to the theme issue. *Philosophical Transactions of the Royal Society, B – Biology*, 362(1485): 1521–9.

Redgrave, P., Prescott, T. J. and Gurney, K. (1999). The basal ganglia: a vertebrate solution to the selection problem? *Neuroscience*, 89: 1009–23.

Rensink, R. A. (2000). The dynamic representation of scenes. *Visual Cognition*, 7: 17–42.

Riesenhuber, M. and Poggio, T. (1999). Hierarchical models of object recognition in cortex. *Nature Neuroscience*, 2: 1019–25.

Rogers, T. T. and McClelland, J. L. (2004). *Semantic cognition: A parallel distributed processing approach*. Cambridge, MA: MIT Press.

Rutte, C. and Taborsky, M. (2007). The influence of social experience on cooperative behaviour of rats (*Rattus norvegicus*): Direct versus generalized reciprocity. *Behavioral Ecology and Sociobiology*, 62(4): 499–505.

Salthouse, T. A. (1986). Perceptual, cognitive, and motoric aspects of transcription typing. *Psychological Bulletin*, 99(3): 303–319.

Seth, A. K. (2007). The ecology of action selection: Insights from artificial life. *Philosophical Transactions of the Royal Society, B – Biology*, 362(1485): 1545–58.

Shanahan, M. P. (2005). Global access, embodiment, and the conscious subject. *Journal of Consciousness Studies*, 12(12): 46–66.

Skinner, B. F. (1935). The generic nature of the concepts of stimulus and response. *Journal of General Psychology*, 12: 40–65.

Spelke, E. S. (2003). What makes us smart? Core knowledge and natural language. In D. Gentner and S. Goldin-Meadow (eds), *Advances in the investigation of language and thought*. Cambridge, MA: MIT Press.

Sperber, D. and Hirschfeld, L. (2006). Culture and modularity. In P. Carruthers, S. Laurence and S. Stich (eds), *The innate mind: Culture and cognition*, vol. 2. Oxford: Oxford University Press, pp. 149–64.

Sterelny, K. (2007). Dawkins versus Gould: Survival of the fittest. Cambridge: Icon.

Teyler, T. J. and Discenna, P. (1986). The hippocampal memory indexing theory. *Behavioral Neuroscience*, 100: 147–54.

Thelen, E. and Smith, L. B. (1994). *A dynamical systems approach to development of cognition and action*. Cambridge, MA: MIT Press.

Thórisson, K. R., Pennock, C., List, T. and DiPirro, J. (2004). Artificial intelligence in computer graphics: A constructionist approach. *ACM SIGGRAPH Computer Graphics*, 38(1): 26–30.

Tinbergen, N. (1951). *The study of instinct*. Oxford: Clarendon Press.

Turing, A. M. (1936). On computable numbers, with an application to the Entscheidungsproblem. *Proceedings of the London Mathematical Society*, series 2, 42(1936–37): 230–65.

Vereijken, B. and Whiting, H. T. A. (1998). Hoist by their own petard: The constraints of hierarchical models. *Brain and Behavioral Sciences*, 21(5): 695.

West, S. A., Griffin, A. S. and Gardner, A. (2007). Evolutionary explanations for cooperation. *Current Biology*, 17: R661–R672.

Winslow, B. B., Takimoto-Kimura, R. and Burke, A. C. (2007). Global patterning of the vertebrate mesoderm. *Developmental Dynamics*, 236(9): 2371–81.

Wolpert, D. H. (1996). The existence of a priori distinctions between learning algorithms. *Neural Computation*, 8(7): 1391–1420.

8
Epistemology, Access and Computational Models

George Luger

8.1 Introduction: the epistemological question

The study of epistemology considers how the human agent knows itself and its world, and in particular whether this agent/world interaction can be considered as an object of scientific study. The empiricist and rationalist traditions have offered their specific answers to this question. I propose a constructivist, model-refinement approach to epistemological issues and offer a Bayesian characterisation of agent/world interactions. I present several Bayesian-based models for diagnostic reasoning and point out epistemological aspects of this approach. I conclude this chapter with some discussion of possible cognitive correlates of this class of computational model.

After finishing a PhD at the University of Pennsylvania in 1973, I accepted a postdoctoral research fellowship at the University of Edinburgh. My research at Penn had focused on models for human problem-solving in the spirit of Allen Newell and Herbert Simon (1972) and Goldin and Luger (1975). There was at that time in the Psychology, Linguistics, Epistemics and Artificial Intelligence departments at the University of Edinburgh a wealth of cognitive tasks under analysis, as well as an exciting research community directing these efforts. These projects included the study of seriation skills in primates and young children (Young, 1976; McGonigle and Chalmers, 1977), object permanence studies with infants (Bower, 1977; Luger et al., 1983, 1984), algebra problem-solving by adults (Luger, 1981; Bundy, 1983), the effects of problem structure on problem-solving behaviour (Luger, 1976; Luger and Bauer, 1978), and even the understanding of perception in moving environments. Besides the normal university teaching and research activities, the fairly regular interdisciplinary seminars in Stewart's Conference Room on Drummond Street gave us the opportunity to understand each other's projects and research, as well as to map out much of the common intellectual territory that supported and mediated our research goals.

In a foundational sense, many of our common problems were epistemological. Whether human or primate, how does an agent work within and manipulate elements of a world that is external to, or, more simply, is *not*, that agent? And, consequently, how can the human agent address the epistemological integration of the agent and its environment? And how is the human agent able to characterise this integration? The roots of these epistemological issues predate Greek philosophy, and across the centuries have had many articulate proponents, including Aristotle, Plato, and in more modern times, Descartes, Locke and the Scots philosopher David Hume, near whose memorial building at the University of Edinburgh many of our experiments and discussions took place.

Epistemological access, or the question of whether/how an agent can understand its own understanding, became a key question for research. In our science at Edinburgh we attempted to assess both how our subjects understood and dealt with their world as well as how we ourselves could understand and characterise that subject–environment interaction. In fact, epistemological access also addresses the foundations for bias and ignorance in ourselves and society, as well as providing clear directives for understanding and creating science itself. Bias and ignorance often follow a lack of knowledge and the subsequent misunderstanding of situations. How can an agent appreciate that its judgements about a situation are simply incorrect? Aberrations often occur because agents do not have the intellectual courage and/or confidence to deconstruct a situation or the accumulated knowledge and commitment to make well-founded judgements. A common example of this is the proposition that creationism, intelligent design and selection-based evolution have equal explanatory power in understanding the state of the natural world.

This chapter offers the author's rapprochement with the issue of epistemology and addresses the deeper problem of epistemological access. The following section takes a philosophical stance, finding in constructivism a plausible integration of the empiricist and rationalist views of the world. The third section, utilising the insights of Bayes' theorem, offers a computational model able to integrate prior expectations with posterior perceptions within the phenomenal world. The fourth section demonstrates this integration with several examples of diagnostic reasoning. I conclude the chapter with more speculative comments on how schema-based diagnosis and model-refinement might be instantiated in the human cortex.

8.2 An integration of rationalism and empiricism

Over the past 60 years, work in artificial intelligence can be understood as an ongoing dialectic between the empiricist and rationalist traditions in philosophy and epistemology. It is only natural that a discipline that as its focus engages in the design and building of artifacts that are intended to capture intelligent activity should intersect with philosophy, and, in particular, with epistemology. I describe this intersection of disciplines in due course, but first we consider philosophy itself.

Perhaps the most influential rationalist philosopher was Rene Descartes (1680), a central figure in the development of modern concepts of the origins of thought and theories of mind. Descartes attempted to find a basis for understanding himself and the world purely through introspection and reflection. Descartes (1680) systematically rejected the validity of the input of his senses and even questioned whether his perception of the physical world was 'trustworthy'. Descartes was left with only the reality of thought; even the reality of his own physical existence had to be re-established. His physical self was established only after making his fundamental assumption, *'Cogito ergo sum'*. Establishing his own existence as a thinking entity, Descartes inferred the existence of a God as an essential creator and sustainer. Finally, the reality of the physical universe was the necessary creation and reflected the veridical trust in this benign God.

Descartes' mind/body dualism was an excellent support for his later creation of mathematical systems, including analytic geometry, where mathematical relationships could provide the constraints for characterising the physical world. It was a natural next step for Newton to describe the orbits of planets around the sun in the language of elliptical relationships of distances and masses. Descartes' clear and distinct ideas themselves became a sine qua non for understanding and describing 'the real'. His physical (*res extensa*) non-physical (*res cogitans*) dualism supports the body/soul or mind/matter biases of much of modern life, literature and religion.

The origins of many of Descartes' ideas can be traced back at least to Plato. The epistemology of Plato supposed that as humans experience life through space and time we gradually come to understand the pure forms of real life separated from material constraints. In his philosophy of reincarnation, the human soul is made to forget its knowledge of truth and perfect forms as it is reborn into a new existence. As life progresses, the human, through experience, gradually comes to remember the pure forms of the disembodied life; learning is remembering. In his narrative of the cave, in *The Republic*, Plato introduces his reader to these pure forms, the perfect sphere, beauty and truth. Mind/body dualism is a very attractive exercise in abstraction, especially for agents confined to a physical embodiment and limited by senses that can mislead, confuse and even fail.

The empiricist tradition, espoused by Locke, Berkeley and Hume, distrusting the abstractions of the rational agent, reminds us that nothing comes into the mind or to understanding except by passing through the sense organs of the agent. On this approach the perfect sphere or *absolute truth* simply do not exist. What the human agent 'perceives' are the things of a physical existence; what it 'knows' are loose associations of these physical stimuli. The extremes of this tradition, expressed through the Scots philosopher David Hume, include a denial of causality and the very existence of an all-powerful God. There is an important distinction here, the foundation of an agnostic position: it is not that a God can't exist, it is that the human agent can't know or prove that he/she *does* exist.

Aristotle was one of the first proponents of the empiricist tradition, although his philosophy also contained the ideas of 'form' and abstraction from a purely material existence. For Aristotle the most fascinating aspect of nature was change. In his *Physics*, he defines his 'philosophy of nature' as the 'study of things that change'. He distinguishes the *matter* from the *form* of things: a sculpture might be 'understood' as the material bronze taking on the form of a specific human. Change occurs when the bronze takes on another form. This matter/form distinction supports the modern computer scientists' notions of symbolic computing and data abstraction, where sets of symbols can represent entities in a world and abstract relationships can describe these entities sharing common characteristics. Abstracting form from a particular material existence supports computation, the manipulation of abstractions, as well as theories for data structures and languages as symbol-based representations.

The modern empiricist has a deep dilemma, however: how can a human come to know/understand new relationships in the physical world? This issue was described more than two thousand years ago by Meno's slave in the Platonic dialogue bearing his name (Plato, 1961):

> And how can you enquire, Socrates, into that which you do not already know? What will you put forth as the subject of the enquiry? And if you find out what you want, how will you ever know that this is what you did not know?

Plato handled this dilemma, as noted above, with his theory of 'learning as remembering' and that experience in the physical world gradually brings the human back to its (pre-birth) understanding of pure forms. But for the modern empiricist the deeper questions remain: What is abstraction, learning, the understanding of new relationships? What is causality, induction and generalisation?

Modern artificial intelligence practitioners have adopted both the empiricist and rationalist views of the world. To offer several simple examples: from the rationalist perspective came the expert system technology in which knowledge was seen as a set of clear and distinct relationships (rules) that could be encoded within a production-system architecture that could then be used to compute decisions in particular situations; traditional robotic planning was seen as presenting the world as a set of explicit constraints that were to be used to accomplish a particular task; and case-based reasoning was a database of collected and clearly specified cases that could then be used to address new and related problems (Luger, 2009, Chapter 7).

From the empiricist perspective of AI there is the creation of semantic networks and conceptual dependencies. These structures, deliberately formed to capture the concept and property associations of the human agent, were used to 'understand' human language and interpret meaning in specific contexts. Neural networks were designed to capture associations in collected data and to interpret and understand patterns in the world. (There were even the obvious – and scientifically useless – claims that neural connectivity was the way intelligent humans performed these tasks). Later approaches to robotics created a 'subsumption' architecture that was supposedly a knowledge-free model for operation in world situations (Luger, 2009, Chapter 7). Artificial life and genetic algorithms were proposed as structures that captured survival of the fittest and thus deserved to be seen as plausible models of the products of evolution.

It is not surprising that these approaches only met with limited success. To give them their due, they have been useful in many of the application domains in which they were designed and deployed. But as models of human cognition, able to generalise to new related situations, or even to generalise or interpret their various results, they were failures. The thought that they could adapt to new categories and problem domains was a delusion. The success of the AI practitioner as the designer of new and important software tools is beyond question; the notion of the cognitive creditability of the products of these tools is simply naive.

We view a constructivist epistemology as a rapprochement between the empiricist and rationalist viewpoints. The constructivist hypothesises that all understanding is the result of an interaction between energy patterns in the world and mental categories imposed on the world by the intelligent agent (Piaget, 1954, 1970; von Glasersfeld, 1978). Using Piaget's descriptions we *assimilate* external phenomena according to our current understanding and *accommodate* our understanding to phenomena that does not meet our prior expectations.

Constructivists use the term 'schemata' to describe the a priori structures used to mediate the experience of the external world. 'Schemata' is taken from the British psychologist Frederic Bartlett (1932); its philosophical roots go back to Kant (1781/1964). On this viewpoint observation is not passive and neutral but active and interpretative.

Perceived information, Kant's a posteriori knowledge, never fits precisely into our preconceived and a priori schemata. From this tension the schema-based biases a subject uses to organise experience are either modified or replaced. The use of accommodation in the context of unsuccessful interactions with the environment drives a process of cognitive *equilibration*. The constructivist epistemology is one of cognitive evolution and continuous refinement. An important consequence of constructivism is that the interpretation of any perception-based situation involves the imposition of the observers (biased) concepts and categories on what is perceived. This constitutes an *inductive bias*.

When Piaget proposed a constructivist approach to understanding the external world, he called it a genetic epistemology. When encountering new phenomena, the lack of a comfortable fit of current schemata to the world 'as it is' creates a cognitive tension. This tension drives a process of schema revision. Schema revision, Piaget's accommodation, is the continued evolution of the agent's understanding towards equilibration.

Schema revision and continued movement toward equilibration is a genetic predisposition of an agent for an accommodation to the structures of society and the world. It combines both these forces and represents an embodied predisposition for survival. Schema modification is both an a priori reflection of our genetics as well as an a posteriori function of society and the world. It reflects the embodiment of a survival-driven agent, of a being in space and time.

There is a blending here of the empiricist and rationalist traditions, mediated by the requirement of agent survival. As embodied, agents can comprehend nothing except that which first passes through their senses. As accommodating, agents survive through learning the general patterns of an external world. What is perceived is mediated by what is expected; what is expected is influenced by what is perceived. These two functions can only be understood in terms of each other. In the following sections I propose several Bayesian models where prior experience conditions current interpretations and current data supports selection of interpretative models.

We, as intelligent agents, are seldom consciously aware of the schemata that support our interactions with the world. As the sources of bias and prejudice both in science and society, we are more often than not unaware of our a priori schemata. These are constitutive of our equilibration with the world and are not usually a perceptible component of our conscious mental life.

Finally, we can ask why a constructivist epistemology might be useful in addressing the problem of understanding intelligence itself. How can an agent within an environment understand its own understanding of that situation? I believe that constructivism also addresses this problem of *epistemological access*. For more than a century there has been a struggle in both philosophy and psychology between two factions: the positivist, who proposes to infer mental phenomena from observable physical behaviour, and a more phenomenological approach which allows the use of first-person reporting to enable the access of cognitive phenomena. This factionalism exists because both modes of access to cognitive phenomena require some form of model construction and inference.

In comparison to physical objects like chairs and doors, which often, naively, seem to be directly accessible, the mental states and dispositions of an agent seem to be particularly difficult to characterise. We contend that this dichotomy between the direct access to physical phenomena and the indirect access to mental phenomena is illusory. The constructivist analysis suggests that no experience of the external (or internal) world is possible without the use of some model or schema for organising that experience. In scientific enquiry, as well as in our normal human cognitive experiences, this implies that *all* access to phenomena is through exploration, approximation and continued model-refinement.

In the following section we consider mathematical (computational) approaches to this exploratory model-refinement process. We begin our analysis with Bayes' methods for probabilistic interpretations and schema building and refine this approach to a form of naive Bayes, which we call the *greatest likelihood* measure, which uses continuous data acquisition to do real-time diagnosis through model-refinement.

8.4 A Bayesian-based and constructivist computational model

We can ask how the computational epistemologist might build a falsifiable model of the constructivist worldview. Historically, an important response to David Hume's scepticism, described briefly in the previous section, was that of the English cleric, Thomas Bayes (1763). When challenged to defend the gospel's and other believers' accounts of Christ's miracles in the light of Hume's demonstrations that such 'accounts' could not attain the credibility of a 'proof', Bayes' genius responded with a mathematical demonstration of how an agent's prior expectations could be related to its current perceptions. Bayes' approach, although it didn't do much for the creditability of miracles, has had an important effect on the design of probabilistic models. In the final section we conjecture that Bayes offers an interesting computational model of epistemological phenomena.

We make a simple start: suppose we have a single symptom or piece of evidence, e, and a single hypothesised disease, h: we want to determine how a bad headache, for example, can be an indicator of a meningitis infection. We can visualise this situation with Figure 8.1, where we see one set, e, containing all the people having bad headaches, and a second set, h, containing all the people that have the disease, meningitis. We want to get a measure of what the probability is of a person having a bad headache also having meningitis.

We now determine the probability that a person having the symptom, e, also has the hypothesised disease, h. This probability can be determined by finding the number of people having both the symptom and the disease divided by the number of people having the disease. (We will concern ourselves with the processes for obtaining these actual numbers later.) Since both these sets of people are normalised by the total number of people, we can represent each number as a probability. We represent the probability of the symptom e given the disease h as $p(e|h)$: $p(e|h) = p(e \cap h) / p(h)$, and $p(e \cap h) = p(e|h) p(h)$.

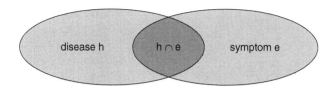

Figure 8.1 A representation of the numbers of people having a symptom, e, and a disease, h. Note that what we want to measure is the probability of a person having the disease, given that they suffer the symptom, **p(h|e)**

We wish to determine the **p(e ∩ h)** value and to do so we have other information from Figure 8.1, including the number of people who have both the symptom and the disease **e ∩ h**, as well as the total number of people who have the symptom, **e**. So we can determine the value for **p(e ∩ h)** with this information: the probability of the disease **h**, given the evidence **e**, **p(h|e)**: **p(h|e) = p(e ∩ h) / p(e)**.

Finally, we have a measure of the probability of the hypothesised disease, **h**, given the evidence, **e**, in terms of the probability of the evidence given the hypothesised disease: **p(h|e) = p(e|h) p(h) / p(e)**.

This last formula is Bayes' law for one piece of evidence and one hypothesised disease. But what have we just accomplished? We have created a relationship between the posterior probability of the disease given the symptom **p(h|e)** and the prior knowledge of the symptom given the disease **p(e|h)**. Our (or in this case the medical doctor's) experience over time supplies the prior knowledge of what should be expected when a new situation – a patient with symptoms – is encountered. The probability of the new person with symptom **e** having the hypothesised disease **h**, is represented in terms of the collected knowledge obtained from previous situations where the diagnosing doctor has seen that a diseased person had a particular symptom **p(e|h)** and how often the disease itself occurred, **p(h)**.

We can make the more general case, along with the same set-theoretic argument, of the probability of a person having a possible disease given two symptoms, say of having meningitis while suffering from both a bad headache and high fever. Again the probability of meningitis given these two symptoms will be a function of the prior knowledge of having the two symptoms when the disease is present along with the probability of the disease.

Next we present the general form of Bayes' law for a particular hypothesis, h_i, from a set of hypotheses, given a set of symptoms (evidence, E). The denominator of Bayes' theorem represents the probability of the set of evidence occurring. With the assumption of the hypotheses being independent, given the evidence, the intersection of each h_i with its piece of the evidence set, forms a partition of the full set of evidence, E, as seen in Figure 8.2.

With the assumption of this partitioning, the earlier equation we presented: $p(e \cap h) = p(e|h)\,p(h)$ can be summed across all the h_i to produce the probability of the set of evidence, $p(E)$, and the denominator of Bayes' relationship for the probability of a particular hypothesis, h_i, given evidence E:

$$p(h_i|E) = \frac{p(E|h_i)p(h_i)}{\sum_{k=1}^{n} p(E|h_k)p(h_k)}$$

$p(h_i|E)$ is the probability that a particular hypothesis, h_i, is true given evidence E.
$p(h_i)$ is the probability that h_i is true overall.
$p(E|h_i)$ is the probability of observing evidence E when h_i is true.
n is the number of possible hypotheses.

The general form of Bayes' theorem offers a functional (computational) description (model) of the probability of a particular situation happening given a set of perceptual clues. Epistemologically, the right-hand side of the equation offers a 'schema' describing how prior accumulated knowledge of occurrences of phenomena can relate to the interpretation of a new situation, the left-hand side of the equation. This relationship can be seen as an example of Piaget's *assimilation* where encountered information fits the accepted pattern created from prior experiences.

To describe further the pieces of Bayes formula: the probability of a hypothesis being true, given a set of evidence, is equal to the probability that the evidence is true given the hypothesis times the probability that the hypothesis occurs. This number is divided by (normalised by) the probability of the evidence itself. The probability of the evidence occurring is seen as the sum over all hypotheses presenting the evidence times the probability of that hypothesis itself.

There are several limitations to using Bayes' theorem as just presented as an epistemological characterisation of the phenomenon of interpreting new (a posteriori) data in the context of (prior) collected knowledge and experience. First, of course, is the fact that the epistemological subject is not

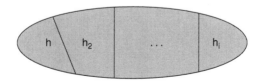

Figure 8.2 The set of evidence, E, is partitioned by the set of possible hypotheses, h_i

a calculating machine. We simply don't have all the prior values for all the hypotheses and evidence that can fit some problem situation. In a complex situation such as medicine where there can be well over a hundred hypothesised diseases and thousands of symptoms, this calculation is intractable (Luger, 2009, Chapter 5).

A second objection is that in most realistic diagnostic situations the sets of evidence are NOT independent, given the set of hypotheses. This makes the mathematical version of full Bayes just presented unjustified. But the rationalisation of the probability of the occurrence of evidence across all hypotheses can also be seen as simply a normalising factor, supporting the calculation of a realistic probability measure for the probability of the hypothesis given the evidence (the left side of Bayes' equation). The same normalising factor is utilised in determining the actual probability of any of the h_i, given the evidence.

Finally, diagnostic reasoning is not just about the calculation of probabilities, it is about the determination of the most likely explanation, given the accumulation of pieces of evidence. Humans are not doing real-time complex mathematics, rather we are looking for the most coherent explanation or possible hypothesis, given the amassed data.

A much more intuitive form of Bayes' rule – often called *naive Bayes* – ignores this $p(E)$ denominator entirely. Naive Bayes simply determines the likelihood of any hypothesis, given the evidence, as the product of the probability of the evidence given the hypothesis times the probability of the hypothesis itself: $p(E|h_i) p(h_i)$.

In most diagnostic situations we are often required to determine which of a set of hypotheses h_i is most likely to be supported. We refer to this as determining the *argmax* across all the set of hypotheses. Thus, if we wish to determine which of all the h_i has the most support we look for the largest $p(E|h_i) p(h_i)$:

$\text{argmax}(h_i) \, p(E|h_i) \, p(h_i)$

In a dynamic interpretation, as sets of evidence themselves change across time, we will call this argmax of hypotheses given a set of evidence at a particular time the *greatest likelihood of that hypothesis at that time*. We show this relationship, an extension of the Bayesian *maximum a posteriori* (or *MAP*) estimate, as a dynamic measure over time **t**:

$gl(h_i|E_t) = \text{argmax}(h_i) \, p(E_t|h_i) \, p(h_i)$

This model is both intuitive and simple: the most likely interpretation of new data, given evidence E at time **t**, is a function of which interpretation is most likely to produce that evidence at time **t** and the probability of that interpretation itself occurring.

We now ask how the argmax specification can produce a computational model of epistemological phenomena. First, we see that the argmax relationship offers a falsifiable approach to explanation. If more data turns up at a particular time an alternative hypothesis can attain a higher argmax value.

Furthermore, when some data suggests an hypothesis, h_i, it is usually only a subset of the full set of data that can support that hypothesis. Going back to our medical hypothesis, a bad headache can be suggestive of meningitis, but there is much more evidence that is also suggestive of this hypothesis, including fever, nausea, and the results of certain blood tests.

We view the evolving greatest likelihood relationship as a continuing tension between possible hypotheses and the accumulating data collected across time. The presence of changing data supports the revision of the greatest likelihood hypothesis, AND, because data sets are not always complete, the possibility of a particular hypothesis motivates the search for data that either supports or falsifies it. Thus, greatest likelihood represents a dynamic equilibrium evolving across time of hypotheses suggesting supporting data and the presence of data combinations supporting particular hypotheses.

When, because of changing data, no new hypothesis is forthcoming, a greedy local search on the data points can suggest (create) new hypotheses. This technique supports *model induction*, the creation of a most likely model to explain the data, an important research component of machine learning (Luger, 2009, Chapter 13).

The following section presents several computational examples from my own research group that utilise this greatest likelihood dynamic equilibration process.

8.5 Computational examples of model-refinement and equilibration

In recent research into the diagnosis of failures in discrete component semiconductors (Stern and Luger, 1997; Chakrabarti et al., 2005) we have an example of creating the greatest likelihood for hypotheses across expanding data sets. Consider the situation of Figure 8.3, where we have two discrete component semiconductor failures.

Figure 8.3 shows examples of two different failures of discrete component semiconductors. This failure type is called an 'open', or the break in a wire connecting a component to others in the system. For the diagnostic expert the presence of a break supports a number of alternative hypotheses. The search for the most likely explanation for the failure broadens the evidence search: How large is the break? Is there any discolouration related to the break? Were there any sounds or smells on its happening? What are the resulting conditions of the components of the system?

Driven by the data search supporting multiple possible hypotheses that can explain the 'open', the expert notes the *bambooing* effect in the disconnected wire, Figure 8.3a. This suggests a revised greatest likelihood hypothesis that explains the open as a break created by metal crystallisation that was most probably caused by a sequence of low-frequency high-current

Epistemology, Access and Computational Models 155

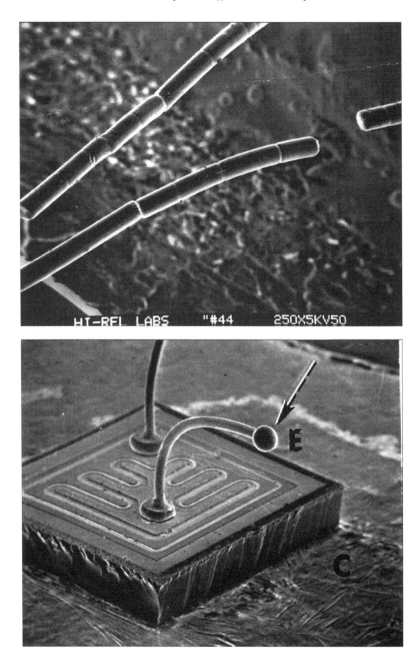

Figure 8.3a.b Examples of two different failures of discrete component semiconductors

pulses. The greatest likely hypothesis for the open of the example of Figure 8.3b, where the break is seen as *balled* is melting due to excessive current. Both of these diagnostic scenarios have been implemented by an expert system-like search through a hypothesis space (Stern and Luger, 1997) as well as with a Bayesian belief net (Chakrabarti et al., 2005). Figure 8.4 presents a Bayesian belief net (BBN) capturing this and other related diagnostic situations.

A Bayesian belief net (Pearl, 1988) is a graph of probabilistic causal relationships that reflects the understanding of an application domain. Because a BBN is a causal model, its graph has two properties: it is directed and without cycles (causes are directed to their effects and no effect can cause itself). These properties also support a further (drastic) reduction of the computational costs of full Bayesian inference: each node in the graph is independent of its non-descendants, given knowledge of its parents.

The BBN, without new data, represents the a priori state of an expert's knowledge of an application domain. In fact, these networks of causal relationships are usually carefully crafted through many hours working with human experts in that application domain. Thus, they can loosely be said to capture expert knowledge implicit in a domain of interest. When new (a posteriori) data are given to the BBN, for example, the wire is 'bambooed', the colour of the copper wire is normal, and so on, the belief network 'infers' the most likely probabilities within its (a priori) model, given this new information. There are many inference rules for doing this (Luger, 2009, Chapter 9); we will describe one of these, loopy belief propagation (Pearl, 1988), later. An important result of using the BBN technology is that as one hypothesis achieves its greatest likelihood other related hypotheses are 'explained away', that is, their likelihood measures decrease.

In the failure of discrete component semiconductors, the discovery of new evidence occurred in a discrete fashion, that is, one piece of information at a time, the first evidence often suggesting consideration of related evidence. In our second example, Chakrabarti et al. (2005), we have a continuous data stream from a set of distributed sensors. In monitoring the 'health' of the transmission of Navy helicopter rotor systems, we receive a steady stream of sensor readings, mainly of temperatures, vibrations and pressure information distributed across the various components of the running system. An example of this data can be seen in the top portion of Figure 8.5, where we have broken this continuous data stream into discrete and partial time slices.

We then use a fast Fourier transform to translate these signals into the frequency domain, as shown in the second row of Figure 8.5. These frequency readings were compared across time cycles to diagnose the running

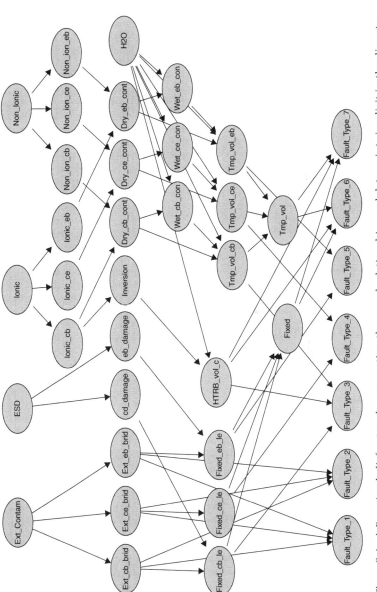

Figure 8.4 A Bayesian belief network representing the causal relationships and data points implicit in the discrete component semiconductor domain. As data is 'discovered' the (a priori) probabilistic values change

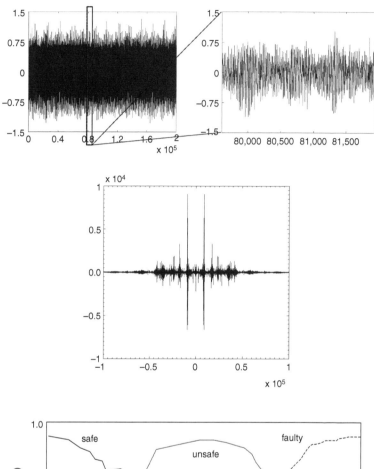

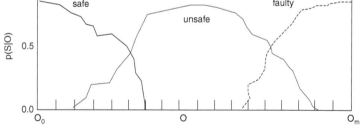

Figure 8.5 Real-time data from the transmission system of a helicopter's rotor. The top component of the figure presents the data stream and an enlarged time slice. The middle level of the figure is the result of the fast Fourier transform of the time slice data (transformed) to the frequency domain. The lower figure represents the hidden states of the rotor system

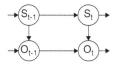

Figure 8.6 The data of Figure 8.5 is processed using an auto-regressive hidden Markov model as in Figure 8.6. States O_t represent the observable values at time t. The S_t states represent the hidden 'health' states of the rotor system, {safe, unsafe, faulty}, at time t

health of the rotor system. The method for diagnosing rotor health is by using the auto-regressive hidden Markov model (A-RHMM) of Figure 8.6. The observable states of the system are made up of the sequences of the segmented signals in the frequency domain while the hidden states are the imputed health values of the helicopter rotor system, as seen in the bottom of Figure 8.5.

The A-RHMM technology is used rather than a simple HMM because the human analysts suggested that the greatest likelihood value of the hidden nodes at any time would have some correlation with the values of these nodes at the previous time period (see the A-RHMM description in Luger, 2009, Chapter 13). Training this system on streams of normal data allows the system to make the correct greatest likelihood measure of the transmission when breakdowns occurred. The Navy supplied data both for the normal running system as well as for transmissions that contained seeded faults (Chakrabarti et al., 2007). Thus the hidden state S_t of the A-RHMM reflects the greatest likelihood hypothesis of the state of the rotor system, given the observed evidence O_t at time **t**.

A final computational example of determining the greatest likelihood measure for hypotheses considers the model-calibration problem itself. What can be done if the data stream cannot be interpreted by the present state's (a priori) model? The problems we have considered to this point simply ask, 'What is the greatest likelihood hypothesis, given a model and a set of data across time?' Now we ask what we can do when there is no interpretation of the model that fits the current data.

Figure 8.7 presents an overview of this situation, where, on the top row, a cognitive model either offers an interpretation of data or it does not. Piaget has described these situations as instances of *assimilation* and *accommodation*. Either the data fits, possibly requiring the model to slightly adjust its (probabilistic) expectations (assimilation), or the model must reconfigure itself, possibly adding new variable relationships (accommodation). The lower part of Figure 8.7 presents our (COSMOS) architecture (Sakanenko and Luger, 2009) that addresses both these tasks.

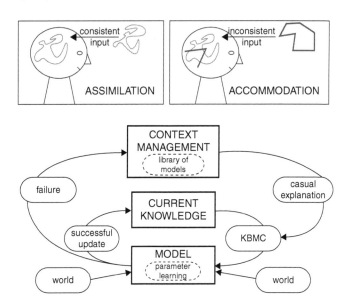

Figure 8.7 Cognitive model use and failure, above; a model-calibration algorithm, below, for assimilation and accommodation of new data

Although this model-calibration algorithm has been tested in complex tasks such as that of pumps, pipes, filters and liquids, complete with real-time measures of pressure, pipe flow, filter clogging, vibrations and alignments, I describe this 'model calibration' idea in a simpler situation of home burglar alarms.

Suppose we have developed a probabilistic home burglar alarm and monitoring system. We then deploy many of these alarm systems in a certain city and test their outputs across multiple situations, in particular monitoring these systems for false positive predictions. Suppose this system is deployed successfully over four winter months where we learn the probabilistic values for the outputs of alarm monitoring systems. The day-to-day deployment produces data that are used to condition the system. After a time of training the new daily data is easily assimilated into the model and the resulting trained system successfully reports both false alarms and actual robberies.

We then find ourselves in the spring months of the year where we encounter multiple fierce desiccating winds that shake the components – those mounted on doors and windows – of the alarm systems and dry out their connections. When our monitoring system sees many more false positive results that no longer fit comfortably into the previously trained system it is necessary to readjust the probabilities of the model and add new parameters reflecting the spring wind conditions. The result will be a new model for the spring monitoring situation.

Furthermore, when our alarm systems are sold in a new city it will need to determine which of its library of models will best fit that new situation. If there are other important variables, such as lots of small earthquake tremors, this variable will probably also need to be modelled. Although the problem of model induction in general is intractable, we feel in these knowledge-intensive situations we can create useful new models, using ideas such as causality and the greedy local search of constraints located near the points of model failure (Sakhanenko et al., 2006; Sakhanenko and Luger, 2009).

There are other examples in the literature of reasoning to the greatest likelihood in diagnostic and prognostic systems, for example, in computer-based speech processing, where, with a series of n-grams, sounds and/or words are tested against the contents of corpora of language data. In these situations techniques such as the probabilistic form of dynamic programming, the Viterbi algorithm (Luger, 2009, Chapter 13), process continuous streams of data to produce greatest likelihood results.

More recent utilisation of the greatest likelihood hypothesis is the study of an agent's active intervention in the world, for example, that agent's direct manipulation of the parameters that change the data stream itself, in an attempt to better understand the most likely model, including possible causality relationships that 'support' an interpretation of the data. This interventionist approach to interpretation was first proposed by Judea Pearl (2000) and his students (Tian and Pearl, 2001), and has been proposed by psychologists as a methodology for an agent's causal understanding of its world (Gopnik et al., 2004). This active interventionist approach for model discovery and refinement is also being explored in my own research group (Rammohan, 2010).

The final section offers some more speculative conjectures about possible cognitive architectures that could support the computational calculation of the greatest likelihood schemas.

8.6 A possible cognitive architecture for greatest likelihood calculation

Neuroscience has seen radical change, based mostly on better imaging technology, over the past 50 years. The neurophysiology and the brain science lectures we all had in the 1950s and 1960s now seem like exercises in the primitive practice of phrenology. Some insights of earlier times still remain quite relevant, however, including Hebbian (1949) learning.

Thanks to new sophistication in fMRI and other neuroimaging techniques, along with the development of sophisticated stochastic tools for model induction, calibration and computational inference, we now know much more about cortical architecture and processing. In particular, prefrontal cortex or Broadmann's areas are often seen as the primary 'location' for model creation and hypothesis generation.

If we consider the Bayesian-based schemas and in particular our formula for calculating a greatest likelihood measure, we note a balance or tension between the left-hand side, the set of hypotheses, and the right-hand side, the scoring of perceptual phenomena, that support these hypotheses. Consider Figure 8.8, a diagram of related components of this situation, where each of a set of hypotheses, h_i, is linked to a number of perceptual indicators, e_i elements of evidence set E that support that hypothesis. It should be noted that even though each hypothesis is assumed to be unique, the pieces of evidence may each support more than one hypothesis.

First, we consider each of these hypotheses and their supporting data sets to be attractor networks (Luger, 2009, Section 11.5). An attractor network, sometimes referred to as a *basin of attraction*, serves as a generalised pattern that can be either partially of totally matched. The insight behind the attractor network model is that stimuli rarely match perfectly to their archetypical patterns, but are messy, partial and usually compromised. Thus the basin of attraction is essential for partial and incomplete matches. But this is a sign of natural intelligence, where perfect information is most likely an unattainable ideal, and yet incomplete partial perceptions are often sufficient.

The second conjecture supporting the diagnostic model of Figure 8.8 is that any perceptual information, given our basins of attraction model, is able to trigger potential hypotheses. That is, as with natural intelligence, data patterns, even partial matches, are able to trigger entire attractor networks. In this sense what we see triggers its explanation, just like appropriate expectations condition our interpretation of perceptual information: the data triggers appropriate models and suggested hypotheses motivate the search for particular pieces of perceptual information.

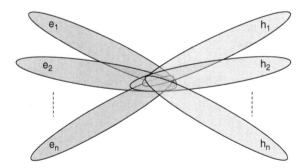

Figure 8.8 A set of hypotheses h_i are linked to multiple (supporting) perceptions, evidence e_j. It is hypothesized that the hypothesis space is located in the pre-frontal cortex (Broadmann's area), while the intersection of the hypothesis space with the various sense modalities is located in the hippocampus

Third, Hebbs' reinforcement rule supports the calculation of the $p(h_i)$ parameter in the greatest likelihood calculation of $p(E_t|h_i) \, p(h_i)$. The likelihood of a particular hypothesis is a measure that is conditioned across time, as the more a particular hypothesis is seen the more it is expected. Similarly, even though a piece of evidence might be indicative of a hypothesis, this fact alone is not sufficient to trigger the greatest likelihood condition for that hypothesis. Thus the importance of the perception of various pieces of data is conditioned on the likelihood of that hypothesis itself, and when the hypothetical situation is extremely rare, the search can continue for other more plausible hypotheses, given the same pieces of evidence.

The discussion to this point has shown how a cognitive *a priori equilibrium* can integrate and interpret a posteriori evidence. Our model also demonstrates that a greatest likelihood hypothesis is supported both by the a posteriori evidence suggestive of particular hypotheses and the likelihood of that hypothesis. This approach to model support and calibration is suggestive of Piaget's *assimilation*.

I would suggest that our architecture is also maximally flexible for the integration of new hypothesis/evidence relationships. New relationships may be learned though an educational process or simply conditioned through observations of a talented diagnostician. Regardless of the method of learning, the new hypothesis/evidence relationships are integrated as further related (Figure 8.8) basins of attraction: new hypotheses are conjoined with their related evidence sets and integrated into the larger diagnostic model. Piaget might refer to this phenomenon as *accommodation*. Similar arguments support the determination of the greatest likelihood for combinations of hypotheses, given sets of evidence. Thus, model calibration and revision is an active process of the intelligent agent interacting with its environment: as it learns, it rewards; when it fails, it revises.

A cortical column–based cognitive architecture (Hawkins, 2004) would support both the representation for and algorithms to compute a greatest likelihood measure. The hypothesis space of Figure 8.8 would be located in the columns of prefrontal cortex or Broadmann's areas. The evidence sets would be linked to the cortical receptors for the human sensory and memory modalities. The critical overlap of the hypothesis space with the sensory modalities is the hippocampus area of cortex. Thus, sense and memory stimulation trigger and support related hypotheses and the possibility of a particular hypothesis would suggest related sensory and memory data.

There exist a number of algorithms created for real-time integration of posterior information and its propagation into (a priori) stochastic models (Luger, 2009, chapters 5, 9, 13). There are also many computational inference rules for calculating the greatest likelihood in probabilistic modelling situations.

For 'cognitive creditability' we propose a form of the *loopy belief propagation* (Pearl, 1989) algorithm as it reflects a system constantly iterating towards equilibrium (or *equilibration*, as Piaget might describe it). A cognitive system is in what can be called *a priori equilibrium* with its continuing states of learned diagnostic knowledge. When presented with the novel information characterising a new diagnostic situation, this a posteriori data perturbs the equilibrium. The (cognitive) system then iterates by sending 'messages' between near-neighbours' prior and posterior components of the model until it finds convergence or equilibrium, usually with support for a greatest likelihood hypothesis.

Iteration of this system can be (intuitively) seen as integrating small perturbations of the values of neighbours in the system, aimed at achieving compatible equilibrating measures, and a stable state of the system is reached. The iteration process itself can be visualised as continuous message passing between near-neighbours ('I've got these values; what are yours? Let's each make slight adjustments moving towards a probabilistic compatibility') attempting to determine the most appropriate set of values for the entire system once the a posteriori information is added to the previous (a priori) equilibrium.

This iteration process can also be seen as a method to account for incomplete or missing information in a situation given a priori equilibrium. The iterative message passing suggests most likely values for missing or lost data, given the state of a priori equilibrium. In our research (Sakhanenko et al., 2008) this is shown to be a form of expectation-maximisation (EM) learning (Dempster et al., 1977). Furthermore, research has demonstrated that the generalised loopy belief propagation algorithm iterates to optimal states of equilibrium when the original a priori belief state is reflected with a non-cyclic-directed graph (Pearl, 1988).

In concluding these discussions we note that we are not claiming the human diagnostic system is *doing* loopy belief propagation on an explicit graphical model when it moves to finding equilibration through interpreting a posteriori information. This type reduction of cognitive phenomena to computational representations and algorithms has long been questioned by researchers, including Anderson (1978, see *representational indeterminacy*) and Luger (1994, Chapter 4).

Rather, the claim of this chapter is that stochastic models coupled with the loopy belief propagation algorithm offer a *sufficient* account of the cortical computation of the *greatest likelihood* measure given a priori cognitive equilibrium and the presentation of novel information. Further, I suggest that this greatest likelihood calculation *is* cognitively plausible, and thus supports an epistemological stance on understanding the phenomena of human diagnostic and prognostic reasoning, as well as addresses the larger question of how agents can come to understand their own acts of interpreting a complex and often ambiguous world.

References

Anderson, J. R. (1978). Arguments concerning representation for mental imagery. *Psychological Review*, 85: 249–77.

Bartlett, F. (1932). *Remembering*. Cambridge: Cambridge University Press.

Bayes, T. (1763). Essay towards solving a problem in the doctrine of chances. *Philosophic Transactions of the Royal Society of London*. London: The Royal Society, pp. 370–418.

Bower, T. G. R. (1977). *A primer of infant development*. San Francisco: W. H. Freeman.

Bundy, A. (1983). *Computer modeling of mathematical reasoning*. New York: Academic Press.

Chakrabarti, C., Rammohan, R. and Luger, G. F. (2005). A first-order stochastic modeling language for diagnosis, in *Proceedings of the 18th International Florida Artificial Intelligence Research Society Conference (FLAIRS-18)*. Palo Alto: AAAI Press.

Chakrabarti, C., Pless, D. J., Rammohan, R. and Luger, G. F. (2007). Diagnosis using a first-order stochastic language that learns. *Expert Systems with Applications*, 32(3). Amsterdam: Elsevier Press.

Dempster, A. P. (1968). A generalization of Bayesian inference. *Journal of the Royal Statistical Society*, 30 (Series B): 1–38.

Descartes, R. (1680). *Six metaphysical meditations, wherein it is proved that there is a god and that man's mind is really distinct from his body*, trans. W. Moltneux. London: Printed for B. Tooke.

Goldin, G. A. and Luger G. F, (1975). Problem Structure and Problem Solving Behavior. In *Proceedings of IJCAI, 1975, Tiblisi, USSR*. Cambridge, MA: MIT Press.

Gopnik, A., Glymour, C., Sobel, D. M., Schulz, L. E., Kushnir, T. and Danks, D. (2004). A theory of causal learning in children: Causal maps and Bayes nets. *Psychological Review*, 111(1), p. 3–32.

Hawkins, J. (2004). *On intelligence*. New York: Times Books.

Hebb, D. O. (1949). *The organization of behavior*, New York: Wiley.

Kant, I. (1781/1964). *Immanuel Kant's critique of pure reason*, trans. N. K. Smith. New York: St. Martin's Press.

Luger, G. F. (2009). *Artificial intelligence: Structures and strategies for complex problem solving*, 6th edn. Boston: Addison-Wesley Pearson Education.

Luger, G. F., (1994). *Cognitive science: The science of intelligent systems*. New York: Academic Press.

Luger, G. F. (1981). Mathematical model building in the solution of mechanics problems: Human protocols and the MECHO Trace. *Cognitive Science* (5), p 55–77.

Luger, G. F. (1978). *Cognitive psychology: Learning and problem solving. unit 28: formal analysis of problem solving behaviour*. Milton Keynes: The Open University Press.

Luger, G. F., (1976). The use of the state space to record the behavioral effects of subproblems and symmetries in the Tower of Hanoi problem. *Int. Journal of Man-Machine Studies*, 8.

Luger, G. F., Lewis, J. A. and Stern, C. (2002). Problem solving as model-refinement: Towards a constructivist epistemology. *Brain, Behavior, and Evolution*, 59: 87–100.

Luger, G. F., Bower, T. G. R. and Wishart, J. G. (1983). A model of the development of the early infant object concept. *Perception*, 12(1): 21–34.

Luger, G. F., Wishart, J. G. and Bower, T. G. R. (1984). Modeling the stages of the identity theory of object-concept development in infancy. *Perception* (13): 97–113.

Luger, G. F. and Bauer, M. A. (1978).Transfer effects in isomorphic problem situations. *Acta-Psychologica* (42) 121–31.

McGonigle, B. O. and Chalmers, M. (1977). Are monkeys logical? *Nature* 267: 694–6.
Newell, A. and Simon, H. A. (1972). *Human problem solving*. Englewood Cliffs, NJ: Prentice-Hall.
Pearl, J. (1988). *Probabilistic reasoning in intelligent systems: Networks of plausible inference*. Los Altos, CA: Morgan Kaufmann.
Pearl, J. (2000). *Causality: Models, reasoning, and inference*. Cambridge: Cambridge University Press.
Peirce, C. S. (1958). *Collected papers 1931–1958*. Cambridge, MA: Harvard University Press.
Piaget, J. (1954). *The construction of reality in the child*. New York: Basic Books.
Piaget, J. (1970). *Structuralism*. New York: Basic Books.
Plato (1961). *The Collected Dialogues of Plato*. E. Hamilton and H. Cairns, (eds). Princeton: Princeton University Press.
Rammohan, R. (2010). *Three algorithms for causal learning*. PhD Dissertation, Computer Science Department, University of New Mexico.
Sakhanenko, N. A., Rammohan, R., Luger, G. F., Stern, C. R. (2008). A new approach to model-based diagnosis using probabilistic logic, in *Proceedings of the 21st International Florida Artificial Intelligence Research Society Conference (FLAIRS-21)*, Palo Alto, CA: AAAI Press, 2008.
Sakhanenko, N. A. and Luger, G. F. (2009). *Model failure and context switching using logic-based stochastic models* (in press; copies available from authors).
Stern, C. R. and Luger, G. F. (1997). Abduction and abstraction in diagnosis: A schema-based account. In *Expertise in Context*, P. J. Feltovitch, et al. (eds). Cambridge, MA: AAAI/MIT Press.
Tian, J. and Pearl, J. (2001). Causal discovery from changes. In *Proceedings of UAI '2001*, San Francisco: Morgan Kaufmann, pp. 512–21.
Von Glaserfeld, E. (1978). An introduction to radical constructivism. In *The Invented Reality*, P. Watzlawick (ed.). New York: Norton, pp. 17–40.
Young, R. M. (1976). *Seriation by children: A production system approach*. Basel: Birkhauser-Verlag.

9
Reasoning about Representations in Autonomous Systems: What Pólya and Lakatos Have to Say

Alan Bundy

9.1 Introduction

Autonomous reasoning systems combine an often logic-based representation of some aspect of the world with rules for manipulating that representation. These representations are usually inherited from the literature or are built manually for a particular reasoning task. They are then regarded as fixed. We have argued that representations should instead be regarded as fluid, that is, their choice, construction and evolution should be under the control of the autonomous agent rather than predetermined and fixed (Bundy and McNeill, 2006).

- Appropriate representation is the key to successful problem-solving. It follows that a successful problem-solver must be able to choose, construct or evolve whatever representation is best suited to solving the current problem.
- Autonomous agents use representations called ontologies. For different agents to communicate they must align their ontologies. In some applications it is not practical to manually pre-align the ontologies of all agent pairs; it must be done dynamically and automatically.
- Persistent agents must be able to cope with a changing world and changing goals. This requires evolving their ontologies as their problem-solving task evolves. W3C call this *ontology evolution*.[1]
- George Pólya has written the classic guide to the art of problem-solving (Pólya, 1945). Imre Lakatos has written a fascinating rational reconstruction of the evolution of mathematical methodology (Lakatos, 1976). Although it was not their intention to do so, both these authors have implicitly provided profound evidence for our thesis that representations should be fluid. In this paper we analyse their work and extract this evidence.

9.2 Pólya's How to solve it

George Pólya (1887–1985) was a world-famous mathematician who did seminal work in series, number theory, mathematical analysis, geometry, algebra, combinatorics and probability. He is, however, best known for his books explaining the art of problem-solving to students of mathematics. The most popular of these books is his *How to solve it* (Pólya, 1945). The backbone of this book is a list of questions[2] that problem-solvers are urged to ask themselves to assist their creativity. Many commentators have encouraged automated reasoning researchers to apply Pólya's questions in the construction of autonomous agents.

Surprisingly, despite widespread knowledge of and enthusiasm for Pólya's ideas, they have had very little influence on practical automated reasoning systems. In this paper we will argue that this is because the nature of the advice has been widely misinterpreted. The main focus of automated reasoning research has been on search control, that is, on mechanisms for choosing which reasoning steps to take among those that are applicable. But Pólya has relatively little to say about this. Rather, most of his advice is about problem representation.[3]

To defend this claim, below I analyse each of Pólya's questions in turn. We will see that while a few of them do relate to proof search, most are about the choice, construction or evolution of problem representations. Pólya organised his questions into four phases: understanding the problem, devising a plan, carrying out the plan and looking back. I've kept these phases, as well as his grouping of several questions into a paragraph.

9.2.1 Understanding the problem

Understanding the problem is the initial phase in Pólya's suggested approach to mathematical problem-solving. As its title suggests, this phase has little to do with proof search and much more to do with representation formation. Few autonomous reasoning systems automatically form a problem representation. Usually, the representation of a problem is manually encoded as an axiomatisation within some logic. Standard sets of such axiomatisations are shared by the research community, which uses them to evaluate and compare automated proof systems. These sets are under constant development, as new axiomatisations are added and old ones amended, but this development is typically done offline by human researchers, not by automated autonomous agents.[4] The main exception to this rule is where an automated reasoning system is part of a question-answering system, and natural language processing is used to translate written or spoken utterances in, say, English, to and from a logical representation. Even here, the underlying formal representation is usually manually encoded, and only additional facts added to it, translated from the input utterances.

Below, each of Pólya's questions is reproduced in italics, followed by some analysis.

What is the unknown? By unknown, Pólya presumably means the value to be discovered as a result of problem-solving. The problem input to a logic-based prover usually comes with a well-labelled goal or conjecture. For instance, it might be the instantiation to be found for x in a problem of the form ∃x. ...So it is not clear what addressing this question would add to the search for a proof. However, in representing an informally stated problem it is vital to identify what is given and what is sought, so we know what can be assumed and what has to be proved.

What are the data? Similar remarks apply. By data, Pólya presumably means the various objects defined by terms in the conjecture. For instance, in a geometry problem, these might form the initial diagram. These are already known in a formally stated problem, but need to be identified when representing the problem. In fact, deciding how to formally represent informally described objects is called *idealisation*, and is the critical issue in representation formation (Bundy, 2006). For instance, in a relative velocity mechanics problem, a ship might be idealised as a point on a horizontal plane, but in a specific gravity problem it might be idealised as a container with volume but indeterminate shape.

What is the condition? Again, similar remarks apply. By condition, Pólya presumably means P in a conjecture of the form P ⇒ Q. Again, such conditions are identified by simple inspection in a formally stated problem. When representing the problem, however, just deciding that a conjecture should take the form P ⇒ Q is a significant decision.

Is it possible to satisfy the condition? Is the condition sufficient to determine the unknown? Or is it insufficient? Or redundant? Or contradictory? Checking that the condition is satisfiable or contradictory does sometimes form part of the proof search; if the condition is contradictory the conjecture is trivially true. Contradictory conditions can arise, for instance, when a conjecture is overgeneralised or when several overlapping case splits are made.

However, checking that the condition is (in)sufficient to determine the unknown is not relevant when a formalised problem is to be solved, unless as a sanity check that you are not being made a fool of. It makes much more sense as a sanity check after formalising a problem to ensure this has been done in a sensible way. All the other checks can also be more usefully interpreted in this way.

Draw a figure. Introduce suitable notation. 'Introduce suitable notation' is a complete giveaway here. That's exactly what you are doing in forming a representation. Unless your proof search includes notational changes or augmentations, it does not make sense as a proof-search hint. Humans often use a figure as a way of clarifying a problem during representation formation, or the diagram may *be* the representation. Figure drawing is not usually a part of automated proof, although there has been some work on the automation of reasoning with diagrams (Jamnik et al., 1997; Winterstein et al., 2002).

Separate the various parts of the condition. Can you write them down? 'Can you write them down?' seems to be an instruction to formalise the condition; prior separation being intended to break this difficult task into more manageable portions. Neither sentence makes much sense from a proof-search viewpoint, but both make sense from a representation-formation viewpoint.

9.2.2 Devising a plan

On the face of it, devising a plan sounds very relevant to proof search. My group has developed a search-control technique we call *proof planning*, in which an outline of a proof is used to guide proof search (Bundy, 1991). Indeed, proof-search advice is part of this grouping of Pólya's advice. However, on closer examination, advice in this group is neither solely nor mainly about proof search. Pólya's plans include representation formulation as well as proof search. Indeed, his idea is to minimise proof search by careful representation. So, again, it is representation that is the central focus.

Have you seen it before? Or have you seen the same problem in a slightly different form? This question suggests using the proof of an analogous theorem as a guide to constructing the proof of the current theorem. There has been periodic work using this technique. Owen (1990) provides a good survey up to 1990. However, such work has seldom been sustained and has not been very influential. So this question is ostensibly a proof-search suggestion.

However, if you listen carefully to uses of analogy in everyday conversation, you will see that it is rarely using an old argument to guide the search for a new one. Rather it is suggesting a way of representing the problem in a way analogous to an existing representation. The solution is thereby immediately suggested, that is, no search is required. Consider, for instance, the following quote from King James I of England and VI of Scotland on taking the English throne in 1604 and arguing, in Parliament, for the union of the two kingdoms.

> I am the Husband and the whole Isle is my lawfull Wife; I am the Head; and it is my Body; I am the Shepherd and it is my flocke; I hope therefore no man will be so unreasonable as to think that I that am a Christian King under the Gospel should be a polygamist and husband to two wives; that I being the Head should have a divided and monstrous Body.

Note that the argument establishes an analogy (or rather three) and then imports the undesirability of properties of the analogy (polygamy, monsterousness) back to properties of the original situation (two kingdoms under one King) to assert their undesirability there. There is no demonstration that arguments (proofs) in the analogy translate into arguments in the original. Rather one is invited to import a representation for the new, two-kingdoms situation by translating an old representation for polygamy and such, complete with a preformed conclusion about the undesirability of this situation.

Do you know a related problem? Do you know a theorem that could be useful? The first question above seems to say the same thing as the previous questions, and similar remarks apply. The instruction to find a useful theorem, however, is a rare example of a genuinely proof-search hint.

Look at the unknown! And try to think of a familiar problem having the same or a similar unknown. Again, similar advice to before and similar remarks about it. This time you are advised to use the unknown as a key and cue for the analogous problem.

Here is a problem related to yours and solved before. Could you use it? Could you use its result? Could you use its method? Should you introduce some auxiliary element in order to make its use possible? More of the same, but now you have the related problem and are invited to recycle its proof. This really does seem to be proof-search advice. You are advised either to use this theorem as a lemma in your current problem or to try to map its proof onto your proof using the kind of analogy mapping discussed in (Owen, 1990). Pólya even draws attention to the need to patch the partial proof arising from this mapping, which is frequently required.

Could you restate the problem? Could you restate it still differently? Go back to definitions. 'Restating the problem' is a clear invitation to revise the problem representation in order to make it simpler to prove. Automated reasoning systems have made no use of such representation variations; they stick with the initial formalisation. The 'definitions', of course, are where the fundamental representational decisions were made, for example, how is this object to be idealised? So going back to definitions is exactly what you have to do to re-formalise the problem.

If you cannot solve the proposed problem, try to solve first some related problem. Could you imagine a more accessible related problem? A more general problem? A more special problem? An analogous problem? Could you solve part of the problem? Keep only part of the condition, drop the other part; how far is the unknown then determined; how can it vary? Could you derive something useful from the data? Could you think of other data appropriate to determine the unknown? Could you change the unknown or the data, or both if necessary, so that the new unknown and the new data are nearer to each other? There's a lot going on in the above advice. Taken as a whole, it seems to suggest forming a related problem with a similar, but different, conjecture. In particular, we are invited to weaken and strengthen the original conjecture in various ways. The advantages of weakening are obvious; this is the classic abstraction approach, whereby a simpler conjecture is formulated and proved, then its proof is used to guide the proof of the original conjecture by analogy or the theorem is used as a key lemma. It's not so obvious why strengthening might help. However, generalisation of a theorem will often throw away unnecessary clutter, simplifying the task.[5] Especially in proofs by mathematical induction, generalisation will strengthen the induction hypothesis, providing a more powerful assumption to prove the similarly strengthened induction conclusion. This advice falls somewhere between representational experimentation and proof search.

Did you use all the data? Did you use the whole condition? Have you taken into account all essential notions involved in the problem? This is pure proof search; suggesting the prioritisation of assumptions and hypotheses.

9.2.3 Carrying out the plan

The heading of this subsection sounds as if the advice in it is about the application of the proof plan to guide the search for a solution to the problem, but closer inspection belies that interpretation.

Carrying out your plan of the solution, check each step. Can you see clearly that each step is correct? Can you prove that it is correct? The steps of logic-based problem-solvers are typically correct by construction, so this advice has little purchase from a proof-search perspective. However, if the solution steps have not have been fully formalised, they may not be correct by construction. Moreover, we will have some additional reasoning to check: that the problem representation we have chosen is faithful to the original problem statement.

9.2.4 Looking back

Looking back is not normally part of automated reasoning. Once a proof is found the job is finished. There has been a limited amount of work on learning from proofs, for example learning new proof plans, tactics or derived rules from example proofs. This sounds related to improving proof search. However, again first appearances are deceptive.

Can you check the result? Can you check the argument? This seems to be the same advice as that given in the last question of subsection 9.2.3. Similar remarks apply.

Can you derive the result differently? Can you see it at a glance? Problem-solvers usually work in a search space of possible solutions. It is technically simple to arrange for this space to be searched for more than one solution, although it is unclear what this gains unless you need a solution with some special property, for example, optimality, efficient execution or reliability. However, if a different *representation* of the problem yields the same result, then this may give us faith in the robustness of the solution to representational variations.

Can you use the result, or the method, for some other problem? This advice relates to that in the section about devising a plan. In particular, it is the flip side of proof by analogy: making solutions available for recycling. Again, I remark that this concerns remembering good representational techniques at least as much as it does proof-search ones.

9.3 Lakatos's Proofs and Refutations

Imre Lakatos (1922–1974) was a philosopher of mathematics and science. At the London School of Economics he worked with Karl Popper, and was influenced to apply to mathematics Popper's ideas on the philosophy of science. His controversial theory was that, as in science, the methodology of mathematics had evolved over time; in particular, that its ability to deal with fallible conjectures and their counter-examples had gradually become more sophisticated. In his book *Proofs and Refutations* he provides an accessible account of this theory via a rational reconstruction of the history of Euler's theorem about polyhedra. Lakatos was also influenced by Pólya. For instance, it was Pólya who suggested using Euler's theorem as the vehicle for Lakatos's work. Some of the methodological techniques described by Lakatos are closely related to Pólya's advice.

Polyhedra are the 3D version of polygons: objects whose faces are polygons. Examples include the five regular Platonic 'solids':[6] tetrahedron, cube, octahedron, dodecahedron and icosahedron, as well as many other semiregular and regular objects. I don't want to be more specific at this stage for reasons that will become apparent.

Lakatos describes a fictional classroom in which a teacher leads a (very bright!) class through this history. The teacher starts by presenting Cauchy's proof of the theorem. The students then confront it with various counter-examples and suggest ways to cope with them. The methodological techniques become more sophisticated as we progress through the book. In particular, there are refinements of definitions, of proofs, and of proof analysis techniques. In her PhD project, Alison Pease implemented Lakatos's methods of surrender, exception-barring, piecemeal withdrawal, monster-barring and lemma-incorporation. These were implemented in her TM (Colton and Pease, 2005) and HRL (Pease et al., 2009) computer systems. We will meet some of these methods in the discussion below.

I recommend anyone who has not read (Lakatos, 1976) to do so. It is short, accessible and educational, but above all fascinating and great fun!

9.3.1 Euler's theorem and Cauchy's 'proof' of it

Euler's theorem is that $V - E + F = 2$, where V is the number of vertices of a polyhedron, E is the number of its edges and F is the number of its faces. The teacher presents the following proof couched as a 'thought experiment', quoted from Lakatos (1976: 7–8).

> **Step 1.** Let us imagine the polyhedron to be hollow, with a surface made of thin rubber. If we cut out one of the faces, we can stretch the remaining surface flat on the blackboard, without tearing it. The faces and edges will be deformed, the edges may become curved, but V and E will not alter, so that if and only if $V - E + F = 2$ for the original polyhedron,

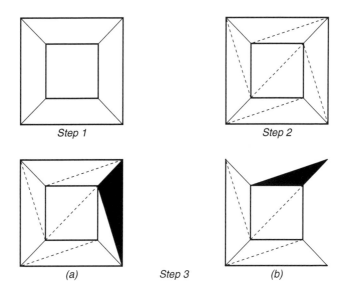

Figure 9.1 A Worked Example of the Proof

Each of the four diagrams shows successive steps in the application of the schematic proof to a cube. First it is stretched onto the plane. Then it is triangulated. Then triangles are removed. The two different types of triangle removal are shown

$V - E + F = 1$ for this flat network – remember we have removed one face. [Figure 9.1, step 1 shows the flat network for the case of a cube.]

Step 2. Now we triangulate our map – it does indeed look like a geographical map. We draw (possibly curvilinear) diagonals in those (possibly curvilinear) polygons which are not already (possibly curvilinear) triangles. By drawing each diagonal we increase both E and F by one, so that the total $V - E + F$ will not be altered. [Figure 9.1, step 2.]

Step 3. From the triangulated network we now remove the triangles one by one. To remove a triangle we either remove an edge – upon which one face and one edge disappear [Figure 9.1, step 3 (a)], or we remove two edges and a vertex – upon which one face, two edges and one vertex disappear [Figure 9.1, step 3 (b)]. Thus if $V - E + F = 1$ before a triangle is removed, it remains so after the triangle is removed. At the end of this procedure we get a single triangle. For this $V - E + F = 1$ holds true. Thus we have proved our conjecture.

Notice that this proof is a procedure, similar to a computer program. Given a particular polyhedron, this procedure will produce a proof of Euler's theorem, customised for this polyhedron. Notice, in particular, that the number and nature of the operations in steps 2 and 3 will vary depending on the polyhedron they are applied to. This kind of proof is called *schematic*. Although schematic proofs are unlike standard logical proofs, they can be automated (see, for instance, the PhD projects of Siani Baker [Baker et al., 1992] and Mateja Jamnik [Jamnik et al., 1997]) using a logical rule called *the constructive omega rule*. These implementations start by working out the steps required in one or more concrete examples, treat these steps as the trace of the required procedure, generalise the trace to induce this general procedure, and then optionally verify that this procedure will produce a correct proof whatever example it is applied to. In his MSc project, Andy Fugard showed that human informal reasoning often takes the form of schematic proof rather than standard logical proof (Fugard, 2005). Schematic proof is not the main topic of this paper, but an overview and discussion of its importance can be found in Bundy et al., (2005).

9.3.2 Definitions can *follow* proofs

Before long (Lakatos, 1976: 13), one of the students has a counter-example. This consists of a solid cube containing a cube-shaped hole, for which $V - E + F = 4$ (Figure 9.2). The first reaction is to just give up the theorem.[7] However, a second reaction soon emerges to disbar the counter-example by ruling it not a polyhedron. Lakatos calls this the method of monster barring.

Two rival definitions of polyhedron are proffered:

1. 'A polyhedron is a solid whose surface consists of polygonal faces.'
2. 'A polyhedron is a surface consisting of a system of polygons.'

Under definition 1 the hollow cube is a polyhedron and under definition 2 it isn't. So by adopting definition 2 the hollow cube is disbarred as a counter-example and the theorem is saved.

Alternative definitions can be proposed because up to now the concept of polyhedron remained undefined. You may ask how it was possible to have a proof about an undefined concept. In logical theories, formal definitions precede proofs. Here, it is the other way around. The use of schematic

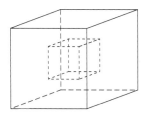

Figure 9.2 The Hollow Cube

A cube containing a cube-shaped hole. Note that V − E + F = 16 − 24 + 12 = 4. Is this a polyhedron?

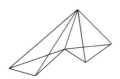

Figure 9.3 Joined Tetrahedra
Each composite polyhedron consists of two tetrahedron joined at an edge (on the left)and a vertex (on the right). For the left-hand one: V − E + F = 6 − 11 + 8 = 3 and for the right-hand one: V − E + F = 5 − 12 + 8 = 1. Are each of these really one polyhedron?

proofs makes this possible. A procedure can be defined without specifying exactly what kind of inputs (in this case polyhedra) it will be given. Only later do we discover that it fails to apply to certain inputs (in this case we can't apply step 1 to the hollow cube). This kind of experience is commonplace in practical computer programming, and is a common source of program bugs.[8]

Note also that definitions 1 and 2 are not totally formal. They call upon other concepts: 'solid', 'surface', 'polygon', 'consists of', 'bounded', 'system' and so on, which must also be defined. If any of these is informal, that is, has grey areas itself, then we have merely postponed the problem. This greyness immediately becomes apparent when definition 2 is challenged in Lakatos (1976: 15) by two counter-examples consisting of two tetrahedron joined at a vertex or edge (Figure 9.3). This is resisted by tightening up the definition of 'system' to exclude such combinations.

This kind of definition refinement was emulated in Pease's HRL system by having a variety of different kinds of definition and moving between them. Initially, for instance, a concept may only be 'defined' in an extensional way as a set of known examples of that concept, for example, a set of known prime numbers. Later, this might be replaced by an intensional definition as a logical formula, for example, prime numbers as numbers with only two divisors. Note that an intensional definition might refer to concepts that are only defined extensionally. During the emulation of the various Lakatos methods, these definitions might change in both directions and to different extensional and intensional definitions.

9.3.3 Alternative idealisations

Some counter-examples do not need to be disbarred, but can be turned into examples by redefining their properties. This method of monster adjustment is proposed in Lakatos (1976: 30). Kepler's star-polyhedron occurs as a counter-example on page 17 of Lakatos (Figure 9.4). It is considered to have 12 faces, each of which is a five-point star. These faces intersect. Under this definition it has 12 vertices and 30 edges, so that $V - E + F = -6$.

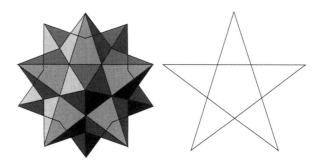

Figure 9.4 Kepler's star-polyhedron

On the left is a small stellated dodecahedron, also called Kepler's star-polyhedron and the 'urchin'. Each of its faces is a pentagram, seen on the right, which has five sides. It has 12 such faces, forming 12 vertices and 30 edges, i.e. V − E + F = 12 − 30 + 12 = − 6. *It can also be seen as containing 60 triangular faces. This time* V − E + F = 32 − 90 + 60 = 2, *as desired. Which is the right interpretation, or are both correct?*

However, on page 31 the faces of this star-polyhedron are redefined to be the triangles at the points of the star-polygons. This gives 60 (5 × 12) faces, 90 edges and 32 vertices, so *V − E + F = 2*, as required.

If we regard the star-polyhedron as being defined as a set of Cartesian coordinates, the definitions of what constitute faces, edges or vertices are somewhat similar to the vision algorithms which find objects in pixel arrays. We should not, therefore, be surprised if minor variants of such definitions return different answers. For instance, does the search for a face stop when we reach an edge or can it carry on through the object to match up with another coplanar face on the other side? This is what determines whether we 'see' 12 star-shaped faces in the Kepler star-polyhedron or 60 triangular ones.

Maybe, as one of the 'students' in Lakatos (1976: 33) claims, it can be both, that is, the same set of points can be viewed as two distinct polyhedra. This forces us to distinguish the object viewed as a set of points (of which there is one) from the object viewed as a polyhedron (of which there are two). This example shows us there are often several ways of idealising an informal object as a formal one. We explored such choices in our Mecho project to solve mechanics problems stated in English (Bundy et al., 1979). For instance, in a lever problem, a man carrying a ladder might play the role of the fulcrum of the lever, whereas the same man standing on a scaffold board might play the role of a weight.

9.3.4 New concepts from repairing faulty conjectures

Another method of dealing with counter-examples is to modify the theorem to incorporate some conditions into it. This is discussed in Lakatos

(1976: 33). Euler's theorem is reformulated as one not about all polyhedra, but only about simple polyhedra. A simple polyhedron is one which can be stretched onto the plane, which is step 1 of the schematic proof in Figure 9.1, that is, those that are topologically equivalent to a sphere. None of the counter-examples is a simple polyhedron, so this excludes them. Lakatos calls this *the method of lemma-incorporation* because he regards steps 1–3 as lemmas and he is ensuring the truth of one of these lemmas by adding a condition to the theorem.

Again this revision requires evolving representations; this time the creation of a new definition for the new concept of simple polyhedra. What is novel is the method of discovering the concept and its definition. The failed proof is analysed with the aid of a counter-example to see at what point the proof fails. A new precondition is then introduced into the faulty conjecture to ensure that this failure does not occur. For instance, step 1 of the proof cannot be carried out for any of the counter-examples, so we restrict the proof to polyhedra for which, by definition, step 1 *can* be carried out. The technique of repairing a faulty conjecture by constructing a new precondition for it by analysing a failed proof attempt is called *precondition analysis*. It has been reinvented and implemented several times in the history of automated reasoning. The earliest occurrence I am aware of is in J. Moore's PhD thesis (Moore, 1974: 147). In her implementation of Lakatos's methods, Alison Pease used Colton's HR concept-formation program to create new concepts to refine faulty conjectures (Colton and Pease, 2005; Pease et al., 2009).

Soon counter-examples are found which fail at other steps of the proof and which require further conditions to be built into the theorem statement, for example, that faces be simply connected in order that step 2 be applicable. Sometimes the assumptions in the proof that are violated are implicit, so that the counter-example first reveals the hidden assumption before supplying the extra condition in the theorem. This situation is maybe analogous to writing a program for integers but implicitly assuming they were positive integers. We only notice when either (a) we apply the program to a negative integer and it goes wrong or (b) we try to verify that the program meets a formal specification and find that the proof attempt breaks down.

One might worry that there is no bottom to this process; that we might go on discovering counter-examples and, hence, hidden assumptions forever; that we will be forced to add new conditions to the theorem, refining our representations indefinitely. One of the students in Lakatos (1976: 40) expresses this worry explicitly as 'an infinite regress in proofs'. In twentieth-century mathematics we seem to have bottomed out in various axiom sets and logical rules of inference, but there is always the potential to unpack these further by deriving existing axioms and rules in terms of yet more primitive ones. The potential for bottoming out is a characteristic of mathematical reasoning which distinguishes it from common-sense reasoning.

It arises because in mathematics we are reasoning about ideal objects for which we can agree on a formal definition. In real life the representation is only a model of reality, a model which we can constantly refine to show more detail.

Notice again how representations are still evolving, with the addition of new concepts, after the initial (faulty) proof has been formulated.

9.3.5 Counter-examples from proof analysis

Not only can counter-examples be used to analyse and modify proofs, but the analysis of proofs can be used to suggest counter-examples. By locating the assumptions in a proof we can design counter-examples which violate these assumptions.

Lakatos calls this interplay between proofs and counter-examples *the method of proofs and refutations*. At its heart is the technique of proof analysis. Proofs are examined in detail to identify concepts which are vague and assumptions which are made. Counter-examples are taken through each proof step in turn to see which ones they violate and so assist proof analysis. The result is a strengthening of definitions and proofs.

9.3.6 New concepts from theorem generalisation

Adding new preconditions to a theorem weakens its content, for example, Euler's theorem covers a narrower class of polyhedra. To counteract this, ways must be found to generalise the theorem.

One way to do this is by proof analysis using examples which violate one of the 'lemmas', that is, proof steps, but for which the theorem is still true. Lakatos calls these local but not global counter-examples. Is there some way to weaken the conditions, for example, by replacing them with disjunctions, and hence allow the theorem to apply to more polyhedra. Lakatos (1976: 57) mentions this possibility, but fails to find any examples of it.

Another route to generalisation is to find a theorem which accounts for the global counter-examples as well. In the case of Euler's theorem a formula is required that gives a value for $V - E + F$ for all polyhedra. Lakatos (1976: 80–1) gives a series of more general and more complex such formulae, culminating in one combining the number of disconnected surfaces, the surfaces of spheroid polyhedra and the number of edges in each face which can be deleted without reducing the number of faces. This formula requires the invention and definition of three new concepts: 'disconnected surface', 'surfaces of spheroid polyhedron' and 'number edges in each face which can be deleted without reducing the number of faces'. Inventing these concepts requires proof analysis using the global counter-examples to identify the properties that cause them to violate the proof and to discover what alternative value the proof gives for $V - E + F$ when these properties are present.

This provides yet another example of the way in which representation evolution, including theorem formation and the introduction of new concepts, is interleaved with problem-solving using the current state of the representation.

9.3.7 Formalisation of informal concepts

In Lakatos (1976, Chapter 2) is a modern formal proof of Euler's theorem. This works by representing polyhedra as sets of vertices, edges and faces together with incidence matrices to say which vertices are in each edge and which edges are in each face. A restricted class of polyhedra is defined using concepts from this representation. The theorem for this restricted class of polyhedra is then turned into a formula of vector algebra and a calculation in this algebra gives the value 2 for $V - E + F$, as required.

This theorem is claimed to be irrefutable, that is, no counter-examples to it are possible. This is because it is proved by a formal derivation from axioms. Proof analysis cannot find any assumptions or vague definitions in the proof, so there is no room for counter-examples to be constructed.

Modulo the possibility of careless slips in applying rules of inference, we can accept this claim. However, the possibility of error has now shifted from the proof to the modelling process. We are asked to accept on trust that the formal definition of polyhedra coincides with our intuitive one; that the restricted class of polyhedra coincides with the intuitively simple polyhedra; and that the axioms and rules of our formal theory are true for polyhedra. This is not a trivial matter. Some of the formal definitions and axioms are quite obtuse and some thought experiments are required to see that they coincide with our intuitions. These thought experiments are not dissimilar to the thought experiment which underlies the concept of schematic proof.

The process of associating a formal representation with an informally stated problem was explored in our Mecho (Bundy et al., 1979), and Eco (Robertson et al., 1991), projects. It is highly non-deterministic. We must choose a formal theory and then find formal counterparts in it for the objects and relations in the informal problem. At each stage we have a lot of choice. Some aspects of the informal problem can be neglected as negligible. Objects and relations can be represented with more or less complexity. In areas like mechanics the ground rules for doing this are well established by centuries of practice. In ecology there are fewer constraints. In all cases the decisions depend on circumstances, for example, how accurate an answer is required; how much calculation are we prepared to do; which aspects of the problem are most important? It is always possible to model the real world in more detail. Do we take account of friction? What about the pressure of the solar wind? Are these particles to be regarded as one object, divided into clumps, or considered individually? This is where the bottomless regress re-enters.

9.4 Conclusion

What lessons can we draw from this analysis?

1. If, as I claim above, Pólya's problem-solving advice is mainly geared to the choice, construction and evolution of representations best suited to efficient problem-solving, and if research on autonomous reasoning agents has been mainly geared to guiding proof search, then this would help explain why the take-up of his advice by the automated reasoning community has been essentially nonexistent.
2. However, the autonomous reasoning agents of the future will need to treat their representations as fluid, as much as they currently derive new conclusions from old ones. In this world, Pólya's advice might prove much more useful.
3. Lakatos's illustrations of mathematical methodology teach us that we need not wait for representation to be complete before reasoning can start. Reasoning can start with an incomplete representation and some of its results may be to develop and refine the representation, as well as reasoning with the current representations.
4. Initially, concepts may be represented in an informal manner which allows some grey area, that is, it may not be clear whether a new object fits the concept or not. An autonomous agent may choose whether or not it does and, hence, refine its concept accordingly.
5. A concept's definition may depend on undefined concepts. These, in their turn, may later be defined in terms of other undefined concepts. In common-sense reasoning this regress may be bottomless, but in mathematics it usually bottoms out in some atomic concepts. This decision may be revisited in some later representation in which definitions are given for previously atomic concepts.
6. New concepts can emerge, for instance, from attempts to patch a failed proof and to generalise a conjecture. For instance, it may be necessary to introduce a new precondition to a conjecture to ensure its correctness. The formal definitions of these new concepts may come later in the process.
7. Part of representation formation is idealisation: the assignment of formal roles and definitions to informal objects. There is often a choice about how this is to be done, a choice that may need to be remade as part of the problem-solving process.
8. By showing that Pólya's advice and Lakatos's analysis are as much, if not more, about representation than about reasoning, we explain why, despite their popularity, they have played such a small role in the automation of reasoning. We also predict that they will play a much bigger role in the automation of fluid representations for autonomous agents.

Acknowledgements

The research reported in this chapter was supported by EPSRC grant EP/E005713/1. I would like to thank the members of the Mathematical Reasoning Group who gave me feedback, especially Michael Chan and Alison Pease. Especial thanks to Predrag Janicic and Alison Pease for drawing the images.

Notes

1. http://www.w3.org/TR/webont-req/#goal-evolution
2. A few are not questions, but injunctions.
3. Which can also require search, but not proof search.
4. Sutcliffe and Puzis (2007) have automated the selection of relevant axioms from a larger set, but their system does not create new axioms.
5. Lakatos also discusses the use of generalisation to patch a faulty conjecture by including some counterexamples to the original conjecture. The TM system implements some Lakatos-inspired methods for correcting faulty conjectures by specialising them to exclude counter-examples [Colton and Pease, 2005], but it does not use specialisation as a route to prove an original correct conjecture.
6. The reason for the scare quotes around 'solid' will soon become apparent.
7. The method of surrender.
8. Note that some programming languages preclude this kind of error by requiring the programmer to formally define the type of inputs to any program.

References

Baker, S., Ireland, A. and Smaill, A. (1992). On the use of the constructive omega rule within automated deduction. In A. Voronkov (ed.), *International Conference on Logic Programming and Automated Reasoning – LPAR 92, St. Petersburg, Lecture Notes in Artificial Intelligence No. 624*. Berlin, Heidelberg: Springer-Verlag, pp. 214–25.

Bundy, A. and McNeill, F. (2006). Representation as a fluent: An AI challenge for the next half century. *IEEE Intelligent Systems*, 21(3): 85–7.

Bundy, A. (1991). A science of reasoning. In J-L. Lassez and G. Plotkin (eds.), *Computational Logic: Essays in Honor of Alan Robinson* MIT Press, pp. 178–98.

Bundy, A. (2006). Constructing, selecting and repairing representations of knowledge. In *Proceedings of ai@50: The Dartmouth Artificial Intelligence Conference: The next fifty years*.

Bundy, A., Byrd, L., Luger, G., Mellish, C., Milne, R. and Palmer, M. (1979). Solving mechanics problems using meta-level inference. In B. G. Buchanan (ed.), *Proceedings of IJCAI-79*. International Joint Conference on Artificial Intelligence, pp. 1017-27.

Bundy, A., Jamnik, M. and Fugard, A. (2005). What is a proof? *Phil. Trans. R. Soc A*, 363(1835): 2377–92.

Colton, S. and Pease, A. (2005). The TM system for repairing non-theorems. *Electronic Notes in Theoretical Computer Science*, 125(3). Amsterdam: Elsevier.

Fugard, A. J. B. (2005). An exploration of the psychology of mathematical intuition. Unpublished M.Sc. thesis, School of Informatics, Edinburgh University.

Jamnik, M., Bundy, A. and Green, I. (1997). Automation of diagrammatic reasoning. In M. E. Pollack (ed.), *Proceedings of the 15th IJCAI*, vol. 1. International Joint Conference on Artificial Intelligence, San Francisco: Morgan Kaufmann Publishers, pp. 528–533.

Lakatos, I. (1976). *Proofs and refutations: The logic of mathematical discovery*. Cambridge: Cambridge University Press.

Moore, J. S. (1974). Computational logic: Structure sharing and proof of program properties, part II. Unpublished Ph.D. thesis, University of Edinburgh.

Owen, S. (1990). *Analogy for automated reasoning*. San Diego: Academic Press Ltd.

Pease, A., Smaill, A., Colton, S. and Lee, J. (2009). Bridging the gap between argumentation theory and the philosophy of mathematics. *Foundations of Science, 14, Special Issue: Mathematics and Argumentation*.

Pólya, G. (1945). *How to solve it*. Princeton: Princeton University Press.

Robertson, D., Bundy, A., Muetzelfeldt, R., Haggith, M. and Uschold, M. (1991). *Eco-logic: Logic-based approaches to ecological modelling*. Cambridge, MA: MIT Press.

Sutcliffe, G. and Puzis, Y. (2007). SRASS: A semantic relevance axiom selection system. In F. Pfenning (ed.), *Procs of CADE-21*, vol. 4603 of *Lecture Notes in Computer Science*. Berlin, Heidelberg: Springer, pp. 295–310.

Winterstein, D., Bundy, A., Gurr, C. and Jamnik, M. (2002). Using animation in diagrammatic theorem proving. In M. Hegarty, B. Meyer and H. N. Narayanan (eds), *Procs of Diagrams 2002*, vol. 2317, *Lecture Notes in Computer Science*, Berlin, Heidelberg: Springer, pp. 46–60.

Part III

Language, Evolution and the Complex Mind

Introduction
Keith Stenning

Just what are the cognitive differences between humans and their closest animal relatives? Was there a 'magic bullet' that transformed humans at some point during what is already a short span of evolutionary time, that paltry five million years separating us from the last common ancestor with chimpanzees? The commonest answer throughout a millennium of philosophical discussion has been: 'Yes, there was a magic bullet – language', even if this was an answer to a rather different question before Darwin. Apes don't have it and humans do, and once they did, humans became capable of reasoning, and a host of other sins.

How does language do that? Well, the story goes, language is a system of signs (originally acoustic, or perhaps gestural, but now also written and reproduced in many media), and because these signs stand for things, operations on them can transform them into other signs, in ways which allow us to arrive at new knowledge about the things they stand for. Operations on 'elephant' are less energetic than operations on elephants, and, besides, the former word is more available than the latter animal. Moreover, 'elephant' can be got into arrangements with other signs which stand for nonexistent but desirable (or undesirable) arrangements of animals ('hallucinated pink elephants'?), which would be impossible to achieve with the animals themselves. Language allows cheap calculation about such arrangements, useful for achieving the former and avoiding the latter. And if you have never wanted to think about elephants, language offers other signs for whatever takes your fancy, existent or fictional. And, oh yes, I nearly forgot, language allows these thoughts to be communicated.

This 'magic bullet' view of language is not such a bad place to start. It is a lot better than no story at all. But it fairly soon should give rise to a host of hard questions: Exactly what is this system of signs called 'language'? Words, sentences, discourses? What cognitive effects does language have, and how? What good would language be without thoughts, or at least something, for it to express? What cognition do we already have to have for language to work its spell? How do we make inferences with it? These are all good questions, and contributing questions is one thing that pre-scientific stories must do. The last language-obsessed half-century of intensive study has shown just how complex and often unintuitive the answers are. But, more troublingly, this magic bullet view also reliably gives rise to a number of distortions.

One distortion is that it makes matters sound as if language is a single device that has been bolted onto the human mind without precursor. Evolution happens by transformation of existing structures and functions, sometimes slow, and sometimes not. Cognitive evolution is not much like plonking new software into a general purpose computer. Let's take a very much more lowly capacity than language – one which is a small part of human language – and use it to illustrate the problems with this bolt-on thinking. Labelling is the function whereby we assign signs (acoustic, sticky, whatever) to things, and we then utter, mutter, mark or otherwise token these labels while performing operations on them, in order to achieve some task relating to the things they stand for. Labelling things is a whole lot less than language; labels are very far from words, even though words can be used as labels. Labelling is (perhaps deceptively) simple enough to look like something that might be bolted onto a mind, yet with enough consequences that it might be interesting to extrapolate its case to language itself. Indeed, as Clark discusses (this section), just teaching chimps a practice of labelling transforms what they are able to do. We have all listened to children muttering labels as they perform some newly acquired hard task which they would have trouble doing silently. Even experts mumble labels. Here we are listening to a device (labelling) which started off 'outside' and has now been installed 'inside' the child's mind? It has been internalised. The fact experimenters can install this same device in a chimp's mind and thereby change its capacities is surely testimony to the bolt-on nature of labelling? A lot less hedged about by proprietary limitations than computer software – fits all primate models!

But what does this device consist of? Take an apparently simple task: remembering a set of objects by uttering their labels. It is tempting to think that the device consists of responding to the sight of the objects by uttering the relevant signs, and continuing to do this when the objects are withdrawn. We can imagine making one of these devices out of a tape recorder and a video camera, and some sealing wax, perhaps. But a moment's thought suggests our prototype won't work very well. For a start it'll go off every time we see anything we know a label for. And how does the sound connect with the thing it stands for? We are going to need to control our device.

So what had to be there in the child's (and the chimp's) mind for the labelling attachment to work? Didn't the child have to have lots of skills surrounding labelling in order to be able to make the attachment work? Labelling, as Clark suggests, even of the simplest kind, is an active skill which is probably best thought of as a way of controlling attention and working memory. This is not to denigrate the rewards of being able to label, but to valorise the complexity of the systems that labelling has to integrate with: working memory, long-term memory, object categorisation, task understanding, strategic decision – most of what are known now as 'executive functions'. The 'articulatory loop', which is one component of human verbal labelling, plays an important role in children's word learning (being able to hold a word-sound before it has a word-meaning [Baddeley, 1998]). But collies are quite good at responding to and learning (if not pronouncing) labels (Kaminski, 2004), and it's not an articulatory loop they have. One human innovation here is the far greater control of production of labels which speech allows. But apes can gain some of this control if the signs are gestural, or the placing of plastic tokens, so they must have a good deal of that necessary background support apparatus in place too, and the question becomes 'what is it there for, and why do they not use it in the way humans do?' Bolt-ons are bad models of the evolution of mind.

All of the papers in this section address, among other things, a question arising from McGonigle and Chalmers' work on relational learning using sustained teaching regimes with squirrel monkeys. These experiments reveal extraordinary analogies between the fine detail of simian and child performance in, for example, transitive choice and seriation abilities. It's not just that after many months of learning 'computer games' for hours per day, these monkeys can perform tasks indicative of transitivity of choice and seriation abilities similar to human six-year-olds. The literature is full of animals learning skills where it turns out that they achieve them in some quite different way than the human model. Much more importantly, after their prolonged learning, McGonigle and Chalmers' monkeys exhibit highly distinctive patterns of speed and error – the 'symbolic distance' effect, for one example – which are hallmarks of human reasoning about dimensions. By no stretch of the imagination are these monkeys taught the symbolic distance effect. The effect consists of reaction times being faster and errors less likely when making judgements of relations between 'distant' pairs of stimuli than close pairs on a dimension. There are no contingencies in the task rewarding these features of performance. They relate to a particular implementation of the reasoning, not to the demands of the task as defined by the experimenter. But this effect is good evidence that the monkeys have a basically similar mental structure for representing dimensions as do children, consisting of a linear ordering device, which they are 'attaching labels' to, and using to respond to relations between them.

One question that entices us immediately: 'Do these sustained and technology-enhanced regimes of learning reveal simian mental structures already there, or do they bring them into being?' Well, they certainly aren't there at the beginning in the sense that the monkeys can't do the target tasks at the beginning. And they certainly aren't brought into existence, in the sense of being bolted on. But there are structures and capacities there at the beginning that are transformable by experience, and what they are transformed into, what we see at the end, is certainly quite novel. So 'both and neither' would seem to be the best answer? As Clark concludes, this may not be a good dichotomy between revealing what is there, and introducing something new. All too often the view that 'something new' is introduced smuggles in the notion that there wasn't much there before (except perhaps some associative can of worms). All too often the idea that 'the structure is already implicitly there' smuggles in the furniture of innate ideas and their non-explanations. Now that biology tells us that much of the structure of the brain is 'learned' in the sense of created by mechanisms recording the contingencies of brain activity (NB this is NOT learning at the psychological level), shouldn't we give up this dichotomy? Even if we should remember that this is NOT nurture winning out.

The same goes for the relation between individual mind and cultural milieu: the idea that we have a mind of a certain nature, that then interacts with our culture to yield our behaviour, and that for any particular behaviour we can apportion its causes between our nature and our culture. Individuals are created by the interactions of their genes and their environment in ways that we now know cannot be factored out from each other in a context-independent manner (Stenning and van Lambalgen, this section). There are no percentages of nature and nurture, or nature and culture. The wiring diagrams of our brains are not produced from blueprints in our DNA. Rather they develop from our conception, by the interaction of DNA with environments: genetic environment, physiological, experiential and cultural environments. Some traits are highly variable and some universally fixed, but they are all fixed or varied by a range of these factors. There are cultural universals, but that does not mean that they have blueprints in the DNA, any more than universals of neuronal organisation mean that; environments can be constant too. There are highly variable cultural features, but then there is variability in the genes as well as the environments.

Students should be warned that a great part of the problems of talking rubbish about nature/nurture/culture is subdisciplinary salesmanship. Cognitive science folk may focus on the mind, on the brain, on the genes, on the culture, on the physical or social environment, or the body, or several other aspects of cognition. Each wants to argue the size of the contribution of their leg of the centipede. Each watches the others ignoring their precious leg. The worst response to such tussles is a ceasefire on agreed percentages. There are no percentages. If the nonsense of such factoring is scientifically clear, perhaps that can serve to dampen the salesmanship and increase the signal/noise ratio.

Colin Allen (Chapter 11) in discussing McGonigle and Chalmers' monkey experiments, expresses scepticism about the conventional dichotomy of associationistic and mentalistic theories of the mind. He notes how the expressive armoury of those who still accept the label 'associationist' has expanded radically since the heyday of S-R learning theory, even expanding to full Turing-power expressiveness if one includes connectionist models within associationism, as is often done. Here the reader may wonder what is left of associationism if it allows itself enough descriptive power to make Chomsky blush? A mini-industry in logic now produces neural (connectionist) models of logics and their fragments. Does this make these logicians into 'associationists'? Ironically, they do this neural implementation to show the inexpressiveness of their formalism, one that can be implemented without a central controller, in a distributed network. The roles of 'abstemious aunt associationism' and 'profligate uncle mentalism' appear to have dissolved in an outbreak of cross-dressing.

Associationism has a curious status in psychology's curious philosophy of its own science. Associationism often serves as little more than a bulwark against the supposed anthropomorphising hordes. But now, with our hugely deeper understanding of computation, do we need this bulwark against anthropomorphism anymore? No one accuses even the most expressive computer learning algorithms of anthropomorphism. Probably the best positive account of associationism's role now in psychology is providing a descriptive language for a descriptive phase of a young science. In a descriptive phase it is crucial to have a language of sufficient expressive power to accommodate what are believed to be the phenomena of record – otherwise they are lost. Hence the expansion of what associationism encompasses. But psychology easily forgets that expressive power is purchased at the expense of explanation. The fact that an expressive language of description is adequate to the phenomena means that that fact itself contributes nothing to the explanation of the data it describes. This is why the last half-century of mentalistic linguistic theory since Chomsky's syntactic theory has sought to limit the expressive power of its formalisms so that their fitting of the data does explain things about human language. Associationism shouldn't be mistaken for some safe haven explanation of the mind to be defended against wild speculations of mentalists. Associationism, as it has lost its parsimony of expression, is not globally an explanation of anything.

So if we can't talk sense about percentages of nature, nurture or culture, and we can't identify the emergence of human cognition with an elevation from associationist to mentalist mind, what can we say? Well, we can talk continuities and discontinuities in evolution, and this book, among other things, seeks to redress what most of its authors perceive as a dominant zeitgeist of discontinuity thinking about human emergence. McGonigle and Chalmers' experiments show that there are notable cognitive continuities between monkeys and humans which, before their work, would have been laughed at as impossible. Monkeys have redoubtable working memories which share some of their important mental structures with human working memory. Their retrievals from their working memories are 'planned' in the same recursive fashion as we plan ours, for example.

But we also need to talk about discontinuities. As Clark (Chapter 10), points out, all this continuity leaves us with the 'glass ceiling problem'. If the cognitive apparatus was already there, what were we waiting for all those millions of years? What are human cognition and language's discontinuities? How does language underpin human rationality, and cognition underpin language? If not as bolt-on pieces of technology, how are precursors altered?

In his chapter, Clark discusses the work of Christoff (2003) who claims a discontinuity (at least in degree) in humans' abilities to evaluate self-generated information such as sub-goals not given in the task specification, a capacity underpinned, they claim, by an enlarged rostrolateral prefrontal cortex (RLPFC). Of course, this interesting speculation is in its infancy. The search must be on for other primates showing capacities for evaluating such self-generated information and their implementation in RLPFC. Until McGonigle and Chalmers' monkeys revealed their planning and execution processes in working memory, monkeys weren't even supposed to internally generate such information, let alone evaluate it. 'Working memory' was conceived for explaining how humans had a 'workspace' for combining task-relevant information, some of which would have just arrived from perception, with information retrieved from long-term memory, and with information self-generated in the process of combination itself. These monkeys clearly themselves produce sub-goals (such as 'do the yellow ones now') as they pick off the icons by groups without repetition, and more generally have something remarkably like a simple working memory.

In this search for discontinuities, Allen, Hinzen (Chapter 12), and Stenning and van Lambalgen (Chapter 13) all emphasise the lack of insightful characterisation of cognition or language required for finding biologically significant phenotypic traits on which to found evolutionary explanations. Allen addresses problems with what one might call 'criterial' tasks. Ones that are set up for establishing all-or-none possession of some mental faculty. He starts out from the 'mirror self-recognition task' which observes the subjects' responses (or lack of them) to a mark on their head which can only be seen in a mirror. As a criterion for self-awareness (conceived as a unitary faculty) the task can be shown to be highly contextually variable and rather too continuous, as well as sensitive to prior experience with mirrors. He moves on to show the same kinds of problems with other tasks, such as imitation. Imitation exhibits some nice ironies in that children at some ages and in some circumstances resolutely display 'surface imitation' tendencies, while other primates tend to imitate in terms of the model's goals, thus displaying 'deeper understanding'. The child may be rescued from the comparison by observing that if one is learning language, one may need to imitate performances that one doesn't yet understand. But this rescue invites questions about whether animals shouldn't also be rescued by a deeper understanding of how their behaviour is embedded in their real environment. Criteria we may need, but we also need contextual sensitivity in their application. Allen argues that developmental characterisation is required to alleviate this problem by enforcing the embedding of behavioural descriptions in the child's and the animal's respective environments. Without such embedding, there's a danger that child and animal are simply being set different tasks.

Hinzen's chapter also proposes re-description of phenotype; in this case, of language semantics, as essentially deictic. Lexical items provide content, but grammar enables the speaker to refer to the same lexical content under different perspectives, and convey that perspective to the hearer, directing his attention to the mode of appearance of that lexical content in the given discourse. 'The king of France is bald' and 'There is a unique king of France and he is bald' may not differ in conventional propositional content, but they differ in the manner of their 'pointing'. The whole of grammar, including recursion, can be re-conceptualised as the control of deixis – an extended sense of deixis in which the pointing is within an abstract structure. Stenning and van Lambalgen propose a re-description of the magic bullet itself, human language. They propose that the process of planning discourse – the defeasible creation of interpretation from non-linguistic environment, general knowledge and sequential linguistic utterance – is the most novel human component capacity for language. Cognitively, language is planned social action, and discourse is the product. Discourse is language-in-operation, rather than language-as-structure, and evolution is above all about what creatures do and the selective pressures on them.

Both 'planning' and 'discourse' have somewhat technical meanings here. Planning covers the cogitative processes we first associate with the word, but also covers the rapid automatic unconscious processes that make up the submerged iceberg of language processing and motor control. Computationally, these processes share more than they differ by, and focusing first on the similarities and backgrounding the differences allows continuities to reveal themselves.

'Discourse' is a word with different connotations in logic and linguistics than in psychology. In psychology, it often means something close to 'dialogue' and emphasises social interchange. In logic and linguistics, discourse is defined in terms of connectedness and contextualisation: what happens next within the context so far created? A discourse is a sequence of utterances in context which progressively create the context in which they are interpreted.

For our present purposes, what probably seems like a picky distinction between dialogue and continuity as the defining feature of discourse, has immediate consequences. Discourse is a kind of planned action. Even if we take private mental process produced solely for controlling one's own action planning (those self-directed mumblings we mentioned before), or McGonigle and Chalmers' monkeys' sequences of tokenings, these are discourse (in the continuity sense) as much as is any dialogue. Primate planning requires private proto-symbols and is already recursive in its character. The monkey's planning is down at the automatic end of the spectrum, but then so is a great deal of the infrastructure of human thought.

Hinzen's chapter takes the nearest path to a 'magic bullet' saltatory account of cognitive and linguistic evolution. He emphasises the differences between human language and the communication systems of apes (trained or untrained). The dimension of difference he emphasises is propositionality or declarativeness. Propositions are bearers of truth-values independent of the actions that one bases on them, or the purposes for which one asserts them. Even the most superficially language-like of the trained 'utterances' of apes do not, he argues, have this property: they are inseparable from contexts of action and reasons for assertion. Words are not like the tokens of these trained ape languages because they are embedded in grammar that can control their focus of reference independently of their lexical content.

Hinzen reads this modern state of human language as implying strong constraints on its evolutionary origins. He associates communication with the code-based model of Shannon and Weaver (and associationistic cognition) with a fixed code book of signals evidenced in involuntary animal signalling systems, and on this basis dismisses communication as the ancestral function that developed into human language. Thus social origins of language in communication are dismissed in favour of a 'saltatory' change which permitted the expression of the pre-existing language of thought, along with other dependent cultural artefacts. In a memorable phrase, he sees language leading to the 'grammaticalisation of the brain' as a result of some single genetic event such as Crow's (2010) proposed change to cortical asymmetries. Whatever 'last genetic change' that happened is seen as the precipitating event of the revolution around 100,000 years ago which gave rise to distinctively modern cognition and language, expressed through art, burial, and the other cultural artefacts we regard as quintessentially human. One kind of evidence for the saltatory nature of the change is cited as the stasis of human cognition for the previous two million years, as evidenced by unchanging stone tool artefacts.

Stenning and van Lambalgen share many but not all of Hinzen's assumptions. These authors all work in the semantic/linguistic/logical traditions, with substantial agreements about the fundamentals of modern human languages (they are propositional, recursive, grammatically structured expressions of thought). They even agree on their first pass answers to the classical question, whether language originates as an expression for thought, or whether thought is an internalisation of public language. Both plump for versions of the former position, impressed by how pointless language would be for a being that did not have propositional thoughts to express. Yet they differ sharply on their accounts of the evolutionary processes at the origin.

Stenning and van Lambalgen see the origin of grammar in the 'grammaticalisation' of discourse distinctions, and the ancestral origin of discourse in primate planning. Hinzen adopts the conventional linguistic position that sentence grammar enables human discourse. But he acknowledges that there is a precise correspondence between grammar and discourse distinctions. Correspondences are symmetrical, and the disagreement is about the direction of causation in evolution, although at a finer grain there are substantial differences about the degree to which language can be separated from the knowledge which is an integral part of its processing. One might see Stenning and van Lambalgen's call to analyse ancestral planning and its exaptation for human discourse planning as proposing to explain how propositionality, declarativeness, recursion, grammar, and so on, arose in the evolutionary process, starting as private 'thought' and eventuating in public language. This is a gradualist account in contrast with Hinzen's saltatory one.

Biology teaches us that to understand evolution requires that we understand the whole creature in its whole environment. Evolutionary developmental biology (evo-devo) further teaches us that evolution is constituted by changes in developmental timing (an excellent non-technical introduction is Carroll, 2005). Allen's and Stenning and van Lambalgen's chapters both call for developmental perspectives to play a larger role in comparing us to our ancestors. Allen upbraids the habit of comparing adult performance of other species with various developmental stages of our children. This may be better than nothing when it is all that is available, but is fraught with difficulty, and depends on a 'recapitulationist' combined with a 'progressivist' view of evolution: that descendants' developmental stages recapitulate the adult forms of ancestral species, and constitute progression towards some goal of perfection.

Of course, there turned out to be a grain of truth in recapitulationism, but it is in general unsatisfactory. As the old joke goes, modern mammals go through a stage when they are entirely dependent on their mother's milk, but we can confidently say that none of mammals' ancestors had an adult stage which shared this property. Recapitulationism was the nineteenth-century progenitor of evo-devo. It bequeathed to evo-devo the core idea that evolution happens by transformations of developmental processes, and evo-devo has been the most important achievement of late-twentieth-century biology. It has shown us how evolution can produce novel structures and processes of great complexity without appealing to the miraculous simultaneous mutation, just thus and so, of the thousands of genes involved. An interesting micro-example is Hinzen's appeal to the candidate genetic change affecting the control of the 'torsion' of the human cerebral hemispheres (Crow 2010), which is the developmental process underlying the unique human asymmetry that both he and Stenning and van Lambalgen see as so important for human language. The critical contribution of evo-devo is that small genetic changes to the control of developmental processes produce large-scale configurational changes to structures and their functions. Language may be an exaptation of planning evolved under pressure for communication, even if the resulting communication has little to do with Shannon and Weaver signalling, and wasn't the pressure that made our ancestors' planning what it was.

Stenning and van Lambalgen argue that to find the human innovatory 'magic bullet', or perhaps the 'magic triggers' would be better, one has to consider the prominent changes in human ontogenetic timings, and this willy-nilly demands understanding human ancestors in their whole environments. Humans display wildly biologically expensive changes in their degree of motoric helplessness at birth, which makes the human infant deal with its physical world in an almost wholly socially mediated way. Acting is getting Mum to act. Evo-devo prompts that we understand what drives this change and with what consequences. Maternal pelvic/neonatal head-size ratios figure as prominently as cognitive factors such as patterns of clausal recursion, and there are more feedback loops than one can knit into a tea cosy. An ape, with all the primate pre-adaptation for complex motor and social planning, with an enlarging brain, a bipedal habit, and a consequent ongoing obstetric crisis, evolved a developmental pattern in which action was dominated by social action, and social and cultural organisation to go with it.

In this development of the arguments, the magic bullet of language has got smeared across a wide range of cognition and, in the smearing, things look a great deal more continuous. Human planning of discourse is continuous with primate planning of action. None of this diminishes the size or the qualitative nature of the differences that remain, whether those turn out to be the evaluation of self-produced information, or the duality of discourse planning, the invasion of grammar, something quite other, or all of the above. None of this continuity requires anthropomorphism to be our view of our ancestors. Brendan McGonigle and Maggie Chalmers were one of very few teams of researchers who combined an evolutionary with a developmental view of cognition, starting way back when evo-devo was in its infancy.

References

Baddeley, A., Gathercole, S. and Papagno, C. (1998). The phonological loop as a language learning device. *Psychological Review*, 105(1): 158–73.

Carroll, S. B. (2005). Endless forms most beautiful: the new science of evo devo and the making of the animal kingdom. London: Norton.

Christoff, K., Ream, J., Geddes, L. and Gabrieli, J. (2003). Evaluating self-generated information: anterior prefrontal contributions to human cognition. *Behavioral Neuroscience*, 117(6): 1161–68.

Crow, T. J. (2010). A theory of the origin of cerebral asymmetry: epigenetic variation superimposed on a fixed right-shift. *Laterality: Asymmetries of Body, Brain and Cognition*, 15(3): 289–303.

Kaminski, J., Call, J. and Fischer, J. (2004). Word learning in a domestic dog: evidence for 'fast mapping'. *Science*, 304(5677): 1682–3.

10
How to Qualify for a Cognitive Upgrade: Executive Control, Glass Ceilings and the Limits of Simian Success

Andy Clark

10.1 Introduction

It is sometimes suggested that words and language form a kind of 'cognitive niche' (Clark, 1998, 2005, 2006, 2008; Chapter 4): an animal-built structure that productively transforms our cognitive capacities. But even if language cognitively empowers us in many deep and unobvious ways, it would be quite wrong to assume that such empowerment occurs in either a neural or an evolutionary vacuum. In evolutionary terms, we need to recognise the various precursors of our own prodigious skills at species-level self-scaffolding. In neural terms, we need to uncover the specific innovations that allow certain kinds of agents to benefit (humans massively, simians somewhat, hamsters not at all) from the empowering effects of exposure to a public linguistic edifice. What we need to understand is thus a delicate balancing act between extra-neural and neural innovation, such that the public material structures of language are enabled (in some beings and not in others) to play significant cognitive roles.

In the present chapter, I first lay out a few of the ways in which language may indeed act as a potent form of cognitive scaffolding. I then briefly rehearse the results of a series of elegant comparative and developmental studies (summarised in McGonigle and Chalmers [2006]) that suggest a surprising amount of evolutionary continuity between human and simian (squirrel monkey) subjects in respect of some of the key 'building block' skills that enable this potent 'mind-tool' (Dennett, 2000) to emerge. I end by asking, 'What then limits simian success?'

10.2 Words as cognitive scaffolding

Language (including speech, gesture and written forms) is sometimes said to play an active role in enabling human thought and reason. Language, that is to say, does not merely communicate thoughts whose genesis and contents are fully language-independent. Rather, language may be part of the apparatus by which thinking and reasoning occur. Such views are associated with the work of a wide and otherwise quite disparate range of theorists, including Vygotsky (1962), Bruner (1966), Dennett (1991), Jackendoff (1996), Donald (1991), Clark (1998, 2006), Goldin-Meadow (2006), McNeill (1992, 2005) and Smith and Gasser (2005). Language, these models variously insist, is best seen as a form of mind-transforming cognitive scaffolding: a culturally heritable, persisting, though never stationary, symbolic edifice that plays a critical role in allowing minds like ours to exist in the natural order.

In this section I examine three distinct but interlocking benefits of this linguistic scaffold. First, the simple act of labelling the world opens up a variety of new computational opportunities and supports the discovery of increasingly abstract patterns in nature. Second, encountering or recalling structured sentences supports the development of otherwise unattainable kinds of expertise. And last, linguistic structures contribute to some of the most important, yet conceptually complex, of all human capacities: our ability to reflect on our own thoughts and characters, and our limited-but-genuine capacity to control and guide the shape and contents of our own thinking.

10.2.1 Augmenting reality

Consider the case of Sheba and the treats, as recounted in Boysen et al. (1996). Sheba (an adult female chimpanzee) has had symbol and numeral training; she knows about numerals. Sheba sits with Sarah (another chimp), and two plates of treats are shown. What Sheba points to, Sarah gets. Sheba always points to the greater pile, thus getting less. She visibly hates this result, but can't seem to improve. However, when the treats arrive in containers with a cover bearing numerals on top, the spell is broken and Sheba points to the lesser number, thus gaining *more* treats.

What seems to be going on here, according to Boysen, is that the material symbols, by being simple and stripped of most treat-signifying physical cues, allow the chimps to sidestep the capture of their own behaviour by ecologically specific fast-and-frugal sub-routines. The material symbol here acts as a manipulable and in some sense merely 'shallowly interpreted' (Clowes, 2007) stand-in able to loosen the bonds between perception and action. Importantly, the presence of the material symbol impacts behaviour not in virtue of being the key to a rich inner mental representation (though it may be this also) but rather by itself, qua material symbol, providing a new target for selective attention and a new fulcrum for the control of action. Such effects, as Clowes (2007) argues, do of course depend on the presence of something akin to a system of interpretation. But it is their ability to provide simple, affect-reduced, perceptual targets that (I want to suggest) explains much of their cognitive potency.

In much the same way the act of labelling creates a new realm of perceptible objects upon which to target basic capacities of statistical and associative learning. The act of labelling thus alters the computational burdens imposed by certain kinds of problems. I have written quite a bit on this elsewhere, so I'll keep this brief. My favourite example (Clark, 1998) begins with the use, by otherwise language-naive chimpanzees, of concrete tags (simple and distinct plastic shapes) for relations such as sameness and difference. Thus, a pair such as cup/cup might be associated with a red triangle (sameness) and cup/shoe with a blue circle (difference). This is not in itself surprising. What is more interesting is that after this training, the tag-trained chimps (and only tag-trained chimps) prove able to learn about the abstract properties of higher-order sameness, that is, they are able to learn to judge of two presented pairs (such as cup/cup and cup/shoe) that the relation between the relations is one of higher-order difference (or, better, lack of higher-order sameness) since the first pair exhibits the sameness relation and the second pair the difference relation (Thompson, Oden and Boysen, 1997). The reason the tag-trained chimps can perform this surprising feat is, so the authors suggest, because by mentally recalling the tags the chimps can reduce the higher-order problem to a lower-order one: all they have to do is spot that the relation of difference describes the pairing of the two recalled tags (red triangle and blue circle).

This is a nice concrete example of what may well be a very general effect (see Clark, 1997; Dennett, 2000). Once fluent in the use of tags, complex properties and relations in the perceptual array are, in effect, artificially reconstituted as simple inspectable wholes. The effect is to reduce the descriptive complexity of the scene. Kirsh (1995) describes the intelligent use of space in just these terms. When, for example, you group your shopping in one bag and mine in another, or when the cook places washed vegetables in one location and unwashed ones in another, the effect is to use spatial organisation to simplify problem-solving, by using spatial proximity to reduce descriptive complexity. It is intuitive that once descriptive complexity is thus reduced, processes of selective attention, and of action-control, can operate on elements of a scene that were previously too 'unmarked' to define such operations over. Experience with tags and labels may be a cheap way of achieving a similar result. Spatial organisation reduces descriptive complexity by means of physical groupings that channel perception and action towards functional or appearance-based equivalence classes. Labels allow us to focus attention on all and only the items belonging to equivalence classes (the red shoes, the green apples, etc.). In this way, both linguistic and physical groupings allow selective attention to dwell on all and only the items belonging to the class. And the two resources were seen to work in close cooperation. Spatial groupings are used in teaching children the meanings of words, and mentally rehearsed words may be used to control activities of spatial grouping.

Simple labelling thus functions as a kind of augmented reality[1] trick by means of which we cheaply and open-endedly project new groupings and structures onto a perceived scene. Labelling is cheap since it avoids the physical effort of actually putting things into piles. And it is open-ended insofar as it can group in ways that defeat simple spatial display; for example, by allowing us to selectively attend to the four corners of a tabletop, an exercise that clearly cannot be performed by physical reorganisation! Linguistic labels, on this view, are tools for grouping, and in this sense act much like real spatial reorganisation. But in addition (and unlike mere physical groupings) they effectively and open-endedly add new 'virtual' items (the recalled labels themselves) to the scene. In this way, experiences with tags and labels warp and reconfigure the problem spaces for the cognitive engine. Relatedly, Lupyan, Rakison and McClelland (in press) show that labelled categories are easier to learn *even when the label is fully redundant*, and adds no information not otherwise available in the stimulus array. They speculate (p. 13) that this is because 'the labels...provide perceptually simple correlates to an otherwise perceptually complex task'.

This suggests a possible answer to an important question raised by Smith and Gasser (2005) who ask why, given that human beings are such experts at grounded, concrete, sensorimotor-driven forms of learning, do the symbol systems of public language take the special and rather rarefied forms that they do?

> One might expect that a multimodal, grounded, sensorimotor sort of learning would favor a more iconic, pantomime-like language in which symbols were similar to referents. But language is decidedly not like this [...] there is no intrinsic similarity between the sounds of most words and their referents: the form of the word dog gives us no hints about the kind of thing to which it refers. And nothing in the similarity of the forms of dig and dog conveys a similarity in meaning. Smith and Gasser (2005: 22)

The question, in short, is 'Why in a so profoundly multimodal sensorimotor agent such as ourselves is language an arbitrary symbol system?' (Smith and Gasser, 2005: 24).

One possible answer, of course, is that language is like that because (biologically basic) *thought* is like that, and the forms and structures of language reflect this fact. But another answer says just the opposite. Language is like that, it might be suggested, because thought (or rather, biologically basic thought) is *not* like that. The computational value of a public system of essentially context-free, arbitrary symbols lies, according to this opposing view, in the way such a system can push, pull, tweak, cajole and eventually cooperate with various non-arbitrary, context-sensitive forms of biologically basic encoding,[2] providing simple inspectable structures in place of complex multimodal representations.

10.2.2 Sculpting attention

The role of structured language as a tool for scaffolding action has been explored in a variety of literatures, ranging from Vygotskyian developmental psychology to cognitive anthropology.[3] Mundane examples of such scaffolding abound, and range from memorised instructions for tying one's shoelaces to mentally rehearsed mantras for crossing the road, such as 'look right, look left, look right again and if all is clear cross with caution' (that's for UK-style left-hand drive roads; don't try that in the USA, folks!). In such cases, the language-using agent is able (once the instructions are memorised, or, in the written case, visually accessed) to engage in a simple kind of behavioural self-scaffolding, using the phonetic or spatial sequence of symbolic encodings to stand proxy for the temporal sequence of acts. Frequent practice then enables the agent to develop genuine expertise, and to dispense with the rehearsal of the helpful mantra.

More interesting than all this, however, is the role of linguistic rehearsal in expert performance itself. In previous work (Clark, 1996) I discussed some ways in which linguaform rehearsal enables experts to temporarily alter their own focus of attention, thus fine-tuning the patterns of inputs that are to be processed by fast, fluent, highly trained sub-personal resources. Experts, I argued, are doubly expert. They are expert at the task in hand, but also expert at using well-chosen linguistic prompts and reminders to maintain performance in the face of adversity. Sometimes inner rehearsal here plays a distinctly affective role, as the expert encourages herself to perform at her peak.[4] But in addition to the important cognitive-affective role of inner dialogue, there may also be cases in which verbal rehearsal supports a kind of perceptual restructuring via the controlled disposition of attention (for a nice example, see the discussion of linguistic rehearsal by expert Tetris players in Kirsh and Maglio, 1992). The key idea, once again, is that the linguistic tools enable us to deliberately and systematically sculpt and modify our own processes of selective attention. In this regard Sutton (in press) describes in some detail the value of 'instructional nudges' (small strings of words, simple maxims). Such nudges, Sutton argues, are often best employed not by the novice but by the expert[5] who can use them to tune and modulate highly learned forms of embodied performance.

Direct cognitive benefits from linguaform encodings are also suggested by recent work by Hermer-Vazquez, Spelke and Katsnelson (1999). In this study, prelinguistic infants were shown the location of a toy or food in a room, then were spun around or otherwise disoriented and required to try to find the desired item. The location was uniquely determinable only by remembering conjoined cues concerning the colour of the wall and its geometry (e.g. the toy might be hidden in the corner between the long wall and the short blue wall). The rooms were designed so that the geometric or colour cues were individually insufficient, and would yield an unambiguous result only when combined together. Prelinguistic infants, though perfectly able to detect and use both kinds of cue, were shown to exploit only the geometric information, searching randomly in each of the two geometrically indistinguishable sites. Yet adults and older children were easily capable of combining the geometric and non-geometric cues to solve the problem. Importantly, success at combining the cues was not predicted by any measure of the children's intelligence or developmental stage except for the child's use of language. Only children who were able to spontaneously conjoin spatial and, for example, colour terms in their speech (who would describe something as, say, to the right of the long green wall) were able to solve the problem. Hermer-Vazquez et al. (1999) then probed the role of language in this task by asking subjects to solve problems requiring the integration of geometric and non-geometric information while performing one of two other tasks. The first task involved shadowing (repeating back) speech played over headphones. The other involved shadowing, with their hands, a rhythm played over the headphones. The working memory demands of the latter task were at least as heavy as those of the former. Yet subjects engaged in speech shadowing were unable to solve the integration-demanding problem, while those shadowing rhythm were unaffected. An agent's linguistic abilities, the researchers concluded, are indeed actively involved in his or her ability to solve problems requiring the integration of geometric and non-geometric information.

The precise nature of this linguistic involvement is, however, still in dispute. Hermer-Vazquez, Spelke and Katsnelson (1999), and following them Carruthers (2002), interpret the results as suggesting that public language provides (or perhaps better, engenders) a unique internal representational medium for the cross-modular integration of information. The linguaform templates of encoded sentences provide, according to Carruthers, special representational vehicles that allow information from otherwise encapsulated resources to interact. This is an attractive and challenging story, and one that I cannot pretend to do justice to here. But it is one that presupposes a specific and quite contentious (see Fodor, 2001) view of the mind as massively (not merely peripherally) modular, requiring linguaform templates to bring multiple knowledge bases into fruitful contact.

Suppose we abandon this presupposition of massive modularity? We may still account (or so I suggest) for the role of language in enabling complex multi-cued problem-solving by depicting the linguistic structures as providing essential scaffolding for the distribution of selective attention to complex (in this case colour/geometry conjunctive) aspects of the scene.

According to this alternative account, linguistic resources enable us better to control the disposition of selective attention to ever-more complex feature combinations.[6] Attention to a complex conjoined cue, I suggest, requires the (possibly unconscious) retrieval of at least some of the relevant lexical items. This explains the shadowing result. And it fits nicely with the earlier account of the cognitive impact of simple labels insofar as linguistic activity (in this case more structured activity) again allows us to target our attentional resources on complex, conjunctive or otherwise elusive elements of the encountered scene. The idea that language enables new forms of selective attention by, in effect, providing new objects for old (i.e. not specifically linguistic) attentive processes, can be further illustrated by the case of arithmetical thought and reason, to which we now turn.

10.2.3 Anchoring thoughts

What is going on when you think the thought that '98 is one more than 97'? According to a familiar model, you must have succeeded (if you managed to think the thought at all) in translating the English sentence into something else. The something else might be a sentence of mentalese (e.g. Fodor, 1987) or a point in some exotic state-space (e.g. Churchland, 1989).

But consider now the account proposed by Stanislas Dehaene and colleagues (see Dehaene, 1997; Dehaene et al., 1999). Dehaene depicts this kind of precise mathematical thought as emerging at the productive intersection of three distinct cognitive contributions. The first involves a basic biological capacity to individuate small quantities: 1-ness, 2-ness, 3-ness and more-than-that-ness, to take the standard set. The second involves another biologically basic capacity, this time for approximate reasoning concerning magnitudes (discriminating, say, arrays of 8 dots from arrays of 16, but not more closely matched arrays). The third, not biologically basic but arguably transformative, is the learned capacity to use the specific number words of a language, and the eventual appreciation that each such number word names a distinct quantity. Notice that this is not the same as fully or 'intuitively' appreciating, in at least one important sense, just what that quantity is. Most of us can't form any clear image of, for example, 98-ness (unlike, say, 2-ness). But we understand nonetheless that the number word '98' names a unique quantity in between 97 and 99.

When we add the use of number words to the more basic biological nexus, Dehaene argues, we acquire an evolutionarily novel capacity to think about an unlimited set of exact quantities.[7] We gain this capacity not because we now have a mental encoding of 98-ness just like our encoding of 2-ness. Rather, the new thoughts depend directly upon our capacity to represent the linguistic tokens (where these may be spoken words or written numbers) of the numerical expressions themselves. They depend, that is, on our capacity to represent the symbol strings of our own public language. The actual numerical thought, on this model, is had courtesy of the *combination* of this tokening (of the symbol string of a given language) and the appropriate activation of the more biologically basic resources mentioned earlier.

Here is some of the evidence for this view, as presented in Dehaene et al. (1999). First, there are the results of studies of Russian–English bilinguals. In these studies, Russian–English bilinguals were trained (quite extensively) on 12 cases involving exact and approximate sums of (the same) pairs of two-digit numbers, presented as words in one or the other language. For example (in English), a subject might be trained on the question 'Four + Five' and asked to select their answer from 'Nine' and 'Seven'. This is called the exact condition, as it requires exact reasoning since the two candidate numbers are close to each other. By contrast, a question like '"Four + Five", select answer from "Eight" and "Three"' belongs to the approximate condition, as it requires only rough reasoning as the candidates are now quite far apart.

After extensive training on the pairs, subjects were later tested on the very same sums in either the original or the other (non-trained) language. After training, performance in the approximation condition was shown to be unaffected by switching the language, whereas in the exact condition, language switching resulted in asymmetric performance, with subjects responding much faster if the test language corresponded to the training language. Crucially, then, there were no switching costs at all for trained approximate sums. Performance was the same regardless of language switching. Training-based speed-up is thus non-language switchable for the exact sums and fully switchable for the inexact ones. Such studies, Dehaene et al. concluded, provide:

> evidence that the arithmetic knowledge acquired during training with exact problems was stored in a language-specific format. For approximate addition, in contrast, performance was equivalent in the two languages providing evidence that the knowledge was stored in a language-independent form. (Dehaene et al., 1999: 973)

A second line of evidence draws on lesion studies in which (to take one example) a patient with severe left-hemisphere damage cannot determine whether 2+2 is 3 or 4, but reliably chooses 3 or 4 over 9, indicating a sparing of the approximation system.

Finally, Dehaene et al. (1999) present neuroimaging data from subjects engaged in exact and approximate numerical tasks. The exact tasks show significant activity in the speed-related areas of the left frontal lobe, while the approximate tasks recruit bilateral areas of the parietal lobes implicated in visuo-spatial reasoning. These results are presented as a demonstration 'that exact calculation is language dependent, whereas approximation relies on nonverbal visuo-spatial cerebral networks' (Dehaene et al., 1999: 970) and that 'even within the small domain of elementary arithmetic, multiple mental representations are used for different tasks' (1999: 973).

Dehaene (1997) also makes some nice points about the need to somehow establish links between the linguistic labels and our innate sense of simple quantities. At first, it seems, children learn language-based numerical facts *without* such appreciation. According to Dehaene, 'for a whole year, children realise that the word 'three' is a number without knowing the precise value it refers to' (Dehaene, 1997: 107). But once the label gets attached to the simple innate number line, the door is open to understanding that *all* numbers refer to precise quantities, even when we lack the intuitive sense of what the quantity is (e.g. my own intuitive sense of 53-ness is not distinct from my intuitive sense of 52-ness, though all such results are variable according to the level of mathematical expertise of the subject).

Typical human mathematical competence, all this suggests, is plausibly seen as a kind of hybrid, whose elements include:

i) images or encodings of actual words in a specific language;
ii) an appreciation of the fact that each distinct number word names a specific and distinct quantity, and;
iii) a rough appreciation of where that quantity lies on a kind of approximate, analogue number line (e.g. 98 is just less than halfway between 1 and 200).

Many of our higher mathematical thoughts rely, if this is correct, on the coordinated action of a heterogeneous swathe of resources, including an internal representation of the numeral, as it occurs in some specific written or spoken code, along with other resources (such as the analogue number line) to which that becomes, via intensive learning routines, roughly keyed so as to yield a sense of relative location. Importantly, the presence of internal representations of the actual number words in a specific public code itself here forms part of the coordinated representational medley whose joint action constitutes and makes possible many kinds of arithmetical knowing.

In this section I explored three ways in which the forms and structures of public codes, such as language and numeral systems, might productively impact human thought and reason. Linking the three ways is a common thread, according to which language provides new objects (Dennett, 1991, 2000) that can act as fulcrums for attention, reason, and categorisation: artificial anchors in the shifting sea of thought. Clearly, this story leaves many crucial questions unanswered. One notable shortfall concerns the vexed question of evolutionary and neural continuity. Just what it is about human brains and/or human history that has enabled structured public codes to get such a comprehensive and multiply empowering grip on minds like ours? It is to this difficult question that I now turn.

10.3 Sculpting simian thought

In an important series of experiments, McGonigle, Chalmers and their colleagues have probed the surprising extent to which human abilities to use, and hence to profit from, culturally inherited structured public codes might be grounded in 'building block' capacities shared with non-human primates.

The starting point for their investigations was a series of experiments (McGonigle and Jones, 1978) on squirrel monkeys that showed clear abilities to code stimulus properties (size and brightness) in a relational, rather than merely in an absolute, fashion and further experiments (McGonigle and Chalmers, 2002) that showed them able to deploy these abilities in a flexible way. In these latter experiments, the monkeys were taught a simple set of five conditional rules such as 'if all the blocks are green, choose the second smallest block', 'if all the blocks are red, choose the smallest block'. The presented colours thus acted as a kind of arbitrary sign indicating which rule to apply. The monkeys proved able to run all five rules concurrently, in random order, and irrespective of details of spatial location and absolute size. Moreover, their performance on this task was highly similar to that of four-year-old human children. Commenting on these results, McGonigle and Chalmers, 2006: 247) suggest that they demonstrate in the squirrel monkey 'an ability to map a set of arbitrary signs onto the relational structure of the task set – an impressive hooking of signs to representations'.

McGonigle and Chalmers (2002), and McGonigle, Chalmers, and Dickinson (2003) extended this work into the domain of high-level serial ('executive') control, as indexed by success on a size seriation task. In classic work on this topic (Inhelder and Piaget, 1969) children were asked to copy a model of multiple rods ordered by ascending or descending size. Young children fail to perform this task for more than a few rods at best, but by the age of five or six are able to roughly copy a model using a trial-and-error strategy. Significantly, around the age of seven, they spontaneously become able to perform the task in a way that is principled (that is, a way that works directly, without relying on trial-and-error placement and replacement) and generative (working for an arbitrary number of rods). At that time they also begin to display sophisticated capacities of 'item insertion': the insertion of any randomly selected rod from the pool into the array. Piaget dubbed this entire set of accomplishments (spontaneous principled ordering, generativity, and item insertion) 'operational seriation' and thought it suggestive of the child's ascent to a new form or level of – in more contemporary parlance – planning, understanding and executive control.

To compensate for important inter-species differences in sheer object manipulation skills when dealing with the real rods, the new experiments used touch-screen technologies that required both child and (in this case) capuchin monkey *(Cebus apella)* simply to touch presented icons, in increasing or decreasing size order, using random (and trial-by-trial varying) spatial layouts. Both groups were also tested for correct item insertion into a presented array, which requires an additional skill of 'ordinal size identification'. Extensive experiment and observation of the human children at different ages and stages of training suggested that success at the size seriation task is probably the precursor skill here, and that:

> as far as the younger child is concerned...the implication was strong that the skill of identifying the ordinal position of every item in a set...originate[s] from serial behaviours dedicated to ordering objects in the world. (McGonigle and Chalmers, 2006: 253)

The important idea here is that it may be the child's experience with the serial ordering task, or with this kind of task, that is responsible for the apparent emergence of a whole new level or form of executive control. The distinction here is between thinking of the experimental setting as merely revealing what is already there, versus seeing it as actively bringing new skills into existence. The child's form of life, replete with language and symbols, counting games, and the like, is more or less certain to offer the kinds of experience that would lead to bouts of size seriation and thence (on this model) to ordinal position skills. But the standard ecology of the monkey is not, superficially at least, similar in that regard. Certainly, it has not been engineered, over cultural-evolutionary time, so as to present the infant monkey with multiple exemplars of structured arrays apt for games of reorganisation according to rules! Nonetheless, the canny use of the touch-screen technology opened up the possibility of performing similar tests on monkeys, thus partially overcoming (in this case at least) a persistent problem in comparative psychology: the problem of distinguishing the multiple and incremental effects of the super-rich, highly engineered, human training environment from effects due solely to the native cognitive apparatus of the species in question. (As Keith Stenning [personal correspondence] notes, there is also the problem of the potential impoverishment of any laboratory environment relative to the natural social and physical environment of the non-human species in question).

The results (McGonigle, Chalmers and Dickinson, 2003) were impressive. Capuchin size seriation (as evidenced by principled model copying) was comparable to that of a human six-year-old. Indeed, '...their performance had all the hallmarks of well-organised executive behaviour, delivered in one smooth production, and very rapidly in the case of successful performance' (McGonigle and Chalmers, 2006: 257). In addition, the monkeys even proved able to classify items into multiple so-called 'reciprocal categories', that is to say ones that share features so that, for example, a blue square belongs in one set, while a blue circle belongs in another. Success in such cases involves recognising that such-and-such an item belongs both to a class and to a series, so as to respect 'a sorting principle based on hierarchical control where one dimension of difference is nested within another' (McGonigle and Chalmers, 2006: 258).

Monkeys, these multiple studies strongly suggest, can be brought to possess several of the key sub-skills implicated in the human's apparently unique ability to master the complex structures of public codes. Starting perhaps with core abilities to represent relational properties in a stimulus array, they can be steered through a sequence of touch-screen training environments that reveal – or perhaps better, that help to induce – capacities of serial ordering, including principled ordering demanding high levels of executive control, as well as multiple concurrent and reciprocal categorisation, again requiring the use of hierarchical and nested control. This learning trajectory represents:

> the most extensive longitudinal monkey study on record (and possibly for any species, including humans) [and displays] considerable momentum...in the manner in which monkeys adapt to tasks of progressive difficulty (both in qualitative as well as quantitative terms) achieving at terminus the most impressive seriation performance ever shown by a nonhuman subject. (McGonigle and Chalmers, 2006: 262)

Mastering these serial executive skills, McGonigle and Chalmers suggest, is the core element in the approach to full 'operational seriation'.

10.4 So what limits simian success?

The spectacular cognitive ascent of the monkey subjects raises an obvious question. The question is, What limits simian success?

10.4.1 Manipulations and manipulanda

In a commentary piece published in 2008, McGonigle and Chalmers offer a robustly optimistic view, suggesting, with reference to the findings just discussed, that:

> new programs of research on serial ordering mechanisms and executive control with primates (McGonigle & Chalmers, 2006; Terrace, 2005) open a window on the emergence of complex behaviors from simple relational primitives. These are already showing the growth of systematicity, compositionality, the deployment of economic strategies with experience which reduce cognitive costs (McGonigle, Chalmers, & Dickinson, 2003), and an upward bound trajectory with a momentum that has not so far been limited by any glass ceiling on simian achievements. (McGonigle and Chalmers, 2008: 143)

The reason for this apparent optimism concerning the potential of simian thought can be traced to their further suggestion that it is the incremental, systematic forms of monkey training experience that the experiments involved that induced (as much as revealed) these new levels of skill. The experiments provided temporally sustained (though at around an hour a day, by human standards still not intensive) teaching, which acted in concert with the vastly enhanced array manipulation capacities made available by the touch-screen technologies. It is this new ecological scaffolding, McGonigle and Chalmers suggest, that allows these simian minds to explore cognitive trajectories not made available by their native cultures. Thus we read that:

> In short, we argue that our learning-based assessments, which could be viewed from one perspective as exposing (i.e. bringing out) basic cognitive competencies, are better viewed as bringing them on by use (McGonigle, 2004). And it is only in a laboratory context that these nurturing conditions can be provided in a principled way. Under natural conditions, by contrast, there is no guarantee that the ecology furnishes systematic challenges, let alone affords opportunities for supervised long-term learning of the sort we describe here, teaching that could match the 'relentless' instruction accorded the child recipient by adult caretakers (Bruner, 1990: 83; McGonigle and Chalmers, 2006: 248)

Nonetheless, none of the monkeys in these experiments progressed beyond principled size seriation. Their skills thus fell short of the full 'operational seriation' package that included item insertion (direct placement of an item of arbitrary size within an existing array). McGonigle and Chalmers speculate, in effect, that the main roadblock on the route to such ever-grander forms of simian cognitive success might be the monkeys' lack of sustained training in an apt and highly self-manipulable environment. By partially remedying this shortfall, surprising levels of executive control, not before shown in any non-human subjects, were enabled. Yet as they go on to stress, the novel ecological scaffoldings they were able to provide still fell far short of allowing the simian what I shall dub 'full manipulative scope'; that is, it fell far short of allowing the monkeys to execute fine array transforming procedures that immediately *externalised* the intermediate results of their actions, and would thus have enabled additional cycles of feedback. Instead, the touch screens gave only transient, buzz-based, right/wrong feedback, rather than anything involving visible, persisting and cumulative changes to the test array. This robbed the simian subjects of the chance to benefit from the externalised consequences of their own actions, and thus deprived them of a major informational resource.

The missing resource is a close cousin of a more pervasive capacity that has recently been described as the capacity to use one's own embodied actions to 'self-structure' an information flow (see Lungarella and Sporns, 2005, and discussion in Clark, 2008, Chapter 1). The bulk of basic information self-structuring involves being able to actively induce time-locked multimodal flows of information that can help us learn (for example, we may learn about object boundaries by actively pushing and prodding objects against a stable backdrop [Fitzpatrick and Arsenio, 2004]). But an important wrinkle is added when the objects to be manipulated, rearranged and re-inspected are either public symbols (as in many of the cases examined in the first section above) or symbol-like, in being perceptually simple objects created expressly for the purpose of training certain forms of abstract thought and reason (as in the cases examined in section 2). In these latter (symbolic, or symbol-like) cases, manipulations of the concrete objects (symbols, or symbol-like items) may create new and temporarily persisting perceptual arrays that are well positioned to support further insights, and/or accelerations of learning. It also provides the subject (the manipulator) with a shareable product that may in turn open up new avenues for peer or caregiver-based social cognitive exploration (as noted by McGonigle and Chalmers, 2006: 263).

Such shortfalls in the ability to create persisting, shareable and self-manipulable externalisations apt to further enhance simian learning are, presumably, remediable. For example, McGonigle and Chalmers mention the use of a 'collection of icons formed at the base of the touchscreen, contingent on icon selection' as a simple means of allowing monkeys to 'view their procedures as externalized arrays', and suggest that such work might enable us to 'evaluate whether a new cycle of causality might be created (McGonigle and Chalmers, 2001) whereby cognitive systems are scaffolded to new heights of achievement, through externalization' (all quotes from McGonigle and Chalmers, 2006: 263).

The vision on offer is thus one in which the major obstacle to further simian success consists in the practicalities of appropriate training with sufficiently potent manipulanda. We can call this the 'manipulation roadblock' story. I want to continue, however, by considering two further key dimensions, each of which might be thought to conceal a roadblock of a different kind. Importantly, none of the resulting dimensions turn out to be orthogonal to any of the others. At that point, our game of 'hunt the roadblock' becomes suspect, revealing a deeper and more important unity underlying the emergence of advanced thought and reason.

10.4.2 Cultural practices

Where McGonigle and Chalmers highlight the dimension of enabled array manipulation, Ed Hutchins (2008, 2011) stresses what he calls 'cultural practices':

> Cultural practices are the things people do and their learned ways of being in the world...a practice will be labelled cultural if it exists in a

cognitive ecology such that it is constrained by or coordinated with the practices of other persons. (Hutchins, 2008: 2012)

Cultural practices, thus construed, are pervasive, and are implicated in every single one of the cases discussed earlier in this chapter. Thus:

> Virtually all external representations are produced by cultural practices. All forms of language are produced by and in cultural practices. Speaking is accomplished via discursive cultural practices. The specifics of each language require its speakers to attend to some distinctions and permit them to ignore others. (Hutchins, 2008: 2012)

Importantly, cultural practices are not simply an agent's internal encodings of sanctioned behaviour patterns and so forth, but include all the constraints arising from bodily mechanics, material setting and embodiment, insofar as these too sculpt the space of socially coordinated practice. Moreover, despite the talk of 'people' in the quoted passages, Hutchins intends his notion of cultural practices to encompass also the practices of non-human animals, such as other primates, and – perhaps most importantly of all for current purposes – to encompass the kinds of novel cross-species cultural practice that result from human attempts to conduct psychological experiments with other animals.

Concerning human intelligence, Hutchins' robust claim is that 'cultural practices account for much of what is needed to account for the organisation of human cognitive systems' (Hutchins, 2011: 440). This is not meant as simply the obvious claim that social experiences impact what we think, and when we think it. Rather, it is the much more interesting and challenging claim that 'Many cognitive outcomes produced by human activity systems are properties of our interactions with material and social settings, but we routinely mistake them for properties of ourselves' (Hutchins, 2008: 2017).

What Hutchins means, I think, is that we often posit far richer forms of neural representation and control than we ought to, and that we do this because we fail to see just how much work is being done by the cultural practices themselves. Hutchins' own work is full of such examples, ranging from practices of Micronesian boat navigation, to Trobriand astronomy, to contemporary ship navigation (for these case studies, see Hutchins, 1980, 1983, 1995, respectively). The key idea is that in each case the problem-solving depends on a complex interplay of multiple heterogeneous elements, spanning brain, body and (social and artefactual) world, but that the bulk of the coordination problem – the problem of how to gather and deploy the right elements at the right time so as to perform the task – is solved *by the cultural practice itself*. Thus we read that the success of a specific Trobriand astronomical endeavour (that uses star patterns to determine the correct timing for key agricultural events), 'depends on [the agent's] brain, of course, but also on his body and his eyes, and on a set of traditional cultural practices that orchestrate the interactions among a complex collection of elements' (Hutchins, 2008: 2010).

These practices include recipes for bodily orientation relative to the pre-dawn sky, that enable further embodied practices of star-pattern following, that enable the identification of the key season-indicating configuration. This complex cultural practice did not originate with the astronomer. Rather, 'the knowledge base (both procedural and declarative)... is the product of millennia of incremental development' (Hutchins, 2008: 2012), and is deeply institutionalised within Trobriand society.

But cultural practices, Hutchins notes, are precisely what get put in place when non-human animals enter certain forms of experimental setting. In the 1997 work by Thompson et al. (also discussed in Clark, 1998, 2008) the settings include practices of using plastic tokens in specific kinds of reward-based interaction: one that ended up scaffolding successful performance on the higher-order task. In the work by McGonigle and Chalmers (2006), the practices included incremental training and testing using the touch-screen technologies. According to Hutchins, then, it is these cultural practices (broadly construed) that most likely produce the 'new cognitive outcomes'. The general thought here is that the laboratory work can also be seen as locating the animals within a new socio-artefactual ecology, where it is the dynamic patterns thus anchored that do much of the eventual cognitive work. In this way, Hutchins seeks to challenge readings (such as Thompson et al.'s own) that depict the token-training regimes as installing the requisite skills and understandings within the chimpanzee. It is the ecology of cultural practices, he claims, that is again doing the key work of orchestration:

> The experimental activity elicits a set of practices that orchestrate the capacities of the chimpanzee in interaction with the material and social world in a way that produces the matching of the within-pair relations of the alternative to the within-pair relations of the sample. The cognitive outcome, performing conceptual match-to-sample, is still not a capacity that belongs to the chimpanzee. If conceptual match-to-sample exists in this case, it belongs to the experiment as a complex system of social practices. (Hutchins, 2008: 2017)

The chimpanzee is no worse off here than the Trobriand astronomer. But in each case, we should be wary, Hutchins is suggesting, of attributing to the individual agent skills and understandings that are in some important sense now distributed across the agent and the social and artefactual setting. For it is only within this larger nexus – that of the cultural practice – that the all-important orchestrations of operations and elements are held in place.

It is not clear, from reading this discussion, to what extent Hutchins would agree or disagree with McGonigle and Chalmers' descriptions of the results of their work with *Cebus apella*. One question might concern the suggestion that the experiments induced principled size seriation as a (newly enabled) capacity *of the monkey* itself. Hutchins might, on the model of his discussion of Thompson et al., seek to question this. While agreeing with McGonigle and Chalmers that the new social ecology of touch-screen manipulanda was indeed crucial to success, Hutchins might suggest that it was so crucial that principled size seriation belongs only to 'the experiment as a complex system of social practices'. Or he might allow that in the latter case (unlike the former) the experimental regime installed within the monkey itself something worth calling a skill of principled size seriation. One way to address this issue might be to explore the transfer of the acquired seriation skills across domains.

In addition, I suspect there are really two issues here that remain somewhat conflated in Hutchins' discussion. One concerns the role of cultural practices as providing 'on the cheap' (i.e. without requiring individual agent insight) effective regimes that enable the orchestration of heterogeneous elements into a successful routine. The other concerns the extent to which we should credit an individual agent with 'knowing how to do X' (where X might be size seriation, or conceptual matching-to-sample) where doing X involves just such culturally scaffolded practices of orchestration: practices not initiated by the agent, and hence in that dimension importantly unlike the case of spontaneous size seriation by the human child. I find the latter 'credit assignment' question somewhat elusive, and will leave it to the epistemologists. If we restrict our attention to the former, then there would seem to be nothing at issue between McGonigle/Chalmers and Hutchins. For the manipulative opportunities that McGonigle and Chalmers foreground clearly form the focus of new sets of (in Hutchins' terms) cultural practices, and these practices support and scaffold displays of ordering and executive control not before elicited from non-human animals.

There is, however, an obvious third dimension to consider. It is the role of the brain itself, and it is to this that I finally turn.

10.4.3 Neural innovations

An obvious (some would say, too obvious) place to look for a glass ceiling to limit any specific form of non-human animal success is inside the kind of head in question. All too obviously, only certain kinds of agents (people, for example, and not hamsters) are apt for the empowering effects of exposure to a rich public linguistic edifice. Once we step beyond what McGonigle and Chalmers nicely dub the old paradigm of 'impoverished learning environments' as our window on the cognitive potential of other animals, we might expect to find all kinds of untapped trajectories: multiple routes to surprising 'cognitive upgrades' but everywhere limited by the specifics of the animals' native neural and, though less decisively (witness the touch-screen ecology), bodily apparatus.

For example, let's return to my much-loved example (see the second section above) of token-trained chimps (*Pan troglodytes*) learning about relations between relations so as to succeed in the relational matching-to-sample task. In this case, so the story goes, the provision of concrete (well, plastic) tokens marking the relations of sameness and difference creates for the learner a new realm of perceptible objects (the associated tokens, tags or linguistic labels) upon which to target more basic capacities of statistical and associative learning. New cultural practices involving (at first) the tags or labels then alter the computational burdens involved in certain kinds of learning and problem-solving, allowing the token-trained chimps (only) to solve more complex problems apparently requiring judgements of higher-order similarity and difference.[8] They do this, it was suggested, by allowing the chimp to internally generate images of the plastic sameness/difference tokens and then to judge *these* to be the same or different, thus reducing the higher-order task to a more tractable lower-order one. (Hutchins [2008] offers a different gloss on the story here, and I refer the reader to his discussion for the details.)

Only language- or token-trained animals (humans, or chimps with the token-training history) seem able to learn to perform the higher-order task. The chimps' experience with concrete tags or tokens thus seems to be the difference that makes a difference. But not all animals are able to thus benefit from token training. Monkeys, unlike chimps, fail at the higher-order task even after successful training with the tokens (Thompson and Oden, 2000). Why might this be so?

One intriguing speculation is that to gain this kind of benefit from the token training requires the presence of neural resources keyed to the processing and evaluation of internally generated information. In particular, there is emerging evidence that the anterior or rostrolateral prefrontal cortex (RLPFC) is centrally involved in a variety of superficially quite different tasks, all of which involve the evaluation of self-generated information (Christoff et al., 2003). Such tasks include the evaluation of possible moves in a Tower of London task (Baker et al., 1996), the processing of self-generated sub-goals during working memory tasks (Braver and Bongiolatti, 2002), and remembering to carry out an intended action after a delay[9] (Burgess et al., 2001, 2007). In general, RLPFC is known to be recruited in a wide variety of tasks involving reasoning, long-term memory retrieval, and working memory. What unites all the cases, according to Christoff et al. (2003) is the need to explicitly (attentively, consciously) evaluate *internally generated information* of various kinds. The relational matching-to-sample task, Christoff et al. believe, requires just this kind of processing: that is to say, it requires the explicit directing of attention to internally generated information[10] concerning (in this case) first-order relations of sameness and difference. The involvement of RLPFC in the inner processing needed to get the most out of the prior experience with concrete tokens for sameness and difference explains, the authors argue, the difference between the monkeys (who fail at the task), the chimps (who succeed), and the human five-year-olds (who seem to be even better at it), for the most relevant comparative brain area (Brodmann Area 10) is twice the relative[11] size in humans as it is in the chimpanzee.

Given the converging behavioural and neuroanatomical evidence, Christoff et al. speculate that:

> [The] explicit processing of self-generated information may exemplify some of the highest orders of transformation in which the prefrontal cortex engages during the perception-action cycle...[and] may also be one of the mental processes that distinguish humans from other primate species. (Christoff et al., 2003: 1166)

Interestingly, while lateral BA10 seems to be engaged during the evaluation of self-generated information of the kinds discussed above, *medial* BA10 has been shown to be activated during judgements of self-generated emotional states (Damasio, 2000; Gusnard et al., 2001). The authors conclude that:

> The ability to become aware of and explicitly process internal mental states – cognitive as well as emotional – may epitomize human mental abilities and may contribute to the enhanced complexity of thought, action, and social interaction observed in humans. (Christoff et al. 2003: 1166)

The speculations concerning RLPFC may or may not turn out to be correct. What matters, for my purposes, is the general picture that here rather concretely emerges. According to this picture there are specific neural innovations that make it possible for some creatures, but not others, to benefit deeply from the ability to associate concrete tokens with abstract relations. To use that ability to leverage further abilities (such as thinking about higher-order relations) requires capacities (such as those involved in the evaluation of internally generated information) that the external scaffolding alone does not provide. Nonetheless, the external scaffolding, in those equipped to make the most of it, can itself play a crucial role, as witnessed by the differences between the token-trained and token-free chimpanzees. The neural innovations and the structured cognitive niche (with its attendant cultural practices) are *both* 'differences that make a difference'.

The proper foci of our cognitive scientific attention are thus multiple and non-exclusive. This is all quite obvious yet bears stating. We need to understand the key neural operations, *and* we need to understand how they conspire with various forms of extra-neural scaffolding to yield the cognitive systems responsible for so much of our problem-solving success.[12]

10.5 Conclusions: chickens, eggs and the free cognitive lunch

We started by reviewing some ways in which linguistic and quasi-linguistic structures, conceived as a kind of 'external scaffolding' for the biological brain, might impact and transform various capacities to think and to reason. When such external scaffoldings really take hold, limited biological brains can embark upon quite remarkable journeys, following trajectories that generate poetry, mathematics, quantum theory, and ultimately perhaps an understanding of their own artifactually enhanced cognitive natures.

But what (we asked in the second section) are the preconditions for such remarkable journeys? It seems clear that a certain sophistication of biologically basic representational capacities is necessary. It simply isn't possible to scaffold a hamster to the heights of cognitive success seen, for example, in the monkeys and the chimpanzees. Yet, even supposing this much to be in place, there are a variety of other, less obvious, stumbling blocks. One such stumbling block, elegantly (if only partially) circumvented in the seminal work by McGonigle, Chalmers and their colleagues using touch-screen technologies, is simply the ability fluently to manipulate arrays of objects or tokens. By removing this brute restriction, it became possible to provide monkeys with training experiences that allowed them incrementally to increase their understanding of key cognitive operations such as size seriation, and to begin to display compositionality and systematicity in their cognitive performances. Provided with apt manipulanda (Dennett, 1991) and the right training routine, these monkeys were able to exhibit forms and levels of understanding never before seen in non-human species.

At this point it might be objected that, interesting and revealing though such results are, they simply highlight what remains most special about the human species: our ability to create and enforce the training routines themselves. There is, you might say, no such thing as a free cognitive lunch. It is here, I think, that Hutchins' anthropological perspective on 'cultural practices' provides a useful corrective. For slowly evolved cultural practices, Hutchins reminds us, are what does (for us) the kind of work that the 'saltationary' provision of the well-chosen laboratory environment does for the monkey or the chimpanzee. Indeed:

> few of the dynamic loops that link people to their environments are invented by the people who exploit them. Rather, the ability to establish and maintain such loops is acquired via participation in culturally organized activities with other people... [..] on average, each individual human's cumulative lifetime contribution to the store of ways of exploiting cognitive environments is negligible. (Hutchins, 2011: 438)

The well-fitted practices, notational schemes and artefactual structures that turbo-charge human cognition are thus themselves the products of gradual cultural evolution, as are the processes of transmission themselves (for an excellent recent account of the emergence of writing that fits this profile, see Dehaene, 2009). As individual modern humans we are thus as much the beneficiaries of some remarkably enabling 'cognitive saltation' as were the monkeys in their touch-screen-populated laboratories.

Hutchins is betting, in effect, that it is the cultural practices themselves that are doing the bulk of the work in making human cognition special. As for the brain, it is best seen (on Hutchins' account) as a powerful organ for cultural learning:

> In this perspective, the brain appears as a special super-flexible medium that can form functional subsystems that establish and maintain dynamic coordination among constraints imposed by the world of cultural activity, by the body, and by the brain's own prior organization. The brain has causal powers, but when it comes to human cognition, most of the causal powers of the brain derive from previous experience in cultural cognition. (Hutchins, 2011: 440)

Quite dizzying vistas of chickens and eggs now vie for our attention. Where I have sometimes (e.g. Clark, 2001, Chapter 8) suggested that it might have been some relatively small neural spark that made possible our deep dovetailing with various forms of external cognitive scaffolding and that thus lit the fires of runaway human cognitive success, Hutchins suggests that it may have been some small cultural spark that opened the door:

> It is equally probable that a series of small changes in cultural practices gave rise to new high-level inter-psychological processes, which in turn shaped certain intra-psychological processes, and these in turn favored certain small neural or neural/bodily differences...which would in turn make possible new cultural practices.... In this account there is no reason to favour changes in the brain over innovations in cultural practices as drivers of primate cognitive development. (Hutchins, 2008: 2018)

My own view remains that a few specific neural innovations – such as an innovation enabling the evaluation of self-generated information – might be playing a crucial enabling role in allowing the kinds of apparently open-ended exploitation of external scaffoldings distinctive of human culture. But Hutchins can accommodate this thought by suggesting that small cultural innovations in the primate lineage nonetheless provided the niche within which such small neural innovations arose and flourished.

Perhaps there is simply no untangling this complex web of possibilities. Indeed, one might reasonably suggest that it really doesn't matter. What we need to understand is, rather, a complex evolving web in which change along any dimension can drive and sculpt the others, and in which every step alters a rubber landscape (see Wheeler and Clark, 2008). Hutchins himself (2011: 440) suggests something like this, depicting the neural and the cultural as 'elements of a single adaptive dynamical system'. This single dynamical system, extended through space and spread across time, is itself the locus of the ratcheting effects that explain much of the shape of human thought and reason. But within this single system we ought never to underestimate the transformative power of cultural practices with their associated material structures. To do so might well be, in the memorable phrase of McGonigle and Chalmers (2008), 'putting Descartes[13] before the horse'.

Acknowledgement

Some of the material in sections 1 and 3 appeared originally in my *Supersizing the Mind: Embodiment, Action, and Cognitive Extension* (Oxford University Press, NY, 2008). Thanks to the Press for permission to re-use this material here. Thanks also to Margaret McGonigle (Margaret Chalmers) and Keith Stenning for invaluable feedback on an earlier draft.

Notes

1. An example of such a display would be the projection, on demand, of green arrows marking the route to a university library onto a glasses-mounted display. The arrows would appear overlaid upon the actual local scene, and would update as the agent moves.
2. This picture fits nicely with Barsalou's account of the relation between public symbols and 'perceptual symbol systems' – see Barsalou (2003), and (for a fuller story about perceptual symbol systems) Barsalou (1999).
3. See, for example, Berk (1994), Hutchins (1995), Donald (2001).
4. Here too, as John Protevi (personal communication) reminds me, the impact is not always positive. We can just as easily derail our own performance by explicit reflection on our own shortcomings.
5. Sutton (in press) *explores* two detailed examples, one involving batting advice for cricketers, the other concerning instructional nudges for piano playing. Concerning the latter he writes that 'The sociologist and jazz pianist David Sudnow [2001] describes how explicit verbal phrases and maxims actually became more useful as his skills in improvised jazz piano increased [...] Sudnow explains his [initial] frustration at his teacher's compressed sayings, such as "sing while you're playing", "go for the jazz", "get the time into the fingers", or especially just "jazz hands". These at first make no sense, as the novice pianist is all too conscious of the embodied insecurity of his playing: but [...] what seemed like just vague words to the novice has now become very detailed practical talk, a shorthand compendium of "caretaking practices" for toning and reshaping the grooved routines'.
6. In partial support of this claim, notice that there is good evidence that children show attentional biases that are sensitive to the language they are learning (or have learnt) – see Bowerman and Choi (2001), Lucy and Gaskins (2001) and Smith (2001). Smith (2001: 113) explicitly suggests that learned linguistic contexts come to 'serve as cues that automatically control attention'.

7. Gordon (2004) presents converging evidence from a tribe in Amazonia that uses only words for one, two and many. Numerical cognition in this tribe was clearly affected, such that 'performance with quantities greater than three was remarkably poor, but showed a constant coefficient of variation, which is suggestive of an analog estimation process' (op cit.: 496). Thanks to Keith Oatley for drawing this to my attention.
8. That is, judging of two pairs of objects that the within-pair relations are the same or different across the two pairs – see 10.2 above.
9. The idea here, which may seem elusive at first, is that such cases involve considering self-generated information about our own prior intentions – see Christoff et al. (2003: 1166).
10. Christoff et al. (2003: 1166) depict the self-generated information as abstract, and contrast this with the case of attending to concrete cues or items. If Thompson et al. (1997) are right, however, the internally generated targets here are shallow imagistic renditions of quite concrete objects (the plastic tokens for sameness and difference) whose *contents* are nonetheless relatively abstract.
11. The relative size of the frontal lobes to the rest of the brain appears to be the same in humans and chimpanzees, but BA10 is twice as large, relative to the rest of the frontal lobe, in the human case – Semendeferi et al. (2002).
12. In the token-training case, the conspiracy is a developmental one, as the external scaffolding eventually drops from view. In other cases, as we saw, there is continued reliance on the external scaffolding. In both cases, however, the picture will be one of key neural innovations combining with cultural ones to yield the capacities we most readily identify with minds like ours.
13. The name 'Descartes' is here used as a kind of shorthand for an image of mind as a fully disembodied engine of reason, one whose key properties are independent of the abilities of manipulation provided by the body, and of the cultural legacy of notations, techniques, and practices.

References

Baker, S. C., Rogers, R. D., Owen, A. M., Frith, C. D., Dolan, R. J., Frackowiak, R. S. J. and Robbins, T. W. (1996). Neural systems engaged by planning: A PET study of the Tower of London task. *Neuropsychologia*, 34(6): 515–26.

Barsalou, L. W. (1999). Perceptual symbol systems. *Behavioral and Brain Sciences*, 22: 577–609.

Barsalou, L. W. (2003). Abstraction in perceptual symbol systems. *Philos. Trans. R. Soc. Lond. B Biol. Sci.* 358: 1177–87.

Berk, L. E. (1994). Why children talk to themselves. *Scientific American*, November: 78–83.

Bowerman, M. and Choi, S. (2001). Shaping meanings for language: Universal and language-specific in the acquisition and shaping of semantic categories. In M. Bowerman and S. Levinson (eds), *Language acquisition and conceptual development*. Cambridge: Cambridge University Press, pp. 475–511

Boysen, S. T., Bernston, G., Hannan, M. and Cacioppo, J. (1996). Quantity-based inference and symbolic representation in chimpanzees (*Pan troglodytes*). *Journal of Experimental Psychology: Animal Behavior Processes*, 22: 76–86.

Braver, T. S. and Bongiolatti, S. R. (2002). The role of frontopolar cortex in subgoal processing during working memory. *NeuroImage*, 15: 523–36.

Bruner, J. (1966). *Toward a theory of instruction*. Cambridge, MA: Harvard University Press.

Bruner, J. (1990). *Acts of meaning.* Cambridge, MA: Harvard University Press.
Burgess, P. W., Quayle, A. and Frith, C. D. (2001). Brain regions involved in prospective memory as determined by positron emission tomography. *Neuropsychologia,* 39(6): 545–55.
Burgess, P. W., Dumontheil, I. and Gilbert, S. J. (2007). The gateway hypothesis of rostral PFC (area 10) function. *Trends in Cognitive Sciences,* 11: 290–8.
Carruthers, P. (2002). The cognitive functions of language. *Behavioral and Brain Sciences,* 25: 657–726.
Christoff, K., Ream, J. M., Geddes, L. P. and Gabrieli, J. D. (2003). Evaluating self-generated information: anterior prefrontal contributions to human cognition. *Behavioral Neuroscience* 117(6): 1161–8.
Churchland, P. M. (1989). *The neurocomputational perspective.* Cambridge, MA: MIT/Bradford Books.
Clark, A. (1996). Connectionism, moral cognition and collaborative problem solving. In L. May, M. Friedman and A. Clark (eds), *Mind and morals.* Cambridge, MA: MIT Press, pp. 109–28.
Clark, A. (1997). *Being there: Putting brain, body and world together again.* Cambridge, MA: MIT Press.
Clark, A. (1998). Magic words: How language augments human computation. In P. Carruthers and J. Boucher (eds), *Language and thought: Interdisciplinary themes.* Cambridge: Cambridge University Press, pp. 162–83.
Clark, A (2001). *Mindware: An introduction to the philosophy of cognitive science.* New York: Oxford University Press.
Clark, A. (2003). *Natural-born cyborgs: Minds, technologies, and the future of human intelligence.* New York: Oxford University Press.
Clark, A (2005). Word, niche and super-niche: How language makes minds matter more. *Theoria* 20(54): 255–68.
Clark, A (2006). Language, embodiment and the cognitive niche. *Trends in Cognitive Sciences.* 10(8): 370–4.
Clark, A (2008). *Supersizing the mind: embodiment, action, and cognitive extension.* New York: Oxford University Press.
Clark, A. and Chalmers, D. (1998). The extended mind. *Analysis,* 58: 7–19.
Clowes, R. W. and Morse, A. F. (2005). Scaffolding cognition with words. In L. Berthouze, F. Kaplan, H. Kozima, Y. Yano, J. Konczak, G. Metta, J. Nadel, G. Sandini, G. Stojanov and C. Balkenius (eds), *Proceedings of 5th International Workshop on Epigenetic Robotics: Modeling Cognitive Development in Robotic Systems* (*Lund University Cognitive Studies,* vol. 123, pp. 101–5). Lund, SE: Lund University Cognitive Studies.
Clowes, R. W. (2007). A self-regulation model of inner speech and its role in the organisation of human conscious experience. *Journal of Consciousness Studies,* 14(7): 59–71.
Colunga, E. and Smith, L. B. (2005). From the lexicon to expectations about kinds: A role for associative learning. *Psychological Review,* 112: 347–82.
Damasio, A. R. (2000). Subcortical and cortical brain activity during the feeling of self-generated emotions. *Nature Neuroscience,* 3: 1049–56.
Dehaene, S. (2009). *Reading in the brain.* New York: Penguin.
Dehaene, S. (1997). *The number sense.* Oxford: Oxford University Press.
Dehaene, S., Spelke, E., Pinel, P., Stanescu, R. and Tviskin, S. (1999). Sources of mathematical thinking: Behavioral and brain imaging evidence. *Science,* 284: 970–4.

Dennett, D. C. (1991). *Consciousness explained*. New York: Little Brown.
Dennett, D.C. (2000). Making tools for thinking. In D. Sperber (ed.), *Metarepresentations: A multidisciplinary perspective* (Oxford: Oxford University Press, 43–62.
Densmore, S. and Dennett, D. C. (1999). The virtues of virtual machines. *Philosophy and Phenomenological Research*, 59: 747–67.
Donald, M. (1991). *Origins of the modern mind*. Cambridge, MA: Harvard University Press.
Donald, M (2001). *A mind so rare*. New York: W. W. Norton.
Fitzpatrick, P. and Arsenio, A. (2004). Feel the beat: using cross-modal rhythm to integrate perception of objects, others, and self. In L. Berthouze, H. Kozima, C. G. Prince, G. Sandini, G. Stojanov, G. Metta and C. Balkenius, (eds), *Proceedings of the Fourth International Workshop on Epigenetic Robotics*, Lund University Cognitive Studies, p. 117.
Fodor, J. (1987). *Psychosemantics: The problem of meaning in the philosophy of mind*. Cambridge, MA: MIT Press.
Fodor, J (2001). *The mind doesn't work that way*. Cambridge, MA: MIT Press.
Goldin-Meadow, S. (2003). *Hearing gesture: How our hands help us think*. Cambridge, MA: Harvard University Press.
Gordon, P. (2004). Numerical cognition without words: evidence from Amazonia. *Science*, 306: 496–9.
Gusnard, D. A., Akbudak, E., Shulman, G. L. and Raichle, M. E. (2001). Medial prefrontal cortex and self-referential mental activity: Relation to a default mode of brain function. *Proceedings of the National Academy of Sciences, USA*, 98: 4259–64.
Hermer-Vazquez, L., Spelke, E. and Katsnelson, A. (1999). Sources of flexibility in human cognition: Dual-task studies of space and language. *Cognitive Psychology*, 39: 3–36.
Hutchins, E. (1980). *Culture and inference: A Trobriand case study*. Cambridge, MA: Harvard University Press.
Hutchins E. (1983) Understanding Micronesian navigation. In D. Gentner and A. Stevens (eds), *Mental models*. Hillsdale, NJ: Lawrence Erlbaum, pp. 191–225.
Hutchins, E (1995) *Cognition in the wild*. Cambridge, MA: MIT Press.
Hutchins, E (2008). The role of cultural practices in the emergence of modern human intelligence. *Phil. Trans. R. Soc. B* 363: 2011–19.
Hutchins, E. (2011). Enculturating the supersized mind. Philosophical Studies 152: 437–46.
Inhelder, B. and Piaget, J. (1969). *The early growth of logic in the child*. Trans. E. A. Lunzer and D. Papert. New York: Norton.
Jackendoff, R. (1996). How language helps us think. *Pragmatics and Cognition* 4(1): 1–34.
Kirsh, D. (1995). The intelligent use of space. *Artificial Intelligence* 73(1–2): 31–68.
Kirsh, D and Maglio, P. (1992). Reaction and reflection in tetris. In J. Hendler (ed.), *Artificial intelligence planning systems: Proceedings of the first annual conference AIPS*. San Mateo, CA: Morgan Kaufman.
Lucy, J. and Gaskins, S. (2001). Grammatical categories and the development of classification preferences: A comparative approach. In M. Bowerman and S. Levinson (eds), *Language acquisition and conceptual development*. Cambridge: Cambridge University Press, pp. 257–83.
Lungarella, M. and Sporns, O. (2005). Information self-structuring: Key principles for learning and development. *Proceedings 2005 IEEE Intern. Conf. Development and Learning*, pp. 25–30.

McGonigle, B. (2004). Cognitive growth as optimised executive control: a learning analysis. Paper presented at Science of Learning Center, University of New Mexico.

McGonigle, B. and Chalmers, M. (2002). Cognitive learning in monkey and man. Specially invited symposium contribution in S. Fountain, M. Bunsey, J. Danks and M. McBeath (eds), *Animal cognition and sequential behavior*. Boston: Kluwer Academic Press, pp. 269–314.

McGonigle, B. and Chalmers, M. (2006). Executive functioning as a window on the evolution and development of cognitive systems. *International Journal of Comparative Psychology* (special issue on comparative, developmental and evolutionary psychology), 19 (2): 241–67.

McGonigle, B and Chalmers, M (2008). Putting Descartes before the horse (again!). *Behavioral and Brain Sciences*, 31: 142–3.

McGonigle, B. O. and Chalmers, M. and Dickinson, A. (2003). Concurrent disjoint and reciprocal classification by *Cebus Apella* in serial ordering tasks: Evidence for hierarchical organization. *Animal Cognition*, 6: 185–97.

McGonigle B O, Jones, B. T., (1978). Levels of stimulus processing by the squirrel monkey: Relative and absolute judgements compared. *Perception* 7(6): 635–59.

McNeill, D. (1992). *Hand and mind*. Chicago: University of Chicago Press.

McNeill, D. (2005). *Gesture and thought*. Chicago: University of Chicago Press.

Schwartz, D. L. and Black, J. B. (1996). Shuttling between depictive models and abstract rules: Induction and fallback. *Cognitive Science*, 20: 457–98.

Schyns, P. G., Goldstone, R. L. and Thibaut, J.-P. (1998). The development of features in object concepts. *Behavioral and Brain Sciences*, 21: 1–54.

Semendeferi, K., Lu, A., Schenker, N. and Damasio, H. (2002). Humans and great apes share a large frontal cortex. *Nature Neuroscience*, 5: 272–6.

Smith, L. (2001). How domain-general processes may create domain-specific biases. In M. Bowerman and S. Levinson (eds), *Language acquisition and conceptual development*. Cambridge: Cambridge University Press.

Smith, L. and Gasser, M. (2005). The development of embodied cognition: six lessons from babies. *Artificial Life*, 11: 13–30.

Sudnow, D. (2001). *Ways of the hand: A rewritten account*. Cambridge, MA: The MIT Press.

Sutton, J. (in press) Batting, Habit, and Memory: the embodied mind and the nature of skill. In Jeremy McKenna (ed.), *The Philosophy of Cricket*, to be published both as a special issue of the journal *Sport in Society* and as a book in the series *Sport in the Global Society* (Taylor and Francis).

Terrace, H. S. (2005). The simultaneous chain: a new approach to serial learning. *Trends in Cognitive Science*, 9(4): 202–10.

Thompson, R. K. R. and Oden, D. L. (2000). Categorical perception and conceptual judgments by nonhuman primates: The paleological monkey and the analogical ape. *Cognitive Science*, 24: 363–96.

Thompson, R. K. R., Oden, D. L. and Boysen, S. T. (1997). Language-naive chimpanzees (*Pan troglodytes*) judge relations between relations in a conceptual matching-to-sample task. *Journal of Experimental Psychology: Animal Behavior Processes*, 23: 31–43.

Vygotsky, L. S. (1962/1986). *Thought and language*. Trans. A. Kozulin [1962]. Cambridge, MA: MIT Press.

Wheeler, M. and Clark, A. 2008. Culture, embodiment and genes: Unravelling the triple helix. *Philosophical Transactions of the Royal Society B*, 363(1509): 3563–75.

11
Private Codes and Public Structures
Colin Allen

> [W]ith new paradigms we can create conditions enabling monkeys to view their cognitive procedures as externalized arrays, without requiring a prior lexical process... (a collection of icons is formed at the base of the touchscreen contingent on icon selection). With these techniques, we are now in a position to evaluate whether a new cycle of causality might be created... whereby cognitive systems are scaffolded to new heights of achievement, through externalization.
>
> – McGonigle and Chalmers (2006: 263)

> *There was a Macaca mulatta*
> *Who learned how to use a computer.*
> *With no need to use ink*
> *She was able to think*
> *And hence she became a lot smarter.*
>
> – this author

11.1 Introduction

Humans externalise cognition in myriad ways. Our tools, marks, trails, speech, writing and dwellings pepper the landscape. The cognitive droppings of our ancestors go back over a million years. More recent cognitive achievements of our species are 'scaffolded' (to repeat the term used by McGonigle and Chalmers in the quotation above) upon those earlier structures. The journey from tallies of grain to the Schrödinger equation was not inevitable, but it would have been impossible without externalised cognition. Our capacity to understand quantum mechanics, such as it is, depends on access to symbols that lie outside the head. And yet we barely understand the process by which such cognitive achievements are scaffolded. It is some part genomic, some part epigenetic and entirely a product of complex developmental processes, 'cycles of causality', whose complex causal strands cannot be simply or linearly separated into genetic factors and environmental factors (Stotz and Allen, 2011).

My starting point in this paper is the pioneering work of Brendan McGonigle and Margaret Chalmers that tries to unravel some of that complexity. As a philosopher of cognitive science with a special history in animal cognition, I am particularly interested in the longstanding debate between 'associationists' and 'cognitivists' about the best way to understand the cognition of the more sophisticated non-human animals, such as in transitive inference (Allen, 2006). As such, I was intrigued by an email message from McGonigle (personal communication, 2006) in which he wrote, 'I don't hold out much scope for associative mechanisms on their own. ... Instead, I favour a "multiple types" of learning approach which targets relationally based mechanisms as qualitatively different and more powerful than those derived from what the late Harvey Carr once described as "the educated salivations of a Russian dog".' With their research on the seriation abilities of monkeys, and into how earlier training experiences supported the development of more sophisticated relational abilities, McGonigle and Chalmers challenge the canonical impulse to constrain scientific understanding of animal cognition within the limits of basic forms of associative learning. It is these relationally based mechanisms – concerned with tracking higher-order and abstract relationships among stimuli – that they probed with the assiduous use of touch screens.

In the opening sentences of 'The Growth of Cognitive Structure in Monkeys and Men', McGonigle and Chalmers (2002: 287) put the importance of understanding development through life history like this:

> There is a widespread view that the sorts of animal learning mechanisms most frequently studied in the laboratory are inductively too weak and unproductive to generate the kinds of behaviours expressed in higher order forms of human cognitive and linguistic adaptation (Chomsky, 1980; Fodor & Pylyshyn, 1988; Piaget, 1971). One reason for this (Harlow, 1949) is that investigations are rarely followed through from one learning episode to another to assess the cumulative benefits (if any) as a function of the agent's task and life history.

The same theme reappears in their 2006 paper, where they claim that, as a result of the prevailing methodology, there is an 'overdependence on a relatively weak inductive mechanism, rejected by cognitive and linguistic researchers alike as one that cannot scale up and deliver teachable cognitive or linguistic skills' and this has created, they say, 'a conceptual vacuum in which language looms as a 'magic bullet' invested with new capabilities of its own and putatively causal to the cognitive abilities unique to humans' (2006: 242).

Their stance against strict behaviourism on the one hand, and against language-centred accounts of cognition on the other, would appear to suit the many people already convinced by their reading of 'cognitive ethology' that the gap between animal minds and human minds has long been exaggerated. Indeed, the implied critique of behaviouristic methodology may seem like old news. But these same people are also likely to believe in the importance of studying animals in ecologically valid contexts. Here, however, McGonigle and Chalmers wield a double-edged sword, for on the reverse swing they argue (2006: 248) that:

> [O]ur learning-based assessments, which could be viewed from one perspective as exposing (i.e., bringing out) basic cognitive competences, are better viewed as bringing them on by use. ... *it is only in a laboratory context* that these nurturing conditions can be provided in a principled way. Under natural conditions, by contrast, there is no guarantee that the ecology furnishes systematic challenges, let alone affords opportunities for supervised long-term learning of the sort we describe here, teaching that could match the 'relentless' instruction accorded the child recipient by adult caretakers. [italics added]

Thus they wade straight into the long and contentious history that divides comparative psychology from ethology, in which the situation of the animals that are the targets of investigation has always been at stake. Ethologists, steeped in a tradition of natural history and ecological validity, have often regarded laboratory animals as artefacts, bearing little resemblance to their uncaged counterparts. Psychologists, having learned the importance of controlled experimentation, have typically regarded wild animals as magnets for anthropomorphically over-interpreted anecdotes, and hardly the proper domain for rigorous scientific work.

My aim in this chapter is not to reheat this long-simmering dispute. In actuality, McGonigle and Chalmers steer an interesting middle course of scepticism about both; questioning, on the one hand, the power of ethological investigation to answer essential questions about animal cognition (without necessarily implying that cognitive ethologists are wrong about what they think they see) and questioning, on the other hand, the tendency of many comparative psychologists to over-generalise from the limitations of their experiments to limitations in the animals themselves. I mention their stance towards animals living under natural conditions because it once again points up the complex causality of development and the difficulties inherent in studying it.

Why might touch screens be an important tool for the comparative developmental psychologist? McGonigle and Chalmers mention the ability of touch screens to circumvent the limitations of monkeys due to the 'serious manipulative restrictions imposed by their motor control systems' (2002: 320). Although they don't use the jargon, the idea that manipulation skills play a significant role in scaffolding cognitive systems aligns with current interest in 'embodied cognition' and 'extended mind'. Human intelligence is partly grounded in our ability to make precise and repeatable notches in materials that include wood, bone, stone, clay and metal. Beyond the jargon lie important questions about the extent to which cognition that operates on external structures is not simply a reflection of pre-existing inner thought, but part of a set of feedback loops between brain, body and environment that have complex combinatoric effects during development.

But if this perspective is correct, how are we to make sense of the further claim by McGonigle and Chalmers that cognitive meaning, both for language and non-linguistic forms of abstract cognition, is grounded in 'private codes'? The claim is made in the context of a discussion of experiments they conducted within a Piagetian framework, investigating the ability of monkeys to sort objects by size. Instead of requiring the monkeys to physically manipulate the actual objects, as Piaget required of the children in his experiments, they allow the monkeys to indicate the correct order by sequentially touching different-sized icons presented on a computer screen. McGonigle and Chalmers maintain that humans externalise these private codes in language but, 'In simians, however, we conclude that these remain as private codes, until their externalization into a public domain is made possible through the vastly improved manipulation skills of humans' (2006: 243). The idea, I take it, is that because human dexterity allows us to build touch screens that monkeys can use, they too can finally externalise their private codes. Of course, one can reasonably go on to point out that even if monkeys can be scaffolded thus far, humans take yet another step of externalisation by using their physically externalised codes on a larger, social scale as media for communication and social cooperation. One might well wonder about the potential for monkeys to do likewise. But I want to take a step in the other direction, towards asking what monkeys bring to the task of using a touch screen, and considering the sense in which the 'private codes' being externalised on-screen would exist at all without the 'supervised long-term learning' that is the hallmark of McGonigle and Chalmers' methodology. Can we say more about the '(nonarbitrary) relationships between physical objects' (2006, 243) that provide the 'objective grounding' for these codes?

One of my objectives in this chapter is to offer some ideas that might help clarify these questions. But I have a larger objective too, which is to help promote a much more developmental approach to animal cognition, particularly to my philosopher colleagues, but also among the comparative psychologists and cognitive ethologists whose work has captured so much more of the recent mindspace of public and academic awareness of animal cognition research than the more difficult path taken by McGonigle and Chalmers. So, before unpacking the notion of private code, a little context.

11.2 State of the nation

We are in the midst of a boom of interest in the scientific study of animal cognition. Barely a week goes by without a new study, and accompanying coverage by science journalists, concerning attempts to show that animals can succeed at various 'high-level' tasks such as self-recognition, imitation, deception, 'theory of mind', tool use and referential communication. Studies of these capacities span a wide range of species, for example, primates, cetaceans, dogs, elephants, parrots and various members of the corvid family. These pursuits have something of the character of trophy hunting by scientists eager to show that their favourite species can (too!) do what another species can do.

The specific tasks investigated usually take human competency as the model to be emulated and it is common to see the cognitive capacities of animals likened to those of human children of various ages, as if to locate the animals with respect to particular benchmarks on the developmental trajectory from neonates to human adult cognitive competency. Although practically everyone acknowledges that these comparisons and the underlying picture of development are too simplistic, it is my view that few scientists study cognitive development in animals adequately. On the one hand, fewer than 5 per cent of the 400-plus articles published in the journal *Animal Cognition* since it was established in 1999 are about cognitive development. A considerable proportion of these 'developmental' studies involve the attempt to establish that adult members of non-human species can succeed in cognitive tasks that are considered to be developmentally significant benchmarks in humans (for a defence of the approach see Parker, 2002). But animals are not humans at an early stage of development, and my view, elaborated in the following sections, is that this approach does not encourage sufficient attention to the developmental and learning processes themselves. On the other hand, while many papers about psychological development in animals get published within the field of developmental psychobiology, nearly all the scientists doing this work operate within the framework of traditional animal learning theory, committed to forms of associationism that McGonigle and Chalmers sought to undermine. However, the scepticism of developmental psychobiologists about the meaning of cognitive concepts drives most of them to actively eschew cognitive vocabulary (see e.g. Wasserman and Blumberg 2006).

In between these poles is a small group of developmentally savvy cognitive comparative psychologists, whose theoretical notions, such as 'emergents' (Rumbaugh et al., 1996; Rumbaugh, 2002) and 'private codes' (McGonigle and Chalmers, 2006), represent an attempt to close the gap between basic associationism that seems insufficient to explain some aspects of animal performance, and anthropocentric models of cognition that are implausibly applied to animals. But some critics might worry that the notion of a 'code' smuggles back into comparative psychology a linguistically inspired notion of representation that should be regarded with suspicion, or that the notion of an emergent labels the phenomenon of cognitive development rather than explains it.

11.3 From self to speech: mirrors, other minds, imitation, tools and talk

The aforementioned 'high-level' tasks have become staples of animal cognition work for a variety of reasons, including that they seem to form a cluster of related capacities in human beings. Take, for instance, mirror self-recognition, which has been investigated via the widely used 'mark test' initially developed by Gallup (1970). In this experiment, anaesthetised chimpanzees were marked (or invisibly sham-marked) on their foreheads, and then observed after recovery from the anaesthetic. Gallup's chimpanzees, who had extensive prior experience with mirrors, showed a significant increase in touching the mark and other apparently self-directed responses when a mirror was present during recovery. Gallup (1979; Gallup et al., 2002) argued that these results show that chimpanzees are self-aware – a trait he believes to be limited to the great apes and humans (Gallup et al., 2002). Gallup et al. also draw a connection between mirror self-recognition and the capacity to attribute mental states to others. They write (2002: 329), 'The rationale for postulating a connection between self-recognition and mental state attribution is simple. If you are self-aware then you are in a position to use your experience to model the existence of comparable experiences in others.' This links mirror self-recognition to so-called 'theory of mind' – a topic that has considerable currency in studies of animal cognition, having originated in that context three decades ago with the work of Premack and Woodruff (1978).

Theory of mind also connects to the topic of imitation, a subject of intensive debate among animal-cognition researchers. At the centre of this debate have been questions about how to define 'imitation': whether to require that imitators understand the motives and goals of the demonstrator, and whether to require that the behaviour of the observer is strictly mimicked. Somewhat ironically, goal-copying has turned out to be easier to demonstrate for non-human animals than has behavioural copying. It seems, for instance, that human children are much more likely than chimpanzees to mimic the actions of their teachers regardless of whether any other goal is served by doing so. For example, after watching a demonstrator retrieve an item of food from a multi-part contraption, chimpanzees skipped the steps of the demonstration that were obviously functionless, whereas children copied those steps faithfully (Horner and Whiten, 2005). Horner and Whiten suggest that the strong tendency of children to recognise the role of teachers and to copy their movements faithfully is a developmentally significant species difference.

Faithful copying has also been implicated in the use of tools and the emergence of cultures. Early hominids used stone tools for over two million years, but changes in the kinds of tools they produced occurred very slowly during this enormous expanse of time. During the past fifty thousand years or less there has been an explosion of innovation in tools and technology, and the accumulation of tools and techniques across multiple generations has been described as due to a cultural 'ratchet' (Tomasello et al., 1993). Tool use in animals has been a hot topic ever since Jane Goodall observed chimpanzees fishing for termites with sticks at Gombe, and there are now very many observations of other kinds of tool use in the wild by chimpanzees, and other more or less disputed claims for tool use in species as diverse as orang-utans, New Caledonian crows and dolphins. The differences in tool usage among wild populations have also led ethologists to argue that these differences are culturally acquired. In an experimental setting, Whiten et al. (2005) also demonstrated what they called conformity to cultural norms of tool use in chimpanzees, where group members copied the behaviour of high-ranking individuals even though they had initially learned to retrieve food from an apparatus using a different technique. Comparative psychologists have also explored the physical understanding of tools in laboratory experiments using a variety of primate species, with conflicting results (Visalberghi et al., 1994, 1995; Povinelli, 2000), and varying interpretations of what such studies show about causal reasoning in animals (Penn and Povinelli, 2007). These studies have been connected back to theory of mind through thinking about the capacity of animals to reason about 'hidden' or invisible causes, with the mental states of others falling into that category.

Questions about symbolic communication have also been extensively studied in a number of high profile studies. These include attempts to teach human-like languages to animals, such as the bonobo Kanzi (Savage-Rumbaugh, 1996; Savage-Rumbaugh et al., 2004) and the African gray parrot Alex (Pepperberg, 1999). But they also include studies of the natural communication systems of animals (see Radick, 2007 for a history), such as the alarm calls of vervet monkeys (Cheney and Seyfarth, 1990), prairie dogs (Slobodchikoff, 2002), chickens (Evans et al., 1993; Evans and Marler, 1995), the social signals of baboons (Cheney and Seyfarth, 2007) and dogs (Bekoff and Allen, 1992). Philosophers familiar with the arguments of Grice (1957; also Dennett, 1983) and Davidson (1982) will immediately understand the potential relevance of such studies to theory of mind, although few scientists have waded far into this philosophical thicket. Instead, those who conduct such studies (and their critics) have been concerned with the capacity of animals to convey environmental information ('functional reference' [Evans and Marler, 1995]) and the signal properties alone and in combination that make this possible (Pepperberg, 1999; Zuberbühler, 2000, 2001; Slobodchikoff, 2002; Savage-Rumbaugh et al., 2004).

The issue of animal communication also connects to culture through the issue of what have been called dialect differences in (for example) birdsong, honeybee dances and whale song. Of course, the relevance to understanding human natural languages, with their rich recursive syntax and compositional semantics, of animal communication, whether using natural or artificial signals, is hotly contested by linguists. For example, Pinker (1994) likened the ape-language studies to an attempt to learn something about elephants' trunks by teaching their nearest living relative, the hyrax, to pick up objects with its rather unremarkable snout. He similarly dismissed the relevance of field studies of animal communication to the origins of language by saying that as a scientific hypothesis it has as much going for it as Lily Tomlin's quip that language was invented by a woman who first exclaimed, 'What a hairy back!' (presumably not the first instance of self-reference). Nevertheless, there are many scientists who assert that the investigation and cautious interpretation of the communicative capacities of non-human animals is a worthwhile endeavour (Hauser, Chomsky and Fitch, 2002).

In this section, I have barely scratched the surface of current research in animal cognition, which has especially burgeoned in the most recent decade. Nevertheless, I hope to have conveyed something about how studies of mirror self-recognition, theory of mind, imitation, tool use and communication are conceived as addressing an interrelated set of questions about the cognitive capacities of animals. Other currently active areas of study can also be assimilated to this nexus. For instance, the experimental work on episodic memory in birds (Clayton and Dickinson, 1998) has engendered discussion of whether it has been or can be established that their memories are genuinely autobiographical (Tulving, 2005), thus connecting to the concept of self-awareness with which Gallup began almost 40 years ago.

11.4 Comparative development

As the preceding section illustrates, studies of the cognitive capacities of animals span a wide range of taxonomic groups. Despite the appearance that such work is strongly comparative, it is rare however to find multiple non-human species compared in a single study. Single species studies of mirror self-recognition have been attempted in elephants (Plotnik et al., 2006), cotton-top tamarin monkeys (Hauser et al., 1995; Hauser et al., 2001 described the failure to replicate the first study), bottlenose dolphins (Reiss and Marino, 2001) and (somewhat notoriously) pigeons (Epstein et al., 1984) – the latter seeking to undermine the interpretation of Gallup's original chimpanzee work. Gallup (e.g. Gallup et al., 2002) has been very critical of several of these studies, and considers only the evidence from orang-utans and bonobos to be on a par with that from chimpanzees.

While Hauser's original studies with cotton-top tamarins are not persuasive, and his research program in general has come under increasing clouds since his academic misconduct was reported in August 2010, his attempt to increase stimulus saliency, by diverging from Gallup's use of a small mark and instead using Day-Glo-coloured hair dye on the white streak that gives cotton-tops their name, raises an important issue about the cross-species validity of the original mirror-mark test. Continuing in this vein, Rajala et al. (2010) provide evidence of mirror-guided self-directed behaviours in rhesus monkeys with cranial implants, including inspection of hinder parts of their anatomy.

Another major methodological problem concerns finding species-appropriate alternatives to the mark-touching behaviour that provided the measure in Gallup's original and subsequent studies – a problem that is especially acute for dolphins, but also requiring modification for members of other species. There is, indeed, something compelling about seeing a dolphin twist and turn in front of a mirror in an apparent attempt to see a mark on its body that cannot be seen directly. However, the contortions that the experimenters themselves go through in order to adapt Gallup's experiment to other species can leave one with the impression that it is human ingenuity in impressing each other that is primarily on display. And since these contortions are not applied equally to different species, including humans, the resulting studies fail to be strictly comparable.

Those carrying out such investigations typically espouse strongly comparative aims, of course. For instance, Reiss and Marino framed their dolphin study in their paper's introduction as follows:

> In humans, MSR [mirror self-recognition] does not emerge reliably until 18–24 months of age and marks the beginning of a developmental process of achieving increasingly abstract psychological levels of self-awareness, including introspection and mental state attribution.... A provocative debate continues to rage about whether self-recognition in great apes implies that they are also capable of more abstract levels of self-awareness. (2001: 5937)

Although the quoted passage is overtly developmental and comparative, the study it frames is really neither. First, with respect to development, the dolphin subjects were not tracked longitudinally as part of the study, and thus we have no idea what prior experiences may or may not have been important to their responsiveness to mirrors or their actual self-conception. (Granting, for the sake of argument, that they have one.) Second, with respect to the comparative aspects, the study itself only involved one species (and only two members of that species), while the other species mentioned are from a lineage that diverged at least 65 million years ago. It is also clear that human development provides the benchmarks. Human children typically engage in mirror-guided, self-directed behaviour before they pass 'theory of mind' tests such as the false-belief task (Wimmer and Perner, 1983), and many psychologists have suggested that the former is a significant step towards the latter (as implied by Reiss and Marino in the quotation above). But whether this is necessarily so for all species is unaddressed by the mirror-recognition-or-bust approach that seems to predominate in this branch of comparative psychology. Furthermore, although the connection of MSR to 'theory of mind' has not been tested directly by experimentation, other putative precursors to theory of mind, such as imitation and pretend play, do not emerge in a strongly correlated way with MSR (Nielson and Dissanayake, 2004).

My intention is not to single out Reiss and Marino for special criticism, nor studies of mirror self-recognition in general. Similar things could be said about the approach taken to any of the other topics introduced in the previous section. Take, for example, the way that theory-of-mind tasks have been tackled. With the invention of the false-belief task, theory-of-mind studies took a developmental turn, at least for human subjects. In the original version of this task, the 'Sally–Anne' task (Wimmer and Perner, 1983), young children were asked where a character named 'Sally' would look for an object that they had seen another character, 'Anne', move and hide while Sally was out of the room. Children under the age of four were generally found to indicate the present location of the object as the place where Sally would look. Somewhere during their fifth year, children normally start to indicate that the Sally character will look in the first location for the hidden item. This is taken as evidence that they have developed the capacity to attribute false beliefs to others – a benchmark for theory of mind. A substantial literature has grown in the human developmental psychology literature about children's attainment of this cognitive benchmark. But even if this was a fruitful way to study cognitive development in humans (not everyone would agree), the Sally–Anne task was unsuitable for comparative work with animals because of its reliance on verbal questions. Eventually, however, non-verbal versions of the false-belief task were developed (Clements and Perner, 1994; Call and Tomasello, 1999), but they seem much more open to interpretation, and have been applied with varying degrees of success to members of various species and much younger children. The approach seems to be one of plucking animals out of their cages to see whether they can do something that is important in human cognitive development, without fully investigating how their histories might have prepared them to succeed or fail.

One response to these complaints is to say that these scientists are simply motivated by different questions. It is one thing to ask *how* subjects come to have the capacities that they do; it is another to ask *what* those capacities are. Defenders of the approach claim that testing animals to see how they match up to developmental benchmarks that are significant in human development provides useful information about the cognitive capacities of different species, regardless of how those capacities were acquired. However, this response assumes that we have a clear conception of the experimental task that the research subject confronts. The problematic nature of this assumption – that the tasks are really the same for members of different species – is highlighted by Susan Jones in her discussion of studies comparing imitation in chimpanzees and children. She writes that: 'Although in all of their comparative work the researchers model the same behaviours for the two species and then measure the same imitative responses, the chimpanzees and children are never really in the same tasks' (Jones, 2005a: 297). She points out that the two kinds of subjects have different histories and are on different developmental trajectories, making the experimental results hard to interpret. Furthermore, Jones points out, children 'learn to imitate in the context of thousands of vocal exchanges with the caretakers...the poor chimpanzee is at a distinct disadvantage – unless someone imitates his vocalizations from an early age' (2005a: 301). The task confronting an individual who has been in imitative social interactions for almost her entire life is not the same as that confronting an individual who must try to figure out what is expected of her today.

The idea that human competencies provide a good benchmark presupposes that we fully understand what those competencies actually are, and how they were acquired. But we are far from understanding this, even for some of the 'highest' achievements of human cognition. For instance, there is evidence that competent symbolic reasoners rely on processes that are sensitive to *unintended* regularities in the symbolic environment. Landy and Goldstone (2007a, 2007b) found that various visual cues of perceptual grouping in symbolic formulae are exploited by competent algebraists, even though these cues are not explicitly taught, and even though their use can lead reasoners away from the intended meanings of the formulae. Likewise, McNeill and Alibali (2004, 2005) present evidence that elementary school children in the United States suffer a large drop in performance in some kinds of arithmetic reasoning problems between first and third grades because the intended meaning of the equality sign is obscured by their implicitly learned responses to spurious regularities in the worksheets used for addition drills, such as the directionality of the perceptual-motor task. What is considered a normative failure by those involved in setting the tasks may in fact be the result of powerful pattern extraction processes at work on objectively measurable structural regularities in the stimuli sets.

The upshot is that many ostensibly 'comparative' and 'developmental' studies fail to be sufficiently comparative because they fail to investigate the actual learning and experience during development of their own research subjects, and because they assume that the experimenters' understanding of the context of the experimental task accurately represents the context in which the subjects are actually making their decisions (see also Stenning and van Lambalgen [2008] for a related critique of the experimental literature on human reasoning 'errors'). Members of different species, and even individuals within the same species, are only in the same experiment if one chooses to ignore a raft of things that might matter to the outcomes of those experiments.

11.5 Development and comparative cognition

The idea that development needs to be taken more seriously is hardly new. Both ethology and behaviouristic psychology were criticised by Lehrman (1953, 1970) for being insufficiently grounded in the biological facts: in the former case for assuming an untenable notion of instinct that ignored or severely underplayed the developmental plasticity of organisms (Lehrman, 1953), and in the latter case for failing to sufficiently address species differences that constrain what animals can be trained to do (Lehrman, 1970). A half-century of rapprochement between 'biological' and 'psychological' approaches to animal behaviour and cognition (e.g. Shettleworth, 1998) have not entirely closed the gap. Critics of cognitive ethology, such as Wynne (2004), Blumberg and Wasserman (1995, 2010), and Shettleworth (in press), still believe they hold the empirical high ground, while defenders, such as Cheney and Seyfarth (2007), question the ability of psychologists to uncover the full power of primate and other animal cognition in the socially and ecologically simplified conditions imposed on laboratory animals.

In the first paragraph of their 2002 paper, quoted above, McGonigle and Chalmers cite Fodor, Chomsky and Piaget as paradigmatic critics of associationist accounts of human cognitive abilities. To philosophers steeped in the folklore of the cognitive revolution, Chomsky's 'Cartesian linguistics' and Fodor's nativism represent the apogee of a rationalistic response to the radical empiricism of Skinnerian psychology. Piaget, however, is much less discussed by philosophers of mind and cognitive science. For instance, if one compares references to the two scientists in *The Stanford Encyclopedia of Philosophy* (SEP) (Zalta, 2010), Chomsky is mentioned or cited in over 50 articles while Piaget appears in fewer than a dozen. The SEP search engine results for 'Piaget' reveal that the articles mentioning him (e.g. 'The Philosophy of Childhood', 'Philosophy for Children', 'The Philosophy of Education', 'Panpsychism') are much less central to the discipline as a whole than those in which 'Chomsky' appears, and the entry on 'Cognitive Science' does not mention Piaget at all.

My point here is not to argue that philosophers should be paying more attention to Piaget specifically (although perhaps they should), but to point out the relative neglect of developmental perspectives in the philosophy of mind and philosophy of cognitive science. I don't deny that philosophers of mind make a lot of hay of the work of some developmental psychologists, but the fact is that methodologically and conceptually the developmental psychologists receiving the most attention from philosophers represent a particularly language-centred perspective on cognitive development. They focus on organisms (human children) for whom the acquisition of language, intentional psychology (or 'theory of mind'), and causal understanding of the world (including 'folk physics'), are (at least to a first approximation) environmentally robust outcomes, and they argue about the extent to which these stable outcomes should be attributed to the possession of innate concepts or knowledge. There has been little attention by philosophers to the role that developmental thinking (or its absence) might play in other debates within the philosophy of mind, such as arguments about the multiple realisability of cognitive capacities (for an exception, see Lloyd, 2004). And while there have been attempts to investigate the developmental effects of raising primates, especially apes, in human-like environments, these efforts suffer both from being examples of the anthropocentric, trophy-hunting approach, and from being shots in the dark with respect to the conditions that actually matter for cognitive development because we don't understand the subtleties of our own cognitive development well enough to know what might work with members of other species.

11.6 Private codes and public structures

It is precisely here, then, that I find the ambitious agenda set out by McGonigle and Chalmers to have the greatest significance. But I also wish to present some friendly amendments.

First, framing the discussion as a debate between associationism and cognitivism may not be entirely helpful (see also, Smith 2000). There are many issues in the mix here, but a general (although not universal) predilection among comparative psychologists for associative accounts of learning provide one major axis. Minimalists accept the label of 'associationism' and seek to reduce all learning to limited set of principles concerning the strengthening and weakening of connections among various combinations of stimuli and responses. McGonigle and Chalmers resist this, of course, maintaining that it leads to the language as 'magic bullet' view.

Today's associationism includes learning mechanisms that are much more powerful than those of the behaviourist heyday and McGonigle and Chalmers might be faulted for not acknowledging the more powerful model of classical conditioning offered by Rescorla and Wagner (1972), as well as subsequent developments of this model (Wagner, 2008). The power of new models of learning makes it less clear whether there's any cognitive capacity, including language learning (although I shall not argue for that here), that cannot be explained in terms of today's more powerful associative mechanisms. Cameron Buckner argues (2011, and dissertation) that recent discussions about 'associationist' versus 'cognitive' explanations of animal behaviour have tended to vacillate between different conceptions of associative mechanisms. For instance, some authors treat connectionist models as associationist. But, under certain idealising assumptions, some connectionist architectures are Turing-equivalent and therefore as powerful as any computational model, the latter being subject also to specific idealising assumptions. If associationists have the full resources of connectionism at their disposal, then associative mechanisms are inductively no weaker than other kinds of computational models. However, with such powerful mechanisms at their disposal, the problem becomes one of understanding how cultural and developmental phenomena provide scaffolding that shapes the learning that is actually observed.

Even if we exclude the most powerful connectionist architectures and newer statistical learning techniques from the class of basic, or minimal, associationist models, and reserve this category for 'the educated salivations of a Russian dog' as supplemented by the tutored bar pressing of a Harvard rat, this was only ever two-thirds of the original associationist package. By the original package, I mean Hume's claim to be the first to enumerate the 'three principles of connexion among ideas, namely, Resemblance, Contiguity in time or place, and Cause or Effect' (Hume, 1748). For Hume, of course, association was between ideas (faint copies of sense impressions) – relata that most comparative psychologists would reject as too subjective (perhaps too much like private codes, in that specific sense) and based too much on a developmentally naive theory of concept learning that was opposed to equally untenable ideas about innate ideas. Setting aside such issues, and without hanging too much on acceptance of Humean associationism here, his three-part taxonomy does nevertheless help us see where behaviourism swept some things under the rug. For while it is arguable that Pavlov was onto contiguity, and Skinner onto cause-effect, a treatment of resemblance was needed for both – which is where notions like 'stimulus generalisation' were deployed. Arguably, although I won't argue it here, 'stimulus generalisation' is more a description of the phenomenon than an explanation of it, and resemblance still resists a completely adequate treatment notwithstanding various theories of similarity among psychologists such as Tversky (1977), Shepard (1987), Markman and Gentner, (2005) and others (but see Vigo 2009a, 2009b; Vigo and Allen, 2009).

By our best accounts, resemblance is a kind of structural relationship. With this in mind, I view McGonigle and Chalmers (2006) as making a move in Hume's direction when they argue that the cognitive competences of monkeys with respect to relations between objects entail the existence of 'private codes'. The notion of a 'code' here introduces the idea that internally represented, prelinguistic, structural relationships underpin such things as the capacity to learn serial orderings. McGonigle and Chalmers intend the availability of such codes equally as a counterweight to the aforementioned 'overdependence on a relatively weak inductive mechanism' among comparative psychologists, and against the assumption that only language makes relational learning possible. The danger is that such 'codes' are themselves just code for the 'ideas' that psychologists became rightly suspicious of during the twentieth century.

The notion that such codes remain private in monkeys until externalised by human artefacts leaves us scrambling for an account of the origins of those codes. McGonigle and Chalmers (2006: 248) hedge on this, suggesting that the intensive, long-term regime in which they investigate monkey learning may be better viewed as bringing *on* rather than bringing *out* the cognitive competencies. This suggests that the relational codes themselves may be a product of development in a particular context. But if that is the case, then it doesn't seem right to suggest that these codes were already present, waiting to be externalised by dint of the superior manipulation skills of humans. The complex interplay between the machines, the monkeys and the experimenters who designed the software (not to mention the comparative psychologists, mathematicians, computer scientists, engineers and philosophers before them) must also be expanded to include the daily experiences of the monkeys *outside* the experimental situation.

However, when McGonigle and Chalmers (2002; see also the chapter by Kusel and Chalmers, this volume) suggest, following Clark (1973), that humans have a biological bias to scan from the ground up which makes the larger objects on the touch screen more salient, and could provide an objective grounding for the relational codes deployed in the learning task. This also suggests an evolutionary origin for the private codes. But that it applies to monkeys (or even to humans) should not be considered as a given. For one thing, the experimenters could perhaps be mistaken about the relevant feature, having introduced spurious correlations between size and other features into the stimulus set, for example, if the icons on a screen are always arranged so that the relative position of the top or bottom edges correlates perfectly with size, then we have no assurance that size is what the monkeys are responding to. Without knowing more about the experiences of the monkeys in the rest of their daily lives, we might have no idea whether they come to the experiment with a bias towards noticing size or edge position, and further experimentation runs the risk of altering those biases since the relevant learning is likely to be context sensitive.

Furthermore, the extent to which experiments focus on sensitivity to manipulations in one feature dimension of simple stimuli, they ignore the more complex relationships among features of sets of stimuli that these animals have to deal with in their daily interactions with each other, their human caretakers (if captive) and the complex ecological context in which resemblance is not a single-dimensional affair. Animals learn how combinations of multiple cues predict various outcomes. A pressing goal for theoretical cognitive science is to describe the structural properties of sets of stimuli that make it possible for animals to coordinate their behaviours around objects and events in their environments without the linguistic ability to say what they mean. Such structure is beginning to be objectively described and quantified (e.g. Vigo, 2009a, 2009b; Vigo and Allen, 2009) although much work remains.

What might a thoroughly developmental comparative psychology look like? Ideally it would consider the entire range of developmental inputs, including but not limited to participation in previous experiments, as potentially relevant to the outcome of any particular experiment, and would control for those inputs. But this is impossible. The sheer complexity and adaptability of organisms to different life histories makes the space intractable. The set of possible sequences of behavioural experiments itself defies enumeration, let alone systematic investigation, not to mention the sheer variety of species that could be studied. Is comparative psychology therefore condemned to be just whistling in the dark? I take a more optimistic view. Given an intractable space, one can proceed by trying to build bridges between parts of the space initially under separate investigation. I think this is the real value of the interdisciplinary work that McGonigle and Chalmers pioneered. In the course of their work they demonstrated that it is possible to investigate the relationship between specific cognitive outcomes and various forms of scaffolding that were not usually in the purview of either strict learning theorists or magic-bullet enthusiasts. They showed how their monkeys could develop cognitive capacities built on prior experience, acquired over the long term. A thoroughly developmental comparative psychology doesn't run away from the problem of the accumulation of capacities by refusing to reuse subjects in different experiments, but treats those prior experiences as a variable. And while it is not feasible to investigate every possible influence, it is possible to be on guard against over-interpreting the results of any given experiment as representing 'the cognitive capacities' of all members of a given species. In the excitement of the chase for particular trophies, this can be hard to remember. So, I believe, comparative developmental psychology stands in need of a new canon of interpretation, not unlike Morgan's (1894). Not a decision procedure, but a useful heuristic against over-interpretation of the results of experiments in comparative cognition: 'In no case may we interpret the action as the outcome of the exercise of a general capacity of the species, if it can be interpreted as the outcome of a developmental process specific to the individual'.

Acknowledgements

I am grateful to Keith Stenning, Margaret McGonigle-Chalmers, Iain Kusel, and the members of my dissertation writers' group at Indiana University, especially Cameron Buckner, for helpful criticism of earlier drafts of this material. I thank the Alexander von Humboldt Foundation and Indiana University for financial support during the writing of this chapter.

References

Allen, C. (2006). Transitive inference in animals: reasoning or conditioned associations? In S. Hurley and M. Nudds (eds) *Rational animals?* Oxford: Oxford University Press, pp. 175–85.
Bekoff, M. and Allen, C. (1992). Intentional icons: towards an evolutionary cognitive ethology. *Ethology*, 91: 1–16.
Blumberg, M. S. and Wasserman, E. A. (1995). Animal mind and the argument from design. *American Psychologist*, 50: 133–144.
Buckner, C. J. (2011). Two Approaches to the Distinction between Cognition and 'Mere Association'. *International Journal of Comparative Psychology*, in press, 24: 314–348.
Buckner (2012). Unpublished doctoral dissertation, Indiana University, Bloomington.
Call, J. and Tomasello, M. (1999). A nonverbal theory of mind test: the performance of children and apes. *Child Development*, 70: 381–95.
Cheney, D. L. and Seyfarth, R. M. (1990). *How monkeys see the world: inside the mind of another species*. Chicago: University of Chicago Press.
Cheney, D. L. and Seyfarth, R. M. (2007). *Baboon metaphysics: the evolution of a social mind*. Chicago: University of Chicago Press.
Chomsky, N. (1980). *Rules and representations*. Oxford: Blackwell.
Clark, H. H. (1973). Space time semantics and the child. In T. E. Moore, (ed.) *Cognitive development and the acquisition of language*. New York: Academic Press, pp. 27–63.
Clayton, N. S. and Dickinson, A. (1998). Episodic-like memory during cache recovery by scrub jays. *Nature*, 395: 272–4.
Clements, W. A. and Perner, J. (1994). Implicit understanding of belief. *Cognitive Development*, 9: 377–95.
Davidson, D. (1982). Rational animals. *Dialectica*, 36: 317–27.
Dennett, D. C. (1983). Intentional systems in cognitive ethology: the 'Panglossian paradigm' defended. *Behavioral and Brain Sciences*, 6: 343–90.
Epstein, R., Lanza, R. P. and Skinner, B. F. (1981). 'Self-awareness' in the pigeon. *Science*, 212: 695–6.
Evans, C. S., Evans, L. and Marler, P. (1993). On the meaning of alarm calls: functional reference in an avian vocal system. *Animal Behaviour*, 46: 23–38.
Evans, C. S. and Marler, P. (1995). Language and animal communication: parallels and contrasts. In H. Roitblat and J. Meyer (eds), *Comparative approaches to cognitive science*. Cambridge, MA: MIT Press, pp. 341–82.
Fodor, J. A. and Pylyshyn, Z. W. (1988). Connectionism and cognitive architecture: a critical analysis. *Cognition*, 28: 3–71.
Gallup, G. G., (1970). Chimpanzees: self-recognition. *Science*, 167: 86–7.
Gallup, G. G. Jr., Anderson, J. R. and Shillito, D. J. (2002). The mirror test. In M. Bekoff, C. Allen and G. Burghardt (eds), *The cognitive animal*. Cambridge, MA: MIT Press, pp. 325–34.
Grice, H. P. (1957). Meaning. *The Philosophical Review*, 66(3): 377–88.
Harlow, H. (1949). The formation of learning sets. *Psychological Review*, 56: 51–65.
Hauser, M. D., Chomsky, N. and Fitch, W. T. (2002). The faculty of language: what is it, who has it, and how did it evolve? *Science*, 298: 1569–79.

Hauser, M. D., Kralik, J., Botto, C., Garrett, M. and Oser, J. (1995). Self-recognition in primates: phylogeny and the salience of species-typical traits. *Proceedings of the National Academy of Sciences,* 92: 10811–14.

Hauser, M. D., Miller, C. T., Liu, K. and Gupta, R. (2001). Cotton-top tamarins (*Saguinus oedipus*) fail to show mirror-guided self-exploration. *American Journal of Primatology,* 53: 131–7.

Horner, V. and Whiten, A. (2005). Causal knowledge and imitation/emulation switching in chimpanzees (*Pan troglodytes*) and children (*Homo sapiens*). *Animal Cognition,* 8: 164–81.

Hume, D. (1748). *An enquiry concerning human understanding.* Oxford: Oxford University Press.

Jones, S. S. (2005). Why don't apes ape more? In S. Hurley and N. Chater (eds), *Perspectives on imitation: from cognitive neuroscience to social science.* Cambridge, MA: MIT Press, pp. 297–301.

Landy, D. and Goldstone, R. L. (2007). Formal notations are diagrams: evidence from a production task. *Memory and cognition,* 35(8): 2033–40.

Landy, D. and Goldstone, R. L. (2007). How abstract is symbolic thought? *Journal of Experimental Psychology: Learning, Memory, and Cognition,* 33(4): 720–33.

Lehrman, D. S. (1953). A critique of Konrad Lorenz's theory of instinctive behavior. *Quarterly Review of Biology,* 28(4): 337–63.

Lehrman, D. S. (1970). Semantic and conceptual issues in the nature-nurture problem. In D. S. Lehrman and J. S. Rosenblatt (eds), *Development and evolution of behavior: essays in memory of T. C. Schneirla.* San Francisco: W. H. Freeman, pp. 17–52.

Lloyd, E. A. (2004). Kanzi, evolution, and language. *Biology and Philosophy,* 19: 577–88.

Markman, A. B. and Gentner, D. (2005). Nonintentional similarity processing. In J. T. Hassin (ed.), *The new unconscious.* New York: Oxford University Press, pp. 107–37.

McGonigle, B. and Chalmers, M. (2006). Ordering and executive functioning as a window on the evolution and development of cognitive systems. *International Journal of Comparative Psychology (2006),* 19: 241–67.

McGonigle, B. and Chalmers, M. (2002). The growth of cognitive structure in monkeys and men. In S. B. Fountain, M. D. Bunsey, J. H. Danks, and M. K. McBeath (eds), *Animal cognition and sequential behavior: behavioral, biological, and computational perspectives.* Dordrecht: Kluwer Academic Publishers, pp. 287–332.

McNeil, N. M. and Alibali, M. W. (2004). You'll see what you mean: students encode equations based on their knowledge of arithmetic. *Cognitive Science,* 28: pp. 451–66.

McNeil, N. M. and Alibali, M. W. (2005). Why won't you change your mind? Knowledge of operational patterns hinders learning and performance on equations. *Child Development,* 76: 883–99.

Morgan, C. L. (1894). *An introduction to comparative psychology.* London: W. Scott, Ltd.

Nielsen, M. and Dissanayake, C. (2004). Pretend play, mirror self-recognition and imitation: a longitudinal investigation through the second year. *Infant Behavior and Development (2004),* 27: 342–65.

Penn, D. C. and Povinelli, D. J. (2007). Causal cognition in human and nonhuman animals: a comparative, critical review. *Annual Review of Psychology,* 58: 97–118.

Pepperberg, I. M. (1999). *The Alex studies: cognitive and communicative abilities of grey parrots.* Cambridge, MA: Harvard University Press.

Piaget, J. (1971). *Biology and knowledge*. Edinburgh: Edinburgh University Press.
Pinker, S. (1994). *The language instinct: how the mind creates language*. New York: William Morrow and Co./HarperCollins.
Plotnik, J. M., de Waal, F. B. M. and Reiss, D. (2006). Self-recognition in an Asian elephant. *Proceedings of the National Academy of Sciences*, 103(45): 17053–7.
Povinelli, D. J. (2000). *Folk physics for apes*. Oxford: Oxford University Press.
Premack, D. and Woodruff, G. (1978). Does the chimpanzee have a theory of mind? *Behavioral and Brain Sciences*, 1: 515–26.
Radick, G. (2007). *The simian tongue: the long debate on animal language*. Chicago: University of Chicago Press.
Rajala, A. Z., Reininger, K. R., Lancaster, K. M. and Populin, L. C. (2010). Rhesus monkeys (*Macaca mulatta*) do recognize themselves in the mirror: implications for the evolution of self-recognition. *PLoS ONE*, 5(9): e12865. doi: 10.1371/journal.pone.0012865
Reiss, D. and Marino, L. (2001). Mirror self-recognition in the bottlenose dolphin: a case of cognitive convergence. *Proceedings of the National Academy of Sciences*, 98: 5937–42.
Rescorla, R. A. and Wagner, A. R. (1972). A theory of Pavlovian conditioning: variations in the effectiveness of reinforcement and nonreinforcement. In A. H. Black and W. F. Prokasy (eds) *Classical conditioning II: Current research and theory*. New York: Appleton Century Crofts, pp. 64–99.
Rumbaugh, D. M. (2002). Emergents and rational behaviorism. *Eye on Psi Chi*, 6: 8–14.
Rumbaugh, D. M., Washburn, D. A. and Hillix, W. A. (1996). Respondents, operants, and emergents: toward an integrated perspective on behavior. In K. Pribram and J. King (eds), *Learning as self-organizing process*. Hillsdale, NJ: Erlbaum, pp. 57–73.
Savage-Rumbaugh, E. S. (1996). *Kanzi: the ape at the brink of the human mind*. New York: John Wiley and Sons.
Savage-Rumbaugh, E. S., Fields, W. M. and Spircu, T. (2004). The emergence of knapping and vocal expression embedded in a *Pan/Homo* culture. *Biology and Philosophy*, 19(4): 541–575.
Shepard, R. N. (1987). Toward a universal law of generalization for psychological science. *Science*, 237(4820): 1317–23.
Shettleworth, S. J. (1998). *Cognition, evolution, and behavior*. Oxford: Oxford University Press.
Shettleworth, S. J. (in press). Clever animals and killjoy explanations in comparative psychology. *Trends in Cognitive Sciences*. doi: 10.1016/j.tics.2010.07.002
Slobodchikoff, C. N. (2002). Cognition and communication in prairie dogs. In M. Bekoff, C. Allen and G. M. Burghardt (eds), *The cognitive animal: empirical and theoretical perspectives on animal cognition*. Cambridge, MA: MIT Press, pp. 257–64.
Smith, L. B. (2000). Avoiding association when it's behaviorism you really hate. In R. Golinkoff and K. Hirsh-Pasek (eds), *Breaking the word learning barrier*. Oxford: Oxford University Press, pp. 169–74.
Stenning, K. and van Lambalgen, M. (2008). *Human reasoning and cognitive science*. Cambridge, MA: MIT Press.
Stotz, K. and Allen, C. (2011) From cell-surface receptors to higher learning: A whole world of experience. In Kathryn S. Plaisance and Thomas A. C. Reydon (eds) *The Philosophy of Behavioral Biology*, Berlin: Springer, pp. 85–123.
Tomasello, M., Kruger, A. and Ratner, H. (1993). Cultural learning. *Behavioral and Brain Sciences*, 16, 495–552.

Tulving, E. (2005). Episodic memory and autonoesis: uniquely human? In H. Terrace and J. Metcalfe (eds), *The missing link in cognition: evolution of self-knowing consciousness*. Oxford: Oxford University Press, pp. 3–56.

Tversky, A. (1977). Features of similarity. *Psychological Review*, 84(4): 327–52.

Vigo, R. (2009). Categorical invariance and structural complexity in human concept learning. *Journal of Mathematical Psychology*, 53: 203–21.

Vigo, R. (2009). Modal similarity. *Journal of Experimental and Theoretical Artificial Intelligence*, 21(3): 181–96.

Vigo, R. and Allen, C. (2009). How to reason without words: inference as categorization. *Cognitive Processing*, 10: 77–88.

Visalberghi, E., Fragaszy, D. M. and Savage-Rumbaugh, S. (1995). Performance in a tool-using task by common chimpanzees (*Pan troglodytes*), bonobos (*Pan paniscus*), an orangutan (*Pongo pygmaus*) and capuchin monkeys (*Cebus apella*). *Journal of Comparative Psychology*, 109: 52–60.

Visalberghi, E. and Limongelli, L. (1994). Lack of comprehension of cause-effect relation in tool-using capuchin monkeys (*Cebus apella*). *Journal of Comparative Psychology*, 108: 15–22.

Wagner, A. R. (2008). Evolution of an elemental theory of Pavlovian conditioning. *Learning and Behavior*, 36(3): 253–365.

Wasserman, E. A., and Blumberg, M. S. (2006). Designing minds. *APS Observer*, 19(10): 25–6.

Whiten, A., Horner, V. and de Waal, F. B. M. (2005). Conformity to cultural norms of tool use in chimpanzees. *Nature*, 437: 737–40.

Wimmer, H. and Perner, J. (1983). Beliefs about beliefs: representation and constraining function of wrong beliefs in young children's understanding of deception. *Cognition*, 13: 103–28.

Wynne, C. D. L. (2004). *Do animals think?*. Princeton, NJ: Princeton University Press.

Zalta, E. N. (2010). *The Stanford Encyclopedia of Philosophy*. Retrieved from: http://plato.stanford.edu/ on 1 August 2010.

Zuberbühler, K. (2000). Referential labeling in Diana monkeys. *Animal Behavior*, 59: 917–27.

Zuberbühler, K. (2001). Predator-specific alarm calls in Campbell's guenons. *Behavioral Ecology and Sociobiology*, 50: 414–22.

12
The Emergence of Complex Language
Wolfram Hinzen

12.1 Introduction

It has now become rather common to accept that any theory of language is constrained by evolutionary and comparative considerations, as much as considerations of the evolution of language should be constrained by linguistic considerations. This is to say: when and as we develop a theory of what language is like and why it is that way, it is useful to look at the archaeological record, as well as at other (non-linguistic) minds, in order to see what makes language special. At the same time, evolutionary hypotheses on the emergence of language benefit from some independent theoretically informed grasp of what the empirical properties of languages are.

The first constraint is illustrated by the prevailing puzzle of what caused the 'Great Leap Forward' (Diamond, 1989) around 80 thousand years ago (kya), which would radically change the evolutionary landscape and the character of mind forever, leading to modern human culture as we know it (McBrearty and Brooks, 2000). While probably not as recent as 35 to 50kya – the period with the richest archaeological evidence found in Europe – this transition remains a very recent and relatively sudden threshold on the evolutionary path to a modern human culture. Section 2 discusses the role that language may have played as a motor of such change, which amounts to a significant shift in the cognitive organisation of early hominids. On some views, the shift depends on the intrusion of grammar into the brain and consequent changes in the organisation of human thoughts. Section 3 argues that at the heart of the emergence of language lies the codification of thought, not communication as such: language structures *what* is to be communicated, and cannot be understood as arising from the constraints of communication alone.

While this account leaves grammar with a biological base, which the term 'Universal Grammar' (UG) traditionally denotes, we nonetheless want to keep the genetic component of the language faculty minimal. Section 4 argues, against frequent arguments for its demise, that UG can nonetheless not be *empty*. This does not mean either, however, that it must be the functionally 'arbitrary' linguistic adaptation that many contemporary UG-free theories combat. On the contrary, Section 5 argues that human grammar is precisely organised around the very units of deixis that must have been central to the reorganisation of the early human communicative setting in which human culture arose. This contention not only provides the basic organisation of grammar with a functional rationale, it also derives the recursivity of complex language, which is often assumed as a primitive in current Minimalist accounts of grammar. Section 6 concludes.

12.2 Revisiting the Great Leap Forward

Experiments with language-trained apes sometimes make one forget that even after the hominoid line split off from our common ancestor with chimpanzees around six to seven million years ago (mya), it would take many million years longer, until *Homo erectus* had evolved into *Homo heidelbergensis*, commonly considered the common ancestor of both modern *Homo sapiens* and the Neanderthals today. The momentous point in human evolution where modern *Home sapiens* branched off is now commonly located in sub-Saharan Africa and dated at around 200kya. A further gap of at least 100kya remains until we encounter clear evidence for behaviourally modern humans in Africa (McBrearty and Brooks, 2000), with the most solid evidence awaiting the Aurignacian revolution around 35kya, after the exodus of these early humans from Africa and their arrival in Europe (Mellars and Stringer, 1989). Human evolution prior to the threshold of 100kya is characterised by slow progress and cultural stagnation, if not the absence of 'culture' altogether. Even to reach the stage at which bipedal early hominids began making crude stone tools would take several million years after the split from the chimpanzee line. These tools begin to appear in the fossil record in Africa from around 2.5mya. This first, 'Oldowan', stone-making industry is to last a million years without any apparent innovation, until the arrival of the 'Acheulian handaxe' in the *erectus* lineage marks a genuine cognitive innovation (Gowlett, 1992). This Acheulian technology would remain the primary way of fabricating tools for at least another million years, until about 100kya.

Also a look at Neanderthal 'culture' in Europe suggests an essentially static stone-tool culture for more than a hundred thousand years. While occupying much of central Europe and western Asia, the absence of cultural variation and progress in time or space, along with a striking lack of art and symbolism for most of this period, appears to be a mark of Neanderthal lifestyle. What is to happen in the same territory between 35 to 50kya suggests a radical break in mental organisation in early human history – so radical that it plausibly led to the wiping out of the Neanderthal species and the sole survival of the invading species. The cultural changes brought along by these invaders are reflected in the emergence of art and symbolic representation, implements made of soft materials (bone, antler), notations, delicate engravings on bone and stone plaques, elegant carvings, music on bone flutes, song and dance, burial of dead in decorated clothing, eyed bone needles marking the arrival of couture, ceramic technology, fish hooks and net sinkers, serious hunting with darts, spears, bows and arrows, and long-distance trade.

Set against the background of millions of years of stagnation or slow progress, one cannot but marvel at the character and the speed of these technological and cultural innovations – effectively, the origin of a modern human culture itself – and wonder as to their cause. It appears as if the changes in question reflect a radically different mode of thought and mental organisation. They present a mind capable of manipulating symbolic representations in the place of natural objects, a mind able to turn back on itself and to reflect on what it sees and experiences. It is unclear whether Neanderthals spoke, but what we do know by now is that the descended larynx, once thought to be a primary innovation for linguistic communication specific to humans, is not specific to humans and widely found in other mammals, from sea lions to deer. As Fitch (2010: 323) points out, evidence for a lowered larynx or a flexible and language-ready vocal tract in Neanderthals, which Krause et al. (2007) claim there is, would not be evidence that they spoke; such anatomy can have different functions, and we see it in animals who don't speak. Moreover, given Neanderthal 'culture', it is by no means obvious what they might have used their vocal tract to say; there is no evidence for *thoughts* that it would have been worthwhile to linguistically express. As Fitch (2010: 336) concludes, 'the venerable hypothesis that limitations of peripheral morphology explain the inability of most animals to speak seems unsustainable'. If so, it must be a neural change – a change in the organisation of thought, or of cognition – which is crucial to the emergence of language.

That said, we want this change, insofar as it is genetically based, to be minimal. Jared Diamond (1989) argues in his classic paper on what he called 'The Great Leap Forward' that the African hominids that would evolve into anatomically modern humans might not merely have shared 98 per cent of their genome with us, as we are sometimes said to do with chimpanzees, but 99.9 per cent. So what accounts for the remainder? It remains plausible, as Diamond contended, that the last genetic change to happen was the change needed to put into place the neural basis for complex language. This is because language exactly has the features that the new form of thought exhibits: it is the paradigm of symbolic representation based on complex, structured expressions, the use of which is not stimulus-bound and independent of the Here and Now. Almost every sentence we utter is a creative exercise, an innovation, a product of the generativity of mind. With language, a creature will have a history and be able to cooperatively plan its future. It will be able to reflect on social values and codify them in art. It will have a means of encoding knowledge in engravings and structured patterns, giving rise to a shared body of knowledge.

This line of thought puts language into the driver's seat of the cultural evolution that we see abruptly emerging in the Aurignacian culture. It is the missing link between our extinct hominid ancestors and ourselves, and indeed a likely cause for their demise. One might contend that culture and language co-evolved, yet the forms of symbolic representation that the new culture so starkly exhibits is centrally what marks out language, which suggests that the two phenomena may be close to indistinguishable. Moreover, by 40kya our African ancestor pioneers had ventured as far as Australia and Northern Eurasia, with no further apparent genetic variability in our linguistic or cognitive endowment since then, or indeed today. In short, by the time of the Great Leap, essentially modern cognitive and linguistic abilities of our species had matured to fixation. It seems as if, once this last genetic change happened, culture could take off in a way that would now not depend on genetic change anymore; it would be more like the evolution of writing, or the invention of science.

12.3 The rise of biolinguistics

So there is, on this view, a biological basis specific to language. Traditionally, this biological basis is the subject of generative grammar as it arose in the 1950s (Chomsky, 1957); structural linguistics as prevailing until then did not have a biolinguistic agenda. Adopting behaviourist premises, internal constraints on cognitive development were largely ignored, as was the question of what made human cognition special and specific; cognition was meant to work the same way in humans, rats and pigeons. As generative grammar evolved into the biolinguistic program (Lenneberg, 1967), the term 'Universal Grammar' (UG) came to be the term for the genetic aspects of language – its 'innate' aspects – understood as one part of what constitutes human linguistic competence. Other such constituent parts are: (i) what is given in experience, and (ii) what is neither given in experience nor genetically based – so-called 'third factor' conditions such as principles of computational efficiency or economy, which are too general to be linguistically specific and which have become central in the description of grammar today (Chomsky, 2005). It appears that with that much, everyone should agree: there are these three factors, and they all enter into language design. Keeping the first, genetic part, minimal, moreover, has become a shared agenda point of both generative linguists in the Minimalist tradition (Hinzen, 2006) and those many others who are unhappy with a substantial innate endowment for language and united in their opposition to UG (Christiansen and Chater, 2008, 2010; Evans and Levinson, 2009; Tomasello, 2008). Whether it could be empty altogether is another question, which I will address in the next section.

The present approach also sides with Christiansen and Chater (2008, 2010) in the rejection of an adaptationist approach to the origins of human language (Pinker and Bloom, 1990; Pinker and Jackendoff, 2005), though the reasons are different. On this adaptationist approach, language is selected for the expression of complex propositional thought: it is a genetic response to the selection pressure acting on organisms that need a means to communicate their thoughts. Thought, that is, imposes conditions that a language system, since it is designed for communication, needs to meet. Its form reflects the contents it is designed to convey. This adaptationist scenario is in contradiction with the approach pursued here. The instigating condition for human culture, on the above scenario, is a neural reorganisation that puts a creative and structured mode of thought into place, and the most plausible cause of this reorganisation is language: what we might call the grammaticalisation of the brain. I return below to how this might have happened. For now note that the adaptationist approach leaves us without an explanation of the origins of the special kind of thoughts that linguistic expressions express. 'Thought' is a highly unspecific term, but what philosophers since antiquity have regarded as the crucial *explanandum* is propositional thought – thought that we can assign truth values to and that is usable in science. Where language is a mere means of 'expressing' such thoughts, and the thoughts are somehow independently there, the emergence of language will not form part of their etiology, and their origin is left unclear.

While the existence of precursors to such thought is not doubtful, no one would claim that the emergence of grammar makes *no* difference to the organisation of thought (even in the absence of such large-scale scepticism as voiced in Penn et al., 2008). Apes, birds and cetaceans think, but it's not that they think the very same kind of thoughts as we do, and they merely fail to express them due to some tragic sensory-motor handicap related to pronunciation. As with the Neanderthals, it's not that, if only they could talk, they would say what we do; their thoughts would come out and look just like ours. As suggested above, the creative potential itself, which goes with human thought and culture, may well be missing.

Again, the hallmark of thought in this sense is symbolic representation, the building of possible worlds through the combination of symbols, and the creative engagement with what we see and experience, as manifested in art. Where the thoughts of apes do 'come out' in sensory-motor externalisation, they do not look like that; communications of apes trained to use a sign language with signing humans do not convey the thoughts that human declarative sentences do. Overwhelmingly, as Mike Tomasello points out, they are requests relating to the performance of very concrete physical actions (such as eating, drinking, or playing games) that, in the communicative settings in the question, the humans controlled and the apes desired (Tomasello, 2008: 251). It is tempting to correlate this feature with the fact that their 'utterances' 'seem to contain basically no relational or grammatical structuring of any kind' (Tomasello, 2008: 249). Relational competence in thought can 'come out' (or be 'brought-out') in a non-communicative setting, as in sequencing tasks of geometrical shapes performed by *cebus appella* monkeys on touch screens (McGonigle, Chalmers and Dickinson, 2003). Yet such relational competence is not externalised in an external communicative medium with a grammar and the inherent deictic potential and propositional character (unrestricted to the Here and Now) that goes with it in the human case. Other questions remain in regard to the character of mind as revealed in such seriation tasks, in relation to the human ability for free and unbounded thought, employed as a 'universal tool', and the set of lexical-conceptual resources it draws on.

Such differences are maybe what we should respect in the simian mind. Why should the character of mind be everywhere the same? Wanting to prove that chimps can talk (express thoughts like us), if only they were given the right sensory-motor equipment, might be like wanting to prove that humans can fly, given some necessary equipment (Chomsky, 2010). Leonardo da Vinci designed relevant appendices attached to human arms. If tried, they might even have worked. Yet what would this have proved in regard to whether humans can fly? Whatever the relational competence that appropriate and longitudinal training regimes can bring on in a monkey (McGonigle and Chalmers, 2008), the fact remains that human infants, even only days from birth if not before, acquire language reflexively when exposed to linguistic stimuli – as reflexively as birds learn to fly; it is in their nature. It is not in the chimp's to do either. Relational thoughts as brought out in sequencing tasks narrows the gap between the human and the non-human since relational structure is so crucial to the human linguistic code (see Hinzen, 2010, for further discussion). Yet a grammar organised around a parts-of-speech system organised hierarchically remains some way off and is, even if 'brought on' in years of training, not an aspect of chimpanzee nature in the way that language is part of human nature.

I have argued that the adaptationist hypothesis on language origins faces problems in explaining the origin of structured thought itself. But the most fundamental problem seems to be that a system designed for communication would not seem to exhibit the features that mark out human language from all other animal communication systems in nature. There is no reason for a communication system to exhibit grammatical organisation, say, if this is not a biological option independently given. Indeed the absence of meaningful syntax remains a crucial design feature of most or all non-human communication systems (Fitch, 2010: 184–5), and non-human animals communicate perfectly well and efficiently without it. Nor does a communication system need to exhibit a creative aspect, as all normal language use does, unlike most non-human communication, which tends to remain situation-specific. Universal aspects of grammatical organisation, such as the sentence/Noun Phrase distinction (Carstairs-McCarthy, 1999), remain unexplained on communicative grounds as well (and saying it is needed to express truth value bearing propositional thoughts would be circular). Nor would we predict the interesting fact that human language is so immensely variable, again unlike typical animal communication systems, while being so uniform in other respects, as in hierarchical organisation, mode of acquisition, independence of sensory modalities, or neural basis. None of this is to argue that human language is not communicatively efficient. It is to say that the *character* of communication changed fundamentally once it became linguistic; what makes human language special is not that it is a communication system, but that it is a linguistic one.

The adaptationist hypothesis will also not predict that human language will have lexical items, given that these seem different from animal vocalisations in both phonological and semantic respects; in semantic respects, because they fail to exhibit the characteristically causal relation that obtains between the animal vocalisation and independently identifiable external stimuli, as Chomsky has argued long since: human use of words is typically not explained by a causal or other physically describable relation between the words and independently identifiable external stimuli. The use of any personal proper name, say, reflects the tracing of an object on the basis of criteria of psychic continuity, rather than the physical features of the relevant person's body. Already the use of words, then, seems to be a feature of human communication strongly discontinuous with the use of symbols in animal communication; it depends on the evolution of a mental lexicon with a special kind of semantics, and that of a generative system generating morphological complexity which can in turn feed into a syntactic combinatorics. All of these are largely absent in non-humans and their communications.

A number of other considerations suggest that language did indeed arise as a system for the free expression and codification of thought rather than arising from communicative constraints. From Aristotle to de Saussure and beyond, language has been looked at as 'sound with meaning'. These come together in the linguistic sign, as two of its sides: every linguistic sign has both a phonological and a semantic analysis. In contemporary terms, every linguistic expression, being highly structured, has to be generated by some procedure, and this procedure, at certain points in the derivation, will have to access two 'interfaces' it forms with outside systems (Chomsky, 2005). At these points, the derivation is assigned a phonetic interpretation (at the sensory-motor interface) and a semantic interpretation (at the semantic interface). Both the semantic interpretability of a linguistic sign and the possibility of articulating and perceiving it impose *constraints* on the syntactic derivation; failing to satisfy these constraints, the syntactic derivation would produce material that performance systems could not use, and linguistic communication would fail. Arguably, however, the sound and the meaning systems may impose *different* constraints, and these may *conflict*. Where this is the case, however, Chomsky (2005, 2010) argues, we see the conflict resolved in favour of meaning rather than sound. This is exactly what we would predict if language evolved as a means for the expression of thought rather than communication as such.

To illustrate, consider that in order to subserve a mode of thought that is creative rather than stimulus-bound and that can be applied to an unbounded set of novel situations, a generative mechanism operating on the elements of an evolved lexicon is needed that generates a discrete infinity of linguistic expressions that can be applied to situations, each with a distinct sound and meaning. Arguably the easiest conceivable operation that does just this is an operation that takes two lexical items, A and B, and puts them in a set {A, B}, leaving the objects themselves unaffected. If the operation can take such sets again as input for a combination with a further lexical item C (or another set), recursive objects of the form {C, {A, B}} will arise. In these sets ordered by the containment-relation we recognise the constituent structures of traditional phrase-structure grammars, in which constituents are organised in a hierarchical fashion, and hierarchy has no bearing on linearity (a subject can be hierarchically ordered in relation to a verb phrase in the exact same way, whether it is pronounced before or after it in speech). Overwhelmingly, now, what matters to grammatical organisation is hierarchy in this sense – not linearity. For example, in the expression 'The man the women adore falls', the subject is 'the man' and it stands in a relation of agreement with its predicate, the verb 'falls'. This is not the predicate that comes first in the linear order, but which comes first in the hierarchical order, giving rise to nesting rather than crossing dependencies:

[The man [that [the women] adore] falls]

If we form a question out of this sentence, we obtain 'Does the man the women adore fall', where 'does' again relates to 'fall', rather than to 'adore'. In comprehension, the gap it leaves in its base position (the position of Tense associated with the predicate 'falls'), from where it moves to the front, will need to be figured out in order to obtain the correct interpretation. The correct interpretative strategy cannot rely on linear order, it will need to reconstruct hierarchical order. It would be greatly helped if 'does' was repeated where it needs to be interpreted, as in 'Does the man the women adored does fall'. But items moved from positions in which they are interpreted are almost never repeated to them. Since repeating them makes no difference to hierarchical order, and hierarchical order matters to interpretation, a computationally more efficient strategy is not to repeat them; the semantic systems do not need to see the moved item in the place of its original position. From the viewpoint of externalisation and communication, on the other hand, the strategy should be to repeat them, and to avoid nesting dependencies and the long-distance relations they create. There is a conflict in interface demands, therefore, and it is resolved in favour of what would be computationally efficient in relation to the semantics. I will later construct an argument that the hierarchical organisation of a linguistic expression in fact exactly matches the thoughts to be expressed, and fully accounts for the basic structures they can have.

In sum, grammar appears to be optimised in relation to the demands of the semantic interface, not the sensory-motor one. This we would predict if language was a system for the free expression of thought rather than arising from communication as such. On the same hypothesis we would equally not be surprised by the finding that language is independent of sensory modality (or mode of externalisation), as it indeed is; whether language is spoken, signed, touched or smelled, makes little difference to its basic organisation, which I here argue is one of the thoughts encoded.

If human language was indeed an adaptation for communication, we would need to find a relevant 'design', in the relevant biological sense; its central features need to be predicted somehow for the use for which it was selected, or 'optimised' in relation to it. This does not seem to be the case: a look at non-human animal communication systems (surveyed in Hauser, 1996) predicts little if any of the features that human language as a communication system uniquely has. To say that it has these unique features because it expresses what none of the other communication systems do, namely human thought, is merely to push back the explanatory problem one step; now the question becomes the origin of thought of that very linguistic sort, which I here argue language (or grammar) is in fact the optimal explanation for. There are other, more conceptual, problems with any adaptationist theorising, but since they are fully general and don't concern the case of language specifically, I will not rehearse them here (see Hinzen, 2006).

12.4 Wiping out UG

While the adaptationist hypothesis closely reflects the gradualism that has gone with the Neo-Darwinian Synthesis, the archaeological record does not suggest that a gradualist scenario is what we should be looking for. Adaptationism has typically gone with a version of biolinguistics on which there is a rich innate endowment specific to language. At the opposite extreme lies the view that there is no UG, indeed no biology of language, at all (Tomasello, 2008). On this view, there is a genetic endowment that is crucially needed for language, but it is not linguistically specific. Finally, there is a middle view, on which UG is minimal and so abstract that it could as well have originated in a short span of time – the view of linguistic Minimalism (Chomsky, 2005). Let us now ask: could the 'first factor' really be empty, as both Tomasello (2008) and Christiansen and Chater (2008,

2010) contend? Not if relevant genetic changes are needed that reorganise the brain in a way that language can be embedded in it. Given that even recombinable lexical items seem to be organised differently at both interfaces (sound and meaning) than animal vocalisations (Hauser et al., 2002), it is not clear where they would come from except from some crucial neural reorganisation. Even if recursion as such is available in non-linguistic domains such as (perhaps) hunting or tool-making, implementing a recursive mechanism in the domain of language that combines lexical items into discretely infinite structures is another massive change. A recent intriguing proposal identifies a gene emerging around six to seven mya, at the origin of our divergence from chimpanzees, related to epigenetic changes around 200kya at the dawn of cognitively modern man (see Crow, 2010). It is argued to have triggered the lateralisation of the human brain, arguably the most asymmetric of brains, and provided a specific neural compartmentalisation of the brain that provides a unique cognitive infrastructure for language not found in any other species or any other cognitive domain. In this sense, the changes in question are specific to language.

Theoretically, UG could be empty if the rise of a modern human culture was premised by a biological adaptation for culture that brought language along as an aspect of that, as Tomasello (2008) argues. Again, no biological change specific to language is needed on this view. My basic objection to this point of view will be that the adaptation for culture in question will explain anything, only if it has language implicitly factored into it. Without grammatically encoded thoughts, a shared intention to communicate or symbolise will not bring such thoughts into place. An appeal to an intention to share a thought, if this thought does not inherently already *correspond* to the kind of meanings that human sentences uniquely have, appears empty. Consider Tomasello's specific argument. Like Evans and Levinson (2009), he regards language as embedded in, and as emerging conventionally from, a prelinguistic 'social-cognitive infrastructure'. Crucial to the latter is, not merely to have intentions and recognise them in other individuals, but to *share* them. Helping and sharing are present in this infrastructure as basic motives for social interaction, which generates a space of joined goals and attention within which communication naturally occurs. So far, in this story, there is no account of the thoughts that such motivated communicators are wanting to share. Put differently, it is not clear what they have to say – or what's in their heads. Without the cognitive changes I discussed above – changes that correlate with the grammaticalisation of thought and the emergence of generativity and creativity – we need to wonder how shared intentions to communicate will give rise to them.

As Tomasello (2008: 321) argues, more specifically, 'the basic skill of shared intentionality is recursive mindreading'. The problem is that recursive mind-reading – higher-order intentionality – is notably *absent* in nonlinguistic animal communication (Fitch, 2010: 191–4). Since recursive mind-reading is paradigmatically what linguistic recursions are about ('He knows that I know that he sees what I see'), the account seems circular: formally encoded, a recursive mind-readout appears as indistinguishable from the semantic representation of a recursive human sentence (or the thought expressed by it). The claim, in short, is that some of the descriptions we are given of the prelinguistic communicative infrastructure are linguistic in disguise. Assuming that some way around this can be found, why is it that the origin of language and grammatical organisation poses no significant problems on this view, of a kind that would require specific biological adaptations?

The reason is that Tomasello, along with a long philosophical tradition, sees language as an 'arbitrary' communicative convention, replacing 'natural' communicative acts such as the pointing gesture to direct the visual attention of others or the use of iconic gestures as in pantomiming. An arbitrary convention carries no epistemological value; all of cognition is there before. Yet what is the evidence, outside of the human linguistic context, that such a cognitive and social infrastructure exists?

A clear connection may well exist between pointing and demonstratives such as *this* and *that* in language, and in turn between iconic gestures and content words (such as nouns and verbs). And yet, pointing in non-linguistic beings is massively different from the use of *this* and *that*, as we see when we observe a child pointing to a toy on the table in a friend's house, and saying, 'I've got that in my house' (Roeper, 2009). The child is pointing to an abstraction, intricately related to different objects or instantiations in different points in time and space. Reference in non-humans appears confined to the Here and Now, and even their pointing to objects in the immediate environment is argued to lack the forms of declarative pointing that naturally replaces the earlier imperative pointing in human infants in the course of ontogenetic maturation (Terrace, 2005). While iconic gestures, in turn, may be related to content words, parts of speech such as nouns and verbs do something that iconic gestures would be hard pressed to do: choosing between a perspective under which we refer to a thing, as when we choose between talking about an event as *Mary's run* (using a noun) or as the event in which *Mary runs* (using a verb in a sentential frame). The two expressions are crucially different deictically, or in the way that we locate an event in time, space, and discourse.

As for the emergence of grammar out of the prelinguistic social-communicative context, Tomasello argues that forms of grammatical organisation arise as ways to 'meet the functional demands of the three basic communicative motives, leading to a grammar of requesting, a grammar of informing, and a grammar of sharing and narrative' (2008: 326). That is, as communicative motives complexify – early humans wanted to communicate a proposition, say – the need for a more complex grammar arose, leading to new linguistic conventions' crystallisation in grammatical constructions that eventually become normatively constrained (2008: 293–9). Yet, neither in biological nor cultural evolution (thinking of how hard-won innovations in stone-tool making were, say), do new functional devices arise in response to functional needs for them, unless the devices are simple extensions of ones already in existence. The mere wish to tell a narrative, where no discourse or grammar exists, will not create one. Referring to the nonexistent, or the ability to string together symbolic expressions so to create a propositional unit with a novel meaning able to tell a truth, is just about as novel as things can get: it defines a new mode of cognition, as argued above.

Christiansen and Chater (2008, 2009), coherent with Tomasello's line of thought, argue that there is further evidence that grammar *could* not in fact have biologically evolved, becoming enshrined in an innate UG that is naturally selected for. In an early human context of ongoing cultural and linguistic evolution, with dispersing human populations and subpopulations, linguistic conventions could not have become biologically fixed though natural selection of genes coding for them, since linguistic environments would have changed too fast and would outdate any fixed UG, making it maladaptive. A genetic language ability would moreover be predicted to adapt to *specific* environments (specific phonologies and syntactic systems), while what we see is that human infants are *universally* adapted to any linguistic environment whose language they can learn. If UG is to contain only abstract and unchanging principles of grammar, nothing would explain why in such a historical setting, just those abstract principles would have genetically fixated that we now ascribe to UG, rather than the more superficial aspects of external linguistic environments.

These problems for a biolinguistic program are circumvented, however, if grammar, rather than the result of hundreds of thousands of years of cultural evolution, was the result of a sudden, perhaps virtually instantaneous (in evolutionary terms), reorganisation of the brain, which *drove* the cultural evolution in question. Indeed, the more minimal the design of UG becomes, as in recent theorising (Chomsky, 2008), the more it becomes virtually impossible to break it into pieces that have gradually evolved (an operation such as Minimalist 'Merge', which creates hierarchical complexity in grammar, cannot evolve gradually; nor does it allow for cross-linguistic differences). The argument also doesn't go through the more principles and parameters of UG as familiar from the 1980s dissolve into 'third-factor conditions', for if so, they start to look like conditions of natural law that cannot evolve gradually and be premised on cultural evolution either. The use of these third-factor conditions in a linguistic domain will nonetheless plausibly require some genetic fixation. What remains of UG in this fashion may then well fall into the broad category of 'internal constraints' in the organism (Fodor and Piattelli-Palmarini, 2010, chapters 2, 3; Hinzen, 2006) that are independent of the external factors operating on phenotypic variations and naturally selecting them, which are the sole focus of Christiansen and Chater's critique: evolution does not, the evo-devo framework in biology suggests, churn out new phenotypic variations at random, with natural selection as the sole or even major factor in their selective retention. On the contrary, the pervasive recent findings of entrenched developmental pathways, master genes, and evolutionary robust regulatory networks may make the search for what human grammar was 'selected for' virtually meaningless, as Fodor and Piattelli-Palmarini (2010) argue. The new biology makes it simply false to expect that, if there is a biology of language in the sense of UG, we need to regard it as the result of the environment shaping the organism 'in its own image', or of the organism recapitulating environmental conditions, an assumption on which the above criticism depends.

There a further important conceptual premise to the argument of Christiansen and Chater (2009) above, which is shared with Tomasello (2008) and Evans and Levinson (2009). Their arguments against UG, Christiansen and Chater (2009) stress, 'only preclude biological adaptations for *arbitrary* features of language' (my emphasis), which they centrally assume traditional UG theories exclusively encode: 'despite the important theoretical differences between current approaches to UG, they all share the central assumption that the core components of UG, whatever their form, are fundamentally arbitrary, from the standpoint of building a system for communication' (2009: 3). Tomasello (2008: 275) similarly directs his critique of UG against the idea that human grammar consists of 'contentless, algebraic "rules"', a conception commonly thought to originate with Chomsky.

However, that generative grammar set out to formally analyse grammatical expressions and find algebraic structure in them does not mean that these expressions are contentless. Generative linguistics has viewed language as 'sound with meaning' from its inception – though it is true that it has *abstracted* from the meaning-side of things for a long while, given that syntactic patterns seemed unexplainable on semantic grounds. It also seems true, as noted above, that *communication as such* does precisely *not* provide the constraints that predict a communication of the human kind, mediated by grammar. Yet, and in direct opposition to the notion of UG used in the criticisms above, the Minimalist version of the generative program in the meantime regards, as described above, grammar as 'perfected' for meaning, or the semantic interface (Chomsky, 2008), directly contradicting claims to the effect that grammatical rules are 'arbitrary' or 'contentless'. Grammatical configurations of specific kinds have direct effect on the organisation of meaning on the other side of the semantic interface; they have inherent interpretive effects.

In the final section, I will attempt to outline what these are – which aspects of the organisation of meaning in language we can trace to grammar, making it instrumental to their existence and the specific mode of thought that we see conveyed in a linguistic medium of communication. For this section,

I conclude that for now, the middle ground that the Minimalist program occupies – assuming that there is a genetic component of language, yet a minimal one – remains the *least* contentious option in the field; the program of finding a minimal basis for getting a system of grammar going appears as one that can and should unite all parties. That there are three factors to language design, genetic, experiential, and third-factor, is arguably the default from which we may start.

12.5 Grammaticalising the mind: the emergence of recursion

As noted above, abstracting from the use of language in discourse and communication was a methodological decision in early generative grammar, when syntax emerged as an apparently autonomous domain of enquiry which one might want to study in its structural aspects and in its own right. The reason for this decision is intelligible in light of the conception of semantics at the time, stemming from logic, which saw language in the role of encoding abstract propositions, semantic reference and truth conditions. Somehow, much of the new riches of syntax uncovered at the time made no sense in the light of these semantic concerns, and they are routinely ignored in standard introductions to formal semantics and in the philosophy of language at large. Logic is substituted for grammar in these domains. Yet, grammar might do something that logic doesn't. Formal analysis of the truth conditions of expressions in classical logic is not designed to encode how speakers, talking to other speakers, encode forms of deixis grammatically in a particular discursive setting. For example, from a formal semantic point of view, the semantics of (1) and (2) might well be argued (and has widely been argued) to be the same:

1. The king of France is bald.
2. There is a unique king of France, and he is bald.

Similarly, we might argue that proper names are generally eliminable (Quine, 1953), and that the semantics of (3) and (4) is in fact identical:

3. Pegasus flies.
4. There is a unique thing which pegasuses and it flies [where 'pegasuses' is a predicate uniquely applying to Pegasus].

Yet, the communicative intentions of a speaker uttering (1) seem misdescribed as making a false claim of existence. Such a speaker refers to the king of France and doesn't assert the existence of anything, for all it appears. He clearly engages, or takes himself to engage in, a referential act. Similarly, referring to Pegasus, the flying horse, is not the same thing as saying, falsely, that something exists that pegasuses; referring when using a proper name is not intuitively making a claim of truth at all. According to Quine, the advantage of (4) is that it brings out that a speaker of (3) is not really referring to anything – which he cannot, if reference is a relation between words and real objects. For Quine, proper name reference is eliminable from logical semantics; it adds nothing to semantics, for which quantification is enough. Language could do without such as deictic devices as demonstratives and proper names. Ipso facto, the special referential behaviour falls outside the scope of semantic theory thus understood (see Hinzen, 2006; Mukherji, 2009, for discussion). Yet much of the *point* of grammar, when added to a prelinguistic system of referential pointing, is that reference will now not be sensitive to 'real existence' in the Here and Now anymore, and hence should be treated independently of that.

Traditionally, the answer to such worries has been: logical semantics is not necessarily *concerned* with communicative intentions; it is about getting truth conditions right, and these may be independent of issues of speaker (as opposed to semantic) reference. Communicative intentions are relegated to the realm of 'pragmatics'. Yet, what if the essence of grammar is to enable forms of deixis that do not exist without it? Could the theory of grammar be reorganised around the fundamental insight that grammar is a device of deixis? Deixis, as Tomasello (2008: 272) argues, always departs from the 'me and you in the here and now – that is, the current joint attentional frame, common ground, Bühler's (1934/1990) deictic center – to ground his acts of reference in what they both perceive or know together'. Pointing, without the help of grammar, is confined to that. Yet, the very point of language is to talk about the news: what is not present in the current joint attentional frame or common ground, and what may not be present in time and space at all. If this is right, the organisation of grammar should reflect this very fact, and as we shall see now, it clearly does.

Take an arbitrary content word from the lexicon, say *milk*. This word, by itself, cannot by itself single out any particular referent – it won't support referential acts to particular kinds of milk or bottles of milk. At best, it sets up a lexical descriptive content that can eventually anchor an act of reference to such things as *skimmed milk*, *milk* (in general, as a kind), *that milk*, *the milk we had together*, or *all of these three bottles of milk*. What we see, then, is that as grammar becomes more complex and compositional, reference becomes more specific. Grammar gives search directions to a hearer for isolating an object of reference and locating it. Every single phrase in grammar that we construct in grammar does precisely that. Starting from a noun such as *milk*, we can expand it to the left, and with each addition, reference changes and becomes systematically more specific:

5a. milk
5b. bottles of milk

5c. three bottles of milk
5d. these three bottles of milk
5e. all (of) these three bottles of milk

Thus, (5a) on its own could only be referred to milk as a natural kind; (5b) could only be used to unspecific sets of individual bottles of milk; (5c) to unspecific sets consisting of three such bottles each; (5d) to a specific such set; (5e) to the sum total of these exact three bottles of milk. At this point the buck stops. Much more a complex Noun Phrase in grammar cannot do; its deictic potential is exhausted. Note that it doesn't even manage to localise any bottles in time; it only localises them in space. Temporal information needs to come from accompanying non-linguistic information, such as an act of pointing, which will add the information: these three bottles *here and now*. Note, crucially, that none of the expansions of phrasal complexity in (5) in any way change the *lexical semantic* content (or 'substantive content') involved; it is always milk we are talking about, no matter how far we expand to the left. Ipso facto, at least in this nominal domain, grammar is a device, not for creating new semantic content, but for deixis; it is a way of *referring* to it. Grammar is thus a device of deixis.

The exact same thing, however – the completion of an act of deixis for a hearer who has to figure out what the speaker is thinking about and referring to – happens in the verbal phrase. Thus, take an arbitrary content verb that is a verb, such as *drink*. Again, we can gradually make the verb more complex, for example by adding inflectional morphology encoding Aspect, as in *drink-ing*, by creating a passive such as *drunk*, or by adding a direct object, as in *drink the milk*, or by creating a more complex and aspectually different VP, as in *drink the milk empty*, which adds a telicity effect. In short, a VP provides grammatical means for encoding an event and locating it in time from the deictic perspective of the speaker; by using such a verbal form, he indicates an event as ongoing or as completed, or as to be completed, in relation to the (temporal) point of speech. So let us call this localisation in time.

None of this, on the other hand, even touches upon what is the most intricate achievement of language: the making of a claim of truth, which depends on configuring a full proposition and assigning it a truth value, which is again a deictic decision; it is a way of presenting it to the hearer as factual, not merely for the speaker and the hearer, but for any third party as well. Neither an NP nor a VP can do this. It is the job of a sentence. Thus take a sentence such as (6):

6. John drank the milk empty

This, unlike any NP or VP, is evaluable for truth and falsehood through the hearer. Now two objects of nominal reference are localised (a particular person and a particular glass or bottle of milk), an event of drinking it to the end is configured, the two are composed in such a way that the two objects figure as participants in this event, and this whole, fully configured event is now predicated of a discourse topic, namely John (it might also have been predicated of the milk, as in a passive). To bring out this topicalisation and the predication involved, we might rephrase (6) as *John is such that [he drank the milk empty]*, where the bracketed constituent is the event, and 'John' is the topic. Again, nothing changes in the lexical content of this event once it becomes part of a sentence. Mode of reference changes, not semantic content.

The change in question can be characterised as a localisation of a proposition in discourse, just as an NP represents localisation in space and the VP localisation in time. Around these exact three dimensions – and apparently no more – language is organised. This, plainly, reflects the nature of the physical world; it so happens that its physical character is such that nothing can be (solely) located in space. Time is needed. And since this is a social world, discourse is needed as well. This explains why the basic units of organisation in human grammar are NPs, VPs and sentences. It also explains why there is nothing more complex in grammar than a sentence. For there is no further major dimension that acts of deixis need to, or could, take into account. So with the sentence, grammar stops. If structure-building goes on, the same maximal form of complexity merely repeats, as when a sentence embeds a sentence embeds a sentence (with the connector 'that' mediating the embeddings in question):

7. [$_S$ Bill believes [$_S$ that Tom hoped [$_S$ that John drank the milk empty]]]

Intriguingly, the exact three units of grammatical organisation, which we have arrived at without any particular reference to grammatical theory, considering pragmatics alone, are also the ones that current Minimalist grammar characterises as the units of grammar: they precisely are what are currently called 'phases'. These are defined as the units of compositional interpretation, though it now seems more appropriate to characterise them as the units of deixis: each of them have a distinctive significance in terms of the integration of lexical semantic information at the syntax-discourse interface. On Chomsky's view, phases are '(close to) functionally headed XPs' (Chomsky, 2001: 14), where XP is a root- or lexical phrase, and functional heads are the ones I have depicted as left expansions above (cf. (5)). No such integration is known to occur in non-linguistic beings. So it is inaccurate to say that UG depicts grammar as 'arbitrary' or 'contentless' and 'algebraic'; it is the enabling factor in the forms of symbolic representation and deictic reference that marks out human communication from all others.

We are now ready to understand the origin of recursion in grammar as well, which in the framework of Hauser et al. (2002) as well as in Minimalism generally is effectively a primitive with no explanation at all (see Arsenijevic and Hinzen, 2012, for a systematic development of this point). Paradigmatic recursions in grammar, as illustrated in (7), are *directional*. Typically, the more complex form of deictic reference embeds the equally or less complex: sentences embed sentences or VPs, VPs embed VPs or NPs. Why is this? Clearly because the semantics of actions in time presupposes that of objects/actors in space, and the semantics of discourse presupposes *both* orientation in space and time (see Casasanto et al., 2010, for some evidence that there is an asymmetry between space and time in the brain, with spatial representation presupposed in temporal representation). In short, social discourse presupposes the physical conditions in which it takes place, and the architecture of the human sentence reflects this: spatial material is most deeply embedded in the clause (internal to the VP), and the VP is embedded deeper than the zone in the clause where discourse-related information is encoded. In this sense, the paradigmatic forms of recursion and constituent hierarchies are a cognitive reflection of the basic design of the natural world and its inherent physical and social dimensions.

Fitting these dimensions together recursively in a single representation requires the iterated and directional use of the same phasal template three times: at the nominal, the verbal and the clausal level, thus crossing two phasal boundaries to reach a third and final one. Once that last level of complexity is reached, the derivation can only cycle, reiterating the same basic structure, as in (7). Recursion is thus not arbitrary, but semantically controlled, at every step. The fundament of the system is the phase, as a unit of deictic significance. Recursion is derived from the limited and inherently semantic ways in which such units need to be fitted into an overall hierarchy, and the new levels of emerging executive control that can hold such hierarchies together.

12.6 Conclusions

Looked at from the viewpoint of the task of implementing novel forms of deixis that are based on the deictic centre of the communicative situation itself as it takes place in time and space, while at the same time allowing reference to be *displaced* from it, human grammar transpires to have an extremely simple design. Independently given as the units of grammatical organisation, the 'phases' of Minimalist theorising now acquire a novel rationale: a rationale deriving from the syntax-discourse interface, and in this sense, pragmatics. All that had to evolve is a fundamentally simple structure-building operation that creates hierarchies, and a structural template, the phase, which will look internally different depending on what lexical content we begin it with. With the intrusion of such forms of grammatical organisation into the thought system of an early hominid (say, *Homo heidelbergensis* 200 to 300kya), and the functional potential of their associated forms of deixis, we are getting a culture-ready brain. Equipped with a new device of reference that expands the horizon away from the Here and Now (while always displacing reference from these as points of departure), early humans came to have a shared past and future.

There appears to be little reason to believe that a brain deprived of such forms of structural organisation would be capable of developing them merely from the pragmatics of a prelinguistic communicative situation as the contemporary critics of UG contend. Human cognition had to evolve predication and propositionality, which are not based on associative cognition, and which create a poverty of the stimulus problem at a massive scale. Yet, at the same time, these elements have a direct rationale in the pragmatics of the communicative situation and are likely to have changed it forever, giving grammar a functional and communicative rationale, in a context where more intricate forms of mind-reading, as per Tomasello's account, may have evolved independently (though the degree to which this is conceivable in the absence of grammar and the specific recursions it creates, remains, as I have argued, unclear). Grammar is nonetheless *not* a 'device for communication', for this misses the essential point: that it is a device for the expression and codification of thought, which itself *changes* the character of communication in essential ways.

Grammar on the present view is essentially synonymous with the emergence of a novel form of thought through the development of a code that we were lucky to be able to send off adequately enough through a sensory-motor channel, using a probably already configured vocal tract. Communication *as such* does not provide a rationale for the thought system entering into such novel linguistic communicative acts. While these conclusions support one of the oldest contentions of Chomskyan generative grammar – that language is not 'for communication' – this traditional contention nonetheless requires a qualification, if my story above is right, for what matters is not a new form of thought as such, either, with a new form of semantics attached to it. As I have argued, substantive semantic content is all there in the lexicon already. What grammar adds is forms of deixis, not of semantic content. And these forms are, inherently, related to acts of reference that take place publicly in the communicative situation in linguistically specific ways. In terms of semantic content and logic, a use of the name 'Pegasus' in discourse may not differ from an existential quantification; deictically, it does. Grammar, as an internal code, is in *this* sense not 'autonomous' from its externalisation or expression in a communicative setting: *both* interfaces are needed, and without its use in discourse, mediated by public phonetic labels attached to similar lexical concepts (see Clark, this volume), the design of grammar cannot be understood.

Notes

1. The Chimpanzee Sequencing and Analysis Consortium 2005 finds about 1.23 per cent of single nucleotide substitutions, so of single letters being different, and about 1.5 per cent difference due to pieces of DNA that have become inserted in a spot in either chimp or human, or because of pieces of DNA that have been deleted in one or the other. The meaningfulness of these figures, though, both then and now, remains obscure. What seems certain is that a genetic and/or developmental change was needed in the last transition to humanity.

2. As Tattersall (2004) points out, the activity of language 'is almost *synonymous* with the symbolic reasoning that marks us off from even our closest relatives in nature' (my emphasis).
3. In particular, as Tattersall (2004) points out, 'Any novelty has to arise spontaneously as an *exaptation*, a structure existing independently of any new function for which it might later be co-opted'. The case that function presupposes form is laid out in Hinzen (2006).
4. The claim just made requires one qualification. Clearly, a sentence can embed an NP, and there can be a recursion of NPs, as in (i):
   i) Bill holds [the belief that Tom has [the knowledge that John drank the milk empty]]
   So sentential recursion can be mediated by non-sentential phasal boundaries. But this doesn't change the essential picture: there could still be no such thing as a sentence with a truth value that is an expansion of an NP (or whose highest syntactic node was categorized as NP) (see Arsenijevic and Hinzen, 2012, for other qualifications and more details).

References

Arsenijevic, B. and Hinzen, W. (2012). On the absence of X-within-X recursion in the design of human grammar. *Linguistic Inquiry* 43:3, to appear summer 2012.

Bühler, K. 1934/1990. *Theory of language: the representational function of language.* Trans. D. F. Goodwin. Amsterdam: John Benjamins.

Carstairs-McCarthy, A. (1999). *The origins of complex language.* Oxford: Oxford University Press.

Casasanto, D., Fotakopoulou, O. and Boroditsky, L. (2010). Space and time in the child's mind: evidence for a cross-dimensional asymmetry. *Cognitive Science* 34: 387–405.

Chomsky, N. (1957). *Syntactic structures.* The Hague: Mouton.

Chomsky, N. (2005). Three factors in language design. *Linguistic Inquiry* 36(1): 1–22.

Chomsky, N. (2008). Approaching UG from below. In U. Sauerland and H.-M. Gärtner, (eds), *Interfaces + Recursion = Language?*, Berlin and New York: de Gruyter, pp. 1–29.

Chomsky, N. (2010). Poverty of stimulus: unfinished business. Lecture presented in the Lecture Series *'Sprache und Gehirn – Zur Sprachfähigkeit des Menschen'*, University of Mainz, Summer 2010.

Christiansen, M. and Chater, N. (2008). Language as shaped by the brain, *Behavioral and Brain Sciences* 31: 489–558.

Christiansen, M. and N. Chater (2009). Language acquisition meets language evolution, *Cognitive Science*: 1–27.

Clark, A. (2010). How to qualify for a cognitive upgrade: executive control, glass ceilings, and the limits of simian success. This volume.

Crow, T. (2009). A theory of the origin of cerebral asymmetry: epigenetic variation superimposed on a fixed right-shift. *Laterality: Asymmetries of Body, Brain and Cognition*, 15(3): 289–303.

Diamond J. (1992). *The third chimpanzee: the evolution and future of the human animal.* New York: HarperCollins.

Evans, N. and Levinson, J. (2009). The myth of language universals: language diversity and its importance for cognitive science. *Behavioral and Brain Sciences* 32: 429–92.

Fitch, W. T. (2010). *Language evolution.* Cambridge: Cambridge University Press.

Fodor, J. A. and Piattelli-Palmarini, M. (2010). *What Darwin got wrong*. London: Profile Books Ltd.
Gowlett, J. A. (1992). Tools: The paleolithic record. In S. Jones, R. D. Martin and D. R. Pilbeam (eds), *Cambridge encyclopedia of human evolution*. Cambridge: Cambridge University Press, pp. 350–60.
Hauser, M. D. (1996). *The evolution of communication*. Cambridge, MA: MIT Press.
Hauser, M. D., Chomsky, N. and Fitch, W. T. (2002). The faculty of language: what is it, who has it, and how did it evolve? *Science* 298, 22 November 2002: 1569–79.
Hinzen, W. (2006). *Mind design and minimal syntax*. Oxford: Oxford University Press.
Hinzen, W. (2010). Emergence of a systemic semantics through minimal and underspecified codes. In C. Boeckx and A.-M. di Sciullo (eds), *The biolinguistic enterprise*. Oxford: Oxford University Press, pp. 417–39.
Krause, J., et al. (2007). The derived FOXP2 variant of modern humans was shared with Neandertals. *Current Biology* 17: 1908–12.
Lenneberg, E. (1967). *Biological foundations of language*. New York: John Wiley and Sons, Inc.
McBrearty, S. and Brooks, A. S. (2000). The revolution that wasn't: a new interpretation of the origin of the origin of modern human behavior. *Journal of Human Evolution* 39: 453–563.
McGonigle, B. O. and Chalmers, M. (2007). Ordering and executive functioning as a window on the evolution and development of cognitive systems. *International Journal of Comparative Psychology* 19(2): 241–67.
McGonigle, B. O., Chalmers, M. and Dickinson, A. (2003). Concurrent disjoint and reciprocal classification by *Cebus Apella* in serial ordering tasks: evidence for hierarchical organization. *Animal Cognition* 6: 185–97.
Mellars, P. and Stringer, C. (eds) (1989). *The human revolution: modelling the early human mind*. Edinburgh: Edinburgh University Press.
Mukherji, N. (2009). *The primacy of grammar*. Cambridge, MA: MIT Press.
Penn, D. C., Holyoak, K. J., Povinelli, D. J. (2008). Darwin's mistake: explaining the discontinuity between human and nonhuman minds. *Behavioral and Brain Sciences* 31(2): 109–30.
Pinker, S. and Bloom, P. (1990). Natural language and natural selection. *The Behavioral and Brain Sciences* 13: 704–84.
Pinker, S. and Jackendoff, R. (2005). The faculty of language: what's special about it?, *Cognition* 95: 201–36.
Quine, W. V. O. (1953). *From a logical point of view*. New York: Harper and Row.
Roeper, T. (2009). *The prism of grammar*. Cambridge, MA: MIT Press.
The Chimpanzee Sequencing and Analysis Consortium 2005. Initial sequence of the chimpanzee genome and comparison with the human genome. *Nature* 437(1), September 2005.
Tattersall, I. (2004). What happened in the origin of human consciousness? *The Anatomical Record* (Part B: New Anat.) 276B: 19–26.
Terrace, H. (2005). Metacognition and the evolution of language. In H. Terrace and P. Metcalfe (eds), *The missing link in cognition*. Oxford: Oxford University Press, 84–115.
Tomasello, M. (2008). *The origins of human communication*. Cambridge, MA: MIT Press.

13
Language Evolution: Enlarging the Picture

Keith Stenning and Michiel van Lambalgen

13.1 Introduction

Contemporary biology understands macro-evolutionary steps as changes in developmental processes: the timing and placement of the expression of genes and their interactions through the environment. This is evo-devo – evolutionary developmental biology. Genes are organised in partially modular control cascades, so a change in a gene far up in a cascade can alter the timing and/or placement of whole complex modular processes, and thus the environment of operation of many other genes, and therefore large-scale coordinated phenotypic features. The classic example is the repeated process that puts pairs of legs on each of an insect's segments, which can, at a single mutation, go on for an extra segment, placing a pair of legs on the insect's head. A large number of genes coordinate to produce a pair of legs, but a single element can control when and where this genetic module is expressed.[1] An excellent introduction to evo-devo for the non-biologist is Carroll (2005).

Human beings differ from their ape ancestors by distinctive, biologically expensive changes in ontogenetic processes. Human infants have radically immature motor systems at birth, slowly develop large brains, have a prolonged period of dependent childhood after weaning, adolescence after sexual maturity, and a prolonged old age. Evo-devo suggests that these (and other) changes in ontogeny may be key to understanding how humans evolved from apes. But what we always first want to know is how human cognitive changes evolved – how did we get so brainy and start speaking? Who cares about this low-level morphology stuff? Well, evolutionary theory cares. These are expensive changes unlikely to be accidental by-products. And it is equally unlikely that such radical changes would be unrelated to the cognitive changes – that they would just happen to take place in the same species' emergence. What we need is a larger picture that can explain how the cognitive innovations are related to the morphological ones. A consequence of evo-devo's new technical understanding of an old perspective on evolution is that upheavals driven by one selective pressure are bound to change many other features of a phenotype, and the selective pressures on a given phenotypic character will often change during the process as the changes in developmental timing expose capacities to new environments. Modern function is only at best an oblique guide to past selected functions. For example, it may be that the only way to grow a large-brained bipedal primate is to extrapolate some ancestral developmental trajectories. Why should this have cognitive effects? And what secondary changes would be required? Language and cognition present formidable obstacles to the identification of insightful phenotypic characters and their changing functions. How should we describe the phenotype of the human capacity for language? And the relevant cognitive phenotypes of our ape ancestors? Many more options are available than with morphology, and our choices will have fundamental implications for what structures and functions are seen as novel, and for the functions and selection pressures considered.

Discussions of language evolution have focused on the structure of sentence codes, and particularly on recursive structure and its creative nature. Sentences are seen as paired with meanings; communication as the swapping of code items; and the novel creativity of recursive sentence structure as the human innovation. But we doubt that cognitive and language evolution can be understood at the level of the highly abstracted meanings of decontextualised sentence-types. It is well known from a half century of psycholinguistics that the meanings of abstracted sentence-types are not what is communicated in discourse (reviewed in Zwaan, 1996). And we doubt that the cognitive basis of recursion in language is novel with humans. Instead we seek to describe the phenotype of the human capacity for discourse – the use of language in context as opposed to the structure of the sets of sentence types (the syntax). We propose that language evolution was driven by the development of the capacity for discourse, and that the structure of sentence codes is a by-product of this evolution. Syntax evolved in the service of discourse. Discourse consists of connected utterances in context. Each utterance engages with and modifies the non-linguistic and linguistic context and creates a highly local but fully interpreted logical language through the operation of defeasible inference on prior knowledge and belief. The constructed context is often subtracted from as well as added to by new utterances. We hear that the cat sat on the mat, and a tabby came to join it. Immediately we know that phrases like 'the first cat' and 'the tabby' are not equivalent, whereas 'the second cat' and 'the tabby' are. These are equivalences only in the local constructed context – absolutely not in English as a universally interpreted language. English quite simply isn't an interpreted language at the level that concerns us here. In context, uttered sentences really do have concretised meanings which are the stuff of communication. The sentences may also be novel, but the driving novelty is the novel mapping of micro-language onto a local world, a mapping which is created anew by the discourse in context. We believe the current evidence about our ape ancestors suggests that this capacity for connected discourse is what is most novel in human language, though not nearly enough attention has been paid to 'discourse capacity precursors' in ape cognition. If connected discourse is what is most novel in language, then its most plausible cognitive precursor is the capacity for planning complex actions. Several authors, notably Greenfield (1991), have noted the close analogy between hierarchically organised motor planning and the planning of sentences. The utterance of sentences is, after all, a case of complex motor planning. Her example was the nesting of a set of cups of increasing size. Her observations of young children showed remarkable analogies between the ordering of difficulty and acquisition of alternative strategies for nesting cups, and for nesting phrases in sentences.

We build on this proposal by observing that planning is involved at several levels in discourse, and that there are several linguistic and cognitive structures that need to be distinguished. Past proposals following Greenfield have emphasised a homology between the structures of complex motor actions and the syntactic structure of sentence productions. We do not disagree that this homology exists, but we see it as a by-product of deeper homologies at semantic and pragmatic levels. Two sets of problems therefore need to be solved. The first is to specify what is novel and what conserved in the development of human planning capacities during language evolution. The second is to understand how these developments of the capacity for discourse are embedded in the larger picture of human phenotypic change. Specifically, we observe that human ontogeny has undergone some biologically radical changes of timing. Childhood and adolescence are essentially new stages interpolated between infancy and juvenility, and between the latter and adulthood, respectively. Evo-devo should ring bells when striking changes in ontogeny show up. The key to understanding selection pressures is understanding the physical and cognitive changes as an integrated process of changing timings in human ontogeny. The next section contrasts the standard code model of linguistics with a perspective on language as discourse – the connected use of language. Section 3 describes discourse as a case of planning and suggests that it is on planning capacities that evolution constructed language. Section 4 takes up the question of what ape planning capacities already existed, and Section 5 what had to be added to yield human discourse planning. Section 6 asks how these changes are related to other prominent innovations in human evolution through changes in the timing of ontogenetic processes.

13.2 The linguist's 'code model' versus the discourse model of communication

The following picture of the function of language is so pervasive that its inadequacies easily go unnoticed: a sentence of a language can be viewed as the code of a message that the speaker wants to transmit, and which must be decoded by the hearer to retrieve the message. This picture presupposes that each sentence carries sufficient information to allow decoding. In one sense this can immediately be seen to be false. The sentence 'The cat sat on the mat', far from being uniquely decodable, invites the questions: What cat? What mat? When? We can suspend these questions, but we still have to imagine the kinds of integrated arrangements of mammals and furniture which are intended by the speaker, and which are not given in the sentences. The root problem with the code model is that the fundamental unit of language is not the sentence, but discourse, connected sequences of sentences in evolving context. Discourses are the units that can be semantically interpreted; sentences can be interpreted in the context of discourse. In formal semantics, these interpretations are called 'discourse models' (Kamp and Reyle, 1993); in the psychological literature, they go by the name of 'situation models' (Zwaan, 1996, for a review). The psycholinguistic studies on situation models show that a remarkable range of causal information is represented in the models. Compare, for example, the following two discourses: John's face brightened when he heard the doorbell ring. A moment later his girlfriend came in. Or: John's face brightened when he heard the doorbell ring. An hour later his girlfriend came in.

Reading times for the second type of discourse, in which a time shift occurs, are significantly longer than those for the first type (Zwaan, 1996). This can be explained by assuming that subjects have a default expectation that the event described in the second sentence of the discourse follows close upon the heels of the event described by the first sentence. This default can be overridden, but at the cost of extra computation. Thus, causal information not verbally encoded plays a role in interpreting the discourse; this will be important when we connect discourse interpretation to planning. One reflection of the primacy of discourse is the role of verb tense in situating events with respect to each other, and the importance of context for determining that order. Every sentence contains a finite verb and so invokes the problem of embedding its interpretation in the discourse's context. It is not always the case that event order corresponds to sentence order, which might reduce the need for context. Here is another mini-discourse: Max fell. John pushed him. The order of the sentences is the reverse of the order of events, because the second sentence is read as explaining the event referred to by the first sentence: Max fell because John pushed him. Moreover, enlarging the discourse can change the order again: Max fell. John pushed him, or rather what was left of him, over the edge, which conjures up a scenario like the following: John does something particularly nasty and bloody to Max which makes him fall, near the edge of a precipice; he then shoves the body over the edge of the precipice. The function of verb tense is much more than locating an event with respect to now; the real difficulty is the incorporation of the event in the discourse model, and this can be done only in the context of the entire discourse and its non-linguistic context. Now we can connect discourse to motor planning. In many ways a discourse model is like a plan (van Lambalgen and Hamm, 2004): it contains actions (the events supplied by verbs) which are embedded in a web of temporal and causal relationships. Such a plan is incrementally (but defeasibly) constructed on the basis of verbal input; each tensed verb contributes a goal of the form: 1. locate the event described by the verb in past, present, or future; 2. mesh the event with events introduced previously.

The second condition requires embedding the new event in the causal web of the discourse model in such a way that its causal preconditions and consequences are faithfully represented. A slightly more formal analysis of the example 'Max fell. John pushed him' will make this clearer. The goals conveyed by this discourse are:

1. 'update discourse with past event $e1 = \text{fall}(m)$ and fit e1 in context'
2. 'update discourse with past event $e2 = \text{push}(j,m)$ and fit e2 in context'

The first part of these goals can be executed immediately, but it is situating e1 with respect to e2 that requires a planning computation. The planning system recruits causal knowledge as well as the principle that causes precede effects. Applied to the case at hand, the planning system scans declarative memory for causal connections between e1 and e2 and finds (roughly) 'e2 is a cause of e1'. This fixes the temporal order of e1 and e2, with e2 preceding e1. If the computations involved in discourse interpretation are formally identical to computation of plans, one may hazard the hypothesis that motor planning was exapted for the purpose of discourse interpretation, a theme that will be developed below.

13.3 The phenotype of 'human discourse capacity'

Discourse is verbal action. And motor action is action. If this is all that they have in common, then this is thin evidence for choosing the latter as the evolutionary origin of the former. There is an interesting line of argument that says that human language started as manually signed language (rather like deaf sign language), and that acoustic implementation of language in speech came later (Corbalis, 2002). This would be a strong connection between language and motor action. But it is not, we think, the most important immediate focus. Speech is also motor action of an exquisitely complex sort, so we can remain agnostic about whether the earlier surface realisation of language was manual/visual or laryngeal/acoustic for our purposes here. Deciding may be important at some stage in interpreting the evidence, but it is not the initial question. The relation we see as fundamental between discourse and motor action is at the level of function – specifically, reasoning function. Motor action is planned with respect to a goal and proceeds on the basis of the actor's best current guess about the context of her action.

We plan to reach out and pick up the coffee cup from the shelf on the basis of immediate visual and proprioceptive information about where we are and where the cup is, along with longer-term knowledge about its likely weight and so forth. It is only when our hand hits the glass that we realise that the cup is a reflection, and we'd better go to hospital instead. What we do not do is consider all logical possibilities, such as reflections, mirages, hallucinations and holograms, before acting. Although action can be dangerous, getting continually caught in infinite proofs is certain death. So we plan with respect to our best current guess, and explore possible alternative obstacles only if our best guess is that they are likely enough to be worth it. Now this is just like discourse. In discourse we plan to create a mutual interpretation of a small fragment of language mapped onto a small fragment of the world, on the basis of our current best guesses about the context we and our interlocutor are in, and the little bit of world we are trying to communicate about. Perhaps I see a child see a coffee cup behind a piece of glass I know the glass is hard to see, and I know that the child is looking for a cup. Being a helpful type, I might say to the child, 'Be careful, the cup is behind glass. If you reach for it you'll cut yourself', or something hopefully less prolix. I have not only a model of the physical context, but a model of the child's likely beliefs ('This is an unobstructed cup I see before me') and of what the child could understand, and much else besides. Planning my warning is just like planning my reaching, in that I have a best-guess model of a very limited part of our context and I plan with respect to that until evidence arises that my best guess is wrong. Clearly it is different from the motor planning example given here in being social. It requires me to coordinate my best-guess model of the physical situation with my best guess of my audience's best-guess model. But some non-language motor behaviour involves similar coordination of models – jointly cooking dinner, or, no doubt, our ancestors' mute teaching of the making of stone tools are examples. So what are the important similarities and differences?

13.3.1 What is ancestral about human planning?

Our first answer is recursion. Primate motor systems are organised hierarchically; not exclusively hierarchically, but hierarchically none the less. Macro movements such as reaching for an object divide into units at different granularities. There is the outward reach, the grasp and the inward retrieval. The grasp is composed of an orientation of the hand, an opening of the thumb and fingers and a closing of the thumb and fingers on the object. And so on right down to the retractions and relaxations of the many muscles involved. And right up to more strategic levels of action – shaking hands or pulling a pint of beer. The units are hierarchical not in the sense that they are discrete in time and properly nested and chained – they aren't. The opening of the grasp starts during the reaching process, and the units are generally smeared considerably in fine execution, just like actual linguistic units. But they are hierarchically organised and coherent to the controller. A reach-and-grasp action is a well-practised unit, but it still breaks down into its components when it has to be re-synthesised with other actions. We can start a reach with an object in our hand, which we have to jettison before grasping another. We can readjust our reach when the object grasped turns out to be much heavier than our best-guess model led us to expect. These are productive structured units of action which are recombined, and bring with them their own fine structures. Hierarchical units generate one kind of recursion – as such recursion is a very primitive principle of organisation of sensorimotor systems. So the question becomes, how were ape sensorimotor planning systems modified to yield human sensorimotor discourse planning systems?

Well, what do we know about the nature of the starting point? The surprise from primate training experiments is just how much can appear in the right supportive context. McGonigle and Chalmers (2006) trained monkeys to exhaustively touch each of a set of icons on a touch screen. In the hardest condition, the icons were randomly repositioned on the screen between touches, requiring working memory of what had already been touched within the current trial. In the most complex conditions, the icons were three shapes of three colours and three sizes – nine distinct icons. The monkeys showed very similar orderings and relative timings as humans show in working memory tasks which invite exploitation of hierarchy. The strategy 'all the green ones, then all the yellow ones,' is highly efficient for memory load. Just try generating random animal names against the clock and you will find hierarchical patterns of recall irrepressible. You will slow down between categories, and speed up within. So did the monkeys, and they are not even apes but monkeys diverged from humans 25 million years ago. They have to spend a lot of time and effort learning this task, but kids too spend a lot of time and effort learning such tasks, and a lot of the learning is the fine-tuning of smearing units for optimal performance efficiency. It is far from clear what natural task, if any, corresponds to this superficially artificial task – perhaps the planning of optimal foraging, or reasoning about pecking order? Our point here is that the working memory capacities required for planning already show recursive structure a very long time before human emergence. Note, we are not claiming that this monkey planning/working memory performance has everything needed for discourse. The motor demands are hugely reduced by the touch screen. There is a great uniformity of simple purpose from trial to trial. Most importantly the task is solitary, not social. It is far from clear to what degree the artificial task calls forth existing capacities, and to what degree the training in the scaffolding environment creates the capacities. It's not even all that clear how to distinguish these possibilities (see Clark's chapter in this book). Nearer to discourse, there has been much attention paid to apes learning artificial communication systems. What is still not apparent from all the ape 'language-learning' experiments is the degree to which apes can acquire the capacity for discourse. They certainly interpret each 'sentence' uttered by the experimenter onto a mutual context, but there seems to be little evidence of the utterance of one sentence creating a new context to which subsequent utterances of the discourse add. Multiple ape utterances are often repetitions, and seem to be more like the English tourist saying it louder, than incremental context construction. This might be partly due to the limitations of the languages taught. If each sentence is essentially an instruction to immediately perform an action, and the animal discharges each instruction before the next is given, there is little scope for connected discourse. In an evolutionary context, it is the 'two-sentence' stage of discourse development that is of paramount importance, rather than the two-word stage of the modern ontogeny of syntax. The development of the 'two-sentence discourse' stage in modern ontogeny has received too little attention. The most plausible evolutionary origin of syntax is the need to structure utterances to indicate the discourse relations between them – 'Max fell because John pushed him' – and it is this structuring which produces recursive sentence embedding: pragmatics first, and syntax second. We should be clear, our argument does not depend on chimps' lacking all discourse skills. The closer the chimp analogues of discourse can be shown to be to human discourse capacities, the happier we will be, because the question of what was missing will thereby be a smaller question (or at least have a shorter answer). Our point is exactly that chimp planning capacities were exapted to produce human discourse capacities, but we need to know what ape ancestor planning capacities were, and rather little attention has been focused on this question (as opposed to the recursion of the sentence code).

13.3.2 What is new about human planning?

So if this is what is old, what then is new in human discourse planning? One candidate is a certain sort of 'duality'. Narratives (and we think narratives are biologically primitive) are specifications of relations between planned actions and their effects – John's shove and Max's fall. This level answers the question 'What happened when?' But narratives themselves have to be planned by the speaker and the plans have to be recognised by the hearer, right down to the sentence level. We have to recognise that 'John pushed him' was planned by the speaker as an explanation, rather than a new following event. This second is the level of the question 'Why am I being told this in this way?' This duality of planning levels is at the heart of human discourse, and it is remarkable, from a processing point of view, because it demands the use of very similar reasoning machinery for simultaneously planning about speaker/listener and characters. Stenning and van Lambalgen (2008) show how easy it is to adapt the logic of discourse planning for theory-of-mind reasoning tasks. Of course children start with simple situations. They continue to have trouble with sequences of narration which reverse sequences of occurrence. They start with 'and then' as their only connective, and so on. And to begin with, adult interpretation makes up for the child's planning inadequacies. Much is often made of how human language can refer to absent objects, but one moral of the chimp-language learning experiments is that chimps in these experiments have little problem with 'talk about absent objects', under the right circumstances. The utterances studied in these experiments are generally dominated by instructions from the experimenter, which can be instructions, for example, to go and get objects which are absent, perhaps in the next room. This is absent reference in a general sense, but in another it is reference to a present requirement. 'Go and fetch a banana now' is not a reference to any banana – it is creation of, and reference to, a present aim.

Liszkowski et al. (2009) report that chimpanzees, unlike children, do not use non-linguistic gestures to refer to absent objects, as children do. This experiment differs from the language-learning experiments in several interesting respects. The apes (and children) are producing gestures rather than comprehending the experimenter's commands. They are introduced to a regularity that desirable items are placed on one platform and undesirable ones on another. In the critical condition, in the absence of a desirable item on the customary platform, it is gestured for by the children but not by the chimps. The experiment nicely analyses what is difficult about absent reference by comparing a control condition where the ape sees the experimenter hide the object beneath the platform on which it is customarily placed. Here the chimp gestures for the desirable object without difficulty, although one could say that it is perceptually absent (though prominent in memory). The problem with absence is that so much is absent. Attending to the right thing can be achieved by the children by a sort of metonymic gesture at the platform usually associated with the object. This metonymy is insufficient for the ape who requires the object to be 'loaded' into working memory by watching it being hidden. So the problem here is not the strictness of the notion of reference used. The successful cases of 'gestural reference' are still not definite reference but rather indefinite goals. The problem is how to manipulate attention to an absent object without the learned association to a symbol that the 'language-learning' experiments provide. What is striking about the child's performance is that it is prelinguistic (these children are 12 months old) and that it hangs on a metonymy established by only a small amount of experience of the placing of kinds of objects. As Tomasello and colleagues have done so much to demonstrate, the ontogeny of child pragmatics appears to recapitulate the only plausible evolution of language in preceding the development of lexicon and syntax.[2]

In fact, most of the research relevant to the development of discourse has been on the development of 'joint attention' in human infants and apes by Tomasello and his collaborators (Tomasello et al., 2005). Human infants are distinctive in developing routines for following what their carers focus on, and drawing their carers' attention towards their own focus of attention, with attendant vocalisations by both parties. All this often without any direct motivation beyond exploration and communication. Discourse is a kind of systematic manipulation and negotiation of attention. It may start out in ontogeny with control of joint attention to perceptually available people and objects, but it very soon encompasses the control of attention to absent objects, shared past events in memory, shared future plans, and to shared imaginary events outside of time. We see these processes as the origin of the linguistic control of deixis, and the propositionality and declarativeness of human language semantics (compare Hinzen's [Chapter 12] discussion of these properties). For these innovations, surely the most constitutive innovations of human language, duality of planning, is essential. To be able to draw attention to absent things with the flexibility of the expert adult, one has to be able to grasp what one's audience already knows, to be able to distinguish your target from all the other things they may confuse, to construct anew the parts of the context they need but do not have. Absence most simply may be physical distance, but if we are to talk about future plans and present desires, and recount past stories, then absence may be nonexistence or even impossibility. Talking about the nonexistent requires even more care with context construction. Even more discourse steps to be connected in the context's construction. Far more potential questions about 'Why am I being told this?' Imagine a creature with an infinite code of sentence/meaning pairs but no discourse capacities – just what would it be able to do with its shiny new grammar? The philosophical origins of our understanding of this reasoning to interpretations introduce a risk of making the processes sound too cogitative. Indeed, the very word 'planning' invokes just such deliberation before action. But these associations must be set aside. The philosophical theories should be taken as specifications of inferences that must somehow be achieved in language processing. How they are made is the psychological issue. The 'planning logic' which we invoke here is naturally thought of as a mathematical model of rapid, automatic and effortless mental processes which mobilise the relevant parts of huge databases of general knowledge, well below the level of cogitative planning. Children begin their discourse careers in highly supportive environments where some simple methods avoid the need for many inferences and shift others onto the burden of the adult listener. Skill in discourse production and comprehension is one area of language where there is very real learning and variation in its success. Which of the two levels of planning came first, and which is more novel? At least some plan recognition of third-party actions is within ape repertoires. The most obvious bet is that it is the level of planning discourse that is a novel redeployment of pre-existing planning capabilities. These abilities are part of what is referred to as 'theory of mind abilities'. One can accept the label for the abilities without taking the metaphor of 'theory-as-implementation' too far (Stenning and van Lambalgen, 2008), or initially taking a stance on whether the development of discourse was instrumental in developing theory-of-mind abilities, or vice versa. Or more likely a whole sequence of chickens and eggs. But what was required was not only the capacity to reason about the audience's knowledge and belief, but to do that while simultaneously reasoning about knowledge and motivations of the participants of the narrative composed. And, of course, the homologous questions in modern development are far from answered yet. For example, Breheny (2006) has asked how children with supposedly no theory-of-mind can speak languages which require a strongly intentional stance. Our bet would be that developing expertise in discourse is one of the activities that drives development of theory-of-mind abilities. On the whole, the study of apes reveals how much of human cognitive machinery was already there, and this is exactly what evo-devo would lead us to expect. The human dual-level planning of the social manipulation of attention is distinctive and is the essence of human discourse and communication. Hare and Tomasello (2004) argued that it may have been as simple an ingredient as contexts of social cooperation which was lacking, and proved the greatest barrier to apes developing these capacities. And one should never underestimate changes in the environment as sources of innovation. Our ancestors upped sticks and went forth, first from the jungle and then from Africa, thus changing their economics considerably. Evolution wasn't of cognition in a vacuum.

13.4 How are innovations in human discourse capacities related to other biological innovations?

If ape planning evolved into human language through a layering of planning about participants and planning about audience, why did that happen? We need to stand back and find a large picture which can integrate other prominent human innovations (enlarged brains, changed life stages, sexual dimorphism, bipedalism, hidden oestrous, enlarged social groups, division of labour, toolmaking, and so forth). We do not pretend to have a complete picture, and we are sure that many of the pieces we consider could be improved on by the experts – they are intended as examples to be replaced by better, or improved ones. One moral of evo-devo is that one should be wary of trying to prematurely construct linear cause-and-effect chains. Functions change and change other functions. There are likely to have been feedbacks and certain to have been non-causal correlations. We are far more confident that there have been interactions between the example factors we will discuss, than in the tentative directions of causality suggested.

13.4.1 Encephalisation

The brain seems somehow implicated in cognition, and the fossil record tells a story of relentless enlargement, continuing a long primate trend, though with a small twist at the end of the plot. From the earliest australopithecines (five million years ago) to Neanderthals, the human brain grew relatively larger, but right at the end of the process, with the emergence of modern Homo sapiens, the brain did a small shrink (Ruff et al., 1997). Just when art flowered and culture as we think of it got under way, our brains got slightly smaller. Does one need a little dumbness to cope with civilisation? Perhaps this shrinking is an artefact of reduced body mass, though the simple corrections do not seem to bear this out. So what exactly does a bigger brain do for us apart from use up enormous metabolic resources? Are big brains really necessary for language? Rather small brains of some kinds of modern human dwarfs can support the use of natural languages of full structural complexity. We conclude that structural complexity of code is not what drove brain enlargement, though this doesn't rule out that what language is used for may have been a driving force for brain enlargement at some stage or stages. And besides, brain enlargement started millions of years before language, and doesn't have to have one cause, or the same cause over time. Human brain enlargement might have been a side effect of changes in growth patterns driven by some other cause or effect, though brains' sheer metabolic expense makes this unlikely. It has been suggested that big human brains are like peacocks' tails, supported by sexual selection. We do not think this particular suggestion works (Stenning and van Lambalgen, 2008), but it serves as reminder that functions can be anything but obvious, and are functions of whole organisms in their social and natural environments. No better from a biological point of view is the pervasive idea that bigger brains bestow greater 'intelligence' and all folk obviously need that. Biological selection pressures need to be specified to a level at which they can be related to concrete behaviour-increasing fitness. It is far from obvious what our big brains were selected for, nor that the selection pressure was constant in kind throughout the last four to five million years. Dunbar (1993) has proposed that primate and human encephalisation was the result of selection pressure to live in larger social groups and has marshalled a good deal of evidence to support this view. Of course there must be something beyond simple size of group involved. Many animals live in much larger social groups than primitive humans did, with rather small brains. Dunbar's argument is that primates need larger brains to live in larger groups, because their social interactions require memory of individuals' history of behaviour, and the planning of individual and group action accordingly. He combines this with the proposal that language's originating function was to take over from the phatic communication of grooming – driven by its greater efficiency when large groups must be kept together. It is not so clear why anything like human language is required for this simple phatic purpose of 'clique maintenance'. Emotion-controlled vocalisations would on the face of it seem simpler and more effective. Though once language has the cognitive capacities of ideational communication, it thereby has many new phatic functions, for example, conveying histories of individual behaviour of relevance to group cohesion. Once more social reasoning is possible, allowing, for example, more elaborate deception, the maintenance of group cohesion will be more difficult. Social reasoning may be an arms race, hugely exacerbated by the advent of language. So perhaps bigger brains evolved for better social reasoning? The large human brain's function may not be obvious, but some of the problems it causes are clear, especially obstetric problems. Bipedality narrowed the human maternal pelvic outlet while human infants' head size increased, with biblical consequences. Several obstetric adaptations followed: broader female pelvises, infant rotations during delivery, and so on, and involved at least two distinct peak selection pressures at different eras, separated by as much as two million years (Bramble and Lieberman, 2004). There is a real possibility of a positive feedback loop here. A social arms race for larger brains leads to birth at increasing immaturity, and increases the degree of investment in our young, increasing the need for group breeding. Larger brains cause more obstetric problems which increase the social dependence of mothers and favours further increases in social reasoning capacities, and therefore bigger brains... The data suggests that the obstetric problems caused by bipedalism are still an active selection pressure today, still evidenced in modern women in the developing world. There is strong evidence that one selective solution to obstetric problems is increased maternal height. There is a correlation within individuals between maternal height and size of pelvic outlet strong enough to make height a clinically useful predictor of difficult births, and cultures with high obstetric stress (as measured by average number of births per mother) are the cultures with the tallest women. Increase of height then has its own implications for timings and/or rates of development – taller individuals mature slightly more slowly.

13.4.2 Toolmaking and its cognitive consequences

Planning is an abstract function that is involved in most domains of action, and it is this abstractness which we appeal to in finding a homology with discourse. But it also means there are many divergences between different kinds of planning. Toolmaking is a complex, planned motor activity. For a theory that complex motor planning was the precursor of language, one important question is, 'What relationship does toolmaking bear to the origin of language?' Was toolmaking an activity which expanded human motor-planning capacities which were later co-opted by other cognition and by language? Or was toolmaking a spin-off of some other expansion of human planning capacities which also fed into language evolution? Perhaps capacities for the division of labour were required for toolmaking, and those capacities were what fed into communication innovations that led to language? Metabolically expensive large brains were at least part of the pressure for a change of diet to meat, facilitated by tools. More feedbacks. The evolution of toolmaking changed the human brain. The making of stone tools was a hominid innovation; a million years of basic flint axe making, with very little further elaboration, preceded the sudden efflorescence of stone tool and weapon technology 40k years ago. It is usually assumed that this late sudden proliferation of tool designs, along with other cultural innovations such as graphic art and burial of the dead, was a result of the innovation of language and culture. What about the earlier long period of stereotyped toolmaking? Conservative it may have been, but this toolmaking required an unprecedented degree of manual skill, and perhaps length of apprenticeship, and probably division of labour. The later Acheulean bifacial tools required more strategic levels of planning, and have been shown to co-opt more frontal regions of the brain in modern practitioners (Stout et al., 2008). Did the brain innovations for toolmaking contribute to subsequent language evolution? The human brain innovates significantly in at least three areas with major involvement in planning: parietal, motor cortical and frontal. The most posterior parietal areas are intimately connected with the fine tactics of flint chipping and tool use (Glover, 2004); next further forward, the language and non-language motor cortical areas are contiguous, and involved in combining actions, for example, sentence-scale planning; and most anterior, the frontal cortex is the strategic planning area (what we perhaps first think of under the label 'planning'). Toolmaking calls forth an asymmetry of hand function between holding and shaping which has even been analogised to the linguistic differentiation between topic and comment (Krifka, 2007; Tallis, 2003). Right-handers hold the worked piece in their left hand and modify it with their right. The development of human patterns of cortical lateralisation is probably at least partly associated with toolmaking and use. Apes show some laterality, but evenly distributed left and right between individuals. Steele and Uomini (2005) marshal archaeological evidence that the human 90:10 distribution of handedness evolved early in stone toolmaking, rather than being coincident with the later complex tool technologies. Perhaps the apprenticeship required even for early toolmaking imposed the near uniform right-handedness displayed by humans (for modern left-handers it is school that still brings down society's controls). Toolmaking was presumably an adult activity, probably male. Males are more lateralised, though generally not as linguistically able. Lateralisation plays some not yet very well specified role in language. Fine motor control of a mid-line organ is clearly implicated. There is some evidence of situation models (not inherently sequential structures) being a right-hemisphere concern, while the sequential construction of language is a left-hemisphere specialisation. All told, it seems that those million years of crude Oldowan toolmaking could have been one strong influence producing the human asymmetrical brain which is plausibly required for the duality of human discourse planning, and the later complexification of tool technology may also have given a boost to strategic levels of planning (compare Hinzen).

13.4.3 Life-stage interpolations – childhood and adolescence

As mentioned in the introduction, the human life-cycle changed considerably between ourselves and apes. Helpless infants require group breeding where mother gets help. Adolescents require social control to delay breeding after puberty. These life-cycle changes are biologically radical, but what are they for and what do they result from? Why tolerate these huge costs and upheavals? If the main selection pressure on brain enlargement is need for social reasoning, one explanation for helpless infants is that the only way for a bipedal ape to grow a larger brain may be to lengthen the period of growth. The best models of the human changes in ontogenetic timing of brain growth involve hypermorphosis – the human brain goes on growing for longer than our ancestors', rather than growing faster or starting sooner. Such evo-devo explanations rely on some parameters of developmental processes being easily modified and others rather rigid. The general biological bet is that primate brain growth rates are rather hard to change and that other processes must fit around such prolongation. Another process that seems to be somewhat inflexible is gestation length: chimps' eight months compared to humans' nine. If the trajectory of human brain growth completion is evenly stretched from the chimp's (around seven years) to the human's (around 20), then this means the human brain has to be born relatively immature, with all the social and cognitive consequences.

One ingredient for explanation of the specific interpolation of childhood in human development is in terms of increased maternal fecundity. Lactation suppresses conception; earlier weaning hastens conception. Having a childhood stage which is weaned but still dependent frees mother for earlier conception. Why increased fecundity was a good thing is not clear. Fecundity is not always selected for, as some crude interpretations of natural selection might suggest. Recent thinking suggests that hominid geographical dispersion may have started a million years earlier than supposed (Dennell and Roebroeks, 2005), and fecundity might be related to colonisation. Or it might be that the extending lactation of helpless infants so reduced fecundity that childhood was required to maintain the population. It is intriguing that adolescence, the other innovation in human life stages, is a delay, operating in the opposite direction, reducing fecundity. The social institution obviously implicated in this change in modern humans is marriage. Human societies delay marriage beyond puberty, and try to limit fecundity to marriage. The social institutions and individual psychology can only be understood in their biological context. But that does not mean that the biology is the cause and the psychology and sociology the effects. The need for longer to learn, or social institutions to control fecundity, could just as well be among the causes of brain growth prolongation. The combined effect of interpolating infancy and adolescence would appear to be females who can reproduce faster but typically begin somewhat later in life. Possibly the multiple changes deliver greater control of varied levels of fecundity contributing to flexibility in multiple environments?

What do these life-cycle changes have to do with the development of discourse capacities? Our helpless infants cannot do much except through communication with others. For them agency is first indirect social action and only later direct physical action. The temporal and spatial contingencies of action through other agents are quite different from those of direct action on the physical world. Human brains at birth are, on some measures, precocious while on others they will not be mature for 20 years. Those brains have been dumped prematurely into the external world, exposing learning mechanisms attuned to the womb, to make what they can of the world. Prolonged helplessness of infants presents mothers with extreme mind reading/mind modification challenges and opportunities which place mothers and infants under novel selective pressures. Child-rearing is a cauldron of social reasoning and social reasoning is a prominent candidate for what drove human brain enlargement. Although conflict is an element of the mother–child relationship, subtly evidenced at all levels from gene expression, through immunology, to behaviour, this relationship also imposes unprecedented levels and duration of cooperation compared to that of our ancestors. This relationship is the origin of discourse as the mutual social manipulation of attention. To adapt an epigram, the proper study of infantkind is mother. For infantkind, planning to achieve anything is planning at two levels: planning for what is to be got to happen and planning for how mother is to be got to do it. Of course this makes it sound very cogitative, and at first mother is the one who does almost all the planning (the infant's frontal lobes barely even functioning until later), but it is from mother's successes and failures of interpretation that the child learns how to plan, at both levels, and at once.

13.5 Conclusions

These are just some of the diverse pieces of the human puzzle. Evo-devo suggests different ways of thinking than have been available before. The emergence of humans from apes clearly does involve significant changes in life cycle – the relative timing of ontogenetic processes. Evo-devo tells us that changes in the relative timing of ontogenetic processes is what generates major evolutionary changes. But it tells us more than just that. It tells us that if the timing of a major developmental process changes – slows, speeds up, lengthens, or truncates – this will change the environment of other developmental processes, and that adjustments may be necessary across a wide range in order to preserve viability. Remember that changes of timings do not have to be uniform. Some processes may start earlier, others later, some accelerate, others decelerate, and so on.

We warned you we had ingredients rather than a cake. Our point is to illustrate that wildly heterogeneous phenotypic features can interact under main selection pressures, obstetrics and brain size altering the environment of learning mechanisms and social dependencies, for example. Feedbacks abound. We have to tame this hodgepodge. A remarkable range of evidence bears on this remarkable range of changes. Wading into this environment with the prejudice that any change from our ancestors is an adaptation whose selected function is transparent is not calculated to enlighten. But it is certainly an environment in which an animal strongly adapted for defeasible action planning is constrained, in infancy, to apply its mental machinery to social manipulation, for want of any other way to act, or, perhaps facetiously, for want of anything else to do. There is no alternative to developing a synoptic picture of how our species changed, even if we need to avoid the idea of a 'grand plan'. There is no shortage of relations between cognitive and morphological changes in developmental timings. A daunting variety of expertise is required, but then a correspondingly large amount of diverse evidence can be brought to bear. Our focus must be on how discourse functioned at different times in different contexts, and how it changed our minds.

13.6 Epilogue

Evolutionarily interesting variation is likely to be variation in ontogenetic processes, along with resulting differences in adults. It is also variation that has significant impact on fitness. We personally have been intrigued how many psychiatric syndromes have an element of changed ontogenetic timing. Two highly prevalent examples are autism, which shows unusual patterns of acceleration and deceleration of perinatal brain growth (Redcay and Courchesne, 2005, Chapter 9; Stenning and van Lambalgen, 2008) and ADHD, which shows a strong slowing of cortical development, particularly of the frontal lobe (Shaw et al., 2007). ADHD is interesting as the most prevalent developmental psychiatric syndrome, and as the one that shows the highest prevalence of 'remission'. The majority of those diagnosed grow out of it. There is evidence that these syndromes are extremes of normal continua, and a good deal is known about them as complex psychological phenotypes. We believe there are rich rewards to be had in conceiving of these phenotypes as beneficial in some 'doses' or environments, and in pursuing their careful functional specification (Pijnacker et al., 2009; van Lambalgen et al., to come). These first forays into logical descriptions of styles of discourse processing are at least suggestive that well-known features of these children's thinking are reflected in their discourse styles. Here might even be human cognitive evolution caught in the act.

Notes

1. This chapter draws heavily on a more extended and more fully referenced treatment in our book (Stenning and van Lambalgen 2008).
2. In terms of Andy Clark's discussion of labelling (Chapter 10), for the 12-month-old child, the non-linguistic control of shared attention is sufficient without the scaffolding of labels: a context in which this is true for the chimp has not yet been found.

References

Bramble, D. and Lieberman, D. (2004). Endurance running and the evolution of homo. *Nature*, 432: 345–52.
Breheny, R. (2006). Communication and folk psychology. *Mind and Language*, 21(1): 74–107.
Carroll, S. B. (2005). *Endless forms most beautiful: the new science of evo devo and the making of the animal kingdom*. London: Norton.
Corbalis, M. (2002). *From hand to mouth*. Princeton: Princeton University Press.
Dennell, R. and Roebroeks, W. (2005). An Asian perspective on early human dispersal from Africa. *Nature*, 438: 1099–1104. doi:10.1038/nature04259.
Dunbar, R. I. M. (1993). Coevolution of neocortical size, group size, and language in humans. *Behavioral and Brain Sciences*, 16(4): 681–94.
Glover, S. (2004). Separate visual representations in the planning and control of action. *Behavioral and Brain Sciences*, 27: 3–78.
Greenfield, P. (1991). Language, tools and the brain: the ontogeny and phylogeny of hierarchically organized sequential behavior. *Behavioral and Brain Sciences*, 14: 531–95.
Hare, B. and Tomasello, M. (2004). Chimpanzees are more skilful in competitive than in cooperative cognitive tasks. *Animal Behaviour*, 68(3): 571–81.
Kamp, H. and Reyle, U. (1993). From discourse to logic: introduction to model-theoretic semantics of natural language, formal logic and discourse representation theory. Volume 42 of *Studies in Linguistics and Philosophy*. Dordrecht, NE: Kluwer Academic Publishers.
Krifka, M. (2007). Functional similarities between bimanual coordination and topic/comment structure. *ISIS: Working Papers of the SFB 632*, 8: 61–96.
Liszkowski, U., Schafer, M., Carpenter, M. and Tomasello, M. (2009). Prelinguistic infants, but not chimpanzees, communicate about absent entities. *Psychological Science*, 20(5): 654–60.
McGonigle, B. and Chalmers, M. (2006). Ordering and executive functioning as a window on the evolution of development of cognitive systems. *International Journal of Comparative Psychology*, 19: 241–67.
Pijnacker, J., Geurts, B., van Lambalgen, M., Buitelaar, J., Kan, C. and Hagoort, P. (2009). Conditional reasoning in high-functioning adults with autism. *Neuropsychologia*, 47(3): 644–651.
Redcay, E. and Courchesne, E. (2005). When is the brain enlarged in autism? A meta-analysis of all brain size reports. *Biological Psychiatry*, 58(1): 1–9.
Ruff, C. B., Trinkhaus, E. and Holliday, T. W. (1997). Body mass and encephalization in pleistocene *homo*. *Nature*, 387: 173–6.
Shaw, P. et al. (2007). Attention-deficit/hyperactivity disorder is characterized by a delay in cortical maturation. *Proceedings of the National Academy of Sciences of the US*, 104: 19649–54. NIMH.

Steele, J. and Uomini, N. (2005). Humans, tools and handedness. In V. Roux and B. Bril, (eds), *Stone knapping: the necessary conditions for a uniquely hominin behaviour*. Cambridge: McDonald Institute for Archaeological Research, pp. 217–39.

Stenning, K. and van Lambalgen, M. (2008). *Human reasoning and cognitive science*. Cambridge, MA: MIT University Press.

Stout, D., Toth, N., Schick, K. D. and Chaminade, T. (2008). Neural correlates of early stone age tool-making: technology, language and cognition in human evolution. *Philosophical Transactions of the Royal Society of London B*, 363: 1939–49.

Tallis, R. (2003). *The hand: a philosophical enquiry into human being*. Edinburgh: Edinburgh University Press.

Tomasello, M., Carpenter, M., Call, J., Behne, T. and Moll, H. (2005). Understanding and sharing intentions: the origins of cultural cognition. *Behavioral and Brain Sciences*, 28: 675–735.

Van Lambalgen, M. and Hamm, F. (2004). *The proper treatment of events*. Oxford and Boston: Blackwell.

van Lambalgen, M., van Kruistum, C. and Parigger, M. (To appear). Discourse processing in attention-deficit hyperactivity disorder (adhd). *Journal of Logic, Language and Information*.

Zwaan, R. (1996). Processing narrative shifts. *Journal of Experimental Psychology: Learning, Memory and Cognition*, 22(5): 1196–1207.

Zwaan, R. and Radvansky, G. (1998). Situation models in language comprehension. *Psychological Bulletin*, 123(2): 162–85.

Brendan McGonigle

Epilogue: Brendan McGonigle

Brendan O. McGonigle received a BA (in 1961) and a PhD (in 1964) from Queen's University, Belfast, Northern Ireland. He arrived at Durham University as a postdoc in 1965, where he became friends with David McFarland. By 1966 they both found themselves at the Institute of Experimental Psychology, Oxford, where they worked together on some problems of common interest.

One of their experiments involved running rats in discrimination learning trials in the Psychology Annexe. This was an old Victorian terraced building, and they had a room on the third floor. The experiments involved daily trials, and they took it in turns to run the trials, each working on alternate days for some weeks. One problem was that the place was overrun with mice. The rats would arrive in the reward box, having made the correct choice, only to be confronted with a mouse stealing the food.

Their experiment was in trouble, so they got together at the scene of the crimes. There the two noticed that the mice habitually ran along the bench and then jumped down onto the floor (they could tell by the mice's tiny footprints). Brendan and David devised a plan: they placed a bucket of water on the spot where the mice jumped onto the floor. Sure enough, the mice jumped into the bucket. Each day, whoever was on duty would empty the bucket of water and mice out of the window before running the experiment.

Unfortunately, they had not realised that the room below had a balcony onto which all the bodies were being deposited. Another scientist inhabited this room and objected, reasonably enough. But they said that they could not change their routine in the middle of the experiment. So the scientist went to the head of department, and they were carpeted.

'You have got to stop,' said the department head.

'No,' said Brendan, 'We can't stop, because we are in the middle of an experiment.'

"I order you to stop.'

'If you do that,' said Brendan, 'we will report the mouse infestation to the health and safety authorities.'

Nothing could come between Brendan and his experiment.

Brendan next went to the United States as an assistant professor and NIH research associate at the Animal Behaviour Lab of Pennsylvania State University. David McFarland visited him there in 1968. Things were not good in the department, due to constant battles between one Professor Hoffman, and one Professor Warren. Brendan needed a break, so Brendan and David set off in the former's car to see the sights. One day they were in Arlington Cemetery and David said to Brendan, 'Look there,' and pointed to two gravestones, side by side, one clearly marked 'HOFFMAN', the other 'WARREN'. Quick as a flash Brendan got out his camera and took a photo. When he eventually got back to work he posted the photo on the department notice board. That ended the departmental bickering.

Following his stint at Pennsylvania State, Brendan became a lecturer in psychology at the University of Edinburgh, in 1969. His primary interest at this point was to account for certain aspects of animal learning in non-associative terms. His first graduate student was Barry Jones. Brendan's mantra at the time was 'If you can't easily explain it to the local bar staff, then you don't properly understand it.' As a consequence, Calvin the barman in The World's End Pub on the Royal Mile soon became an expert on the problems of seeing and learning in the monkey and rat. Towards the end of Barry's studentship, the Science Research Council (a funding organisation) arranged a site visit. The leader of the visit was the late Stuart Sutherland from Sussex University. During the site visit he said somewhat grumpily that he was 'disagreeably surprised' at Barry's turning down his offer of a studentship (some years earlier) but was very pleased to see that he had struck gold in Brendan's lab. The venue of the site visit interview was changed at the last minute to Barry's room which, unfortunately, was used by the lab's personnel to brew beer using kit from the lab. This did not go unnoticed by Professor Sutherland. Fortunately, he was 'agreeably surprised' at the quality of the bitter but thought the stout a bit sweet. Totally true to form, Brendan said he was right about the bitter but quite wrong about the stout and explained why. Unfortunately Barry was unable to hear the explanation as the conversing pair moved out of earshot.

Another of Brendan's early graduate students was Margaret Chalmers. In 1977 she became his postdoc, and they started a life-long scientific collaboration. They were married in 1996.

George Luger came to Edinburgh as a postdoctoral research fellow in the Department of Artificial Intelligence in 1974. He came to work with the students and faculty of the Department of Psychology; here he met Brendan and started collaborative research which was to last for three decades. After accepting a professorial position (1979) in Computer Science at the University of New Mexico (which eventually included appointments to the UNM Psychology and Linguistics Departments) George continued his collaborative efforts with the University of Edinburgh. The collaborative research was supported by MRC, NATO, the Royal Society and NSF grants over the next thirty years. The collaboration included parts of three sabbatical years George spent in Edinburgh and several visits by Brendan to his research group in the US. When George worked in Brendan's labs in the Appleton Tower, their discussions continued in many of the more interesting restaurants and pubs of Edinburgh, and included frequent delightful dinners at the McGonigle house in Denholm. George Luger's favourite Brendan quote is 'The problem with ordering a half-pint near the end of the evening is that you just might toss it over your shoulder.'

Alan Bundy first met Brendan at an epistemics conference in Edinburgh in the early 1980s. Brendan was presenting work on transitive reasoning in monkeys. Alan and Brendan had a long discussion about whether the situation could be modelled with logic. Brendan suggested they recruit a PhD student to explore the possibilities. In 1983, they managed to persuade Mitch Harris to do this study. What followed were many happy hours of arguing and modelling with Mitch, Brendan and Maggie, culminating in a model that was a very good fit to the experimental data. When Mitch later moved to the robot group in the AI department, Brendan got involved in this research too, which proved to be a very productive collaboration.

Keith Stenning arrived in Edinburgh in 1983 to take up a Lecturership in Psychology and Cogntiive Science. Keith had taken tutorials from Brendan in Oxford in 1967. As a joint philosophy/psychology student, fired with enthusiasm for Chomsky's then current onslaught on associationism, he was guaranteed to be a hard market for anything about animal learning. Once he had calmed down enough to appreciate their shared assumptions, the tutorials turned out a feast. Brendan was a wonderful teacher of the complexities of a position. The tutee exited, somewhat toned down, but with a lot more ammunition, and much the richer for the experience.

In 1983, Edinburgh psychology department was still stalked by more than one un-dead professor, locked in combat, and there didn't seem to be the right cemetery to repeat Brendan's Arlington trick. Regular sandwich lunch with Brendan in the monkey-lab office was a huge intellectual stimulus. Translating Brendan's long developed personal language into any of the several languages of Cognitive Science could be a challenge, but one infinitely worthwhile. All sorts of personal asides emerged. It turned out, for example, that Brendan's career as a student of animal behaviour started at a tender age, with rescue of a levret, which must truly be one of the shyest animal subjects, but one which reveals much about the man.

In 1989, having been awarded a scholarship from the Italian Ministry of University and Scientific and Technological Research for postgraduate studies abroad, Carlo De Lillo spent the summer in Edinburgh to prepare for a master's programme at a prestigious British university. However, he met

Brendan during a fortuitous visit to his laboratory, and it did not take long for Brendan to convince Carlo to join his lab and do a PhD instead of doing a masters at another university. As Carlo says, 'Brendan's original personality and enthusiasm proved immediately compelling. Brendan's reputation among senior colleagues in Rome and his fascinating *Nature* papers, which I subsequently read, convinced me that I made the right decision. In the four years that I spent at Brendan's lab, I experienced a formative mixture of dazzling thinking and passionate theorising about major theoretical issues in cognitive science, which often extended to gargantuan dinners and pub discussions. My current research is still profoundly rooted in what I learned there. To say that Brendan was a larger-than-life character would be a gross understatement. Indeed it would be impossible to do justice with words to Brendan's exceptional and unconventional character, as any person who has known him would know.'

In 1991 Marcos Rodriguez finished his PhD at the University of Wales, Aberystwyth and he and his supervisor, Mark Lee, were looking for funding to develop a self-organising neural model that was able to generate adaptive animal-like behaviour. Mark Lee had a long association with Chris Malcolm of the Robotics Lab in the AI department at Edinburgh and during a one of his visits had a memorable evening in a city pub with Chris Malcolm, Brendan and Tim Smithers. It soon became very clear that this group had a very different view of intelligence from the prevailing approach. The subsumption-based architecture of Rodney Brooks was then overturning the dominant symbolic paradigm but the Malcolm-McGonigle-Smithers team knew that insect models would not be sufficient to account for human cognition and were full of stimulating ideas and novel insights. Over many happy dinners, Brendan and Mark formulated a joint grant proposal, and during one of these dinners, Mark was alarmed when Brendan ordered 'a bottle of your best champagne' in celebration of the grant they were 'about to win'. This seemed like tempting fate far too much, but was typical of Brendan, and of course they did win the grant!

In 1992, Joanna Bryson was also pursuing a conversion MSc in Edinburgh's Department of Artificial Intelligence. Brendan was then collaborating with Alan Bundy in Mathematical Reasoning and with Ulrich Nehmzow in Robotics, and also giving occasional lectures in the Cognitive Science courses open to AI students. At some point, Joanna encountered Brendan giving a talk about monkey rationality, and was sufficiently intrigued that she arranged to meet him in his office in George Square. 'Honestly, I think part of what fascinated me was his perspective on intelligence, and the other part was that I found him slightly incomprehensible.' She left Edinburgh to pursue a PhD in the MIT AI Laboratory, but after a few years became disillusioned with the strong emphasis on engineering and weak emphasis on science of her group. She asked Brendan whether she could come for a 6- to 12-month research visit, but at the time Brendan had no available research assistant funding. Undaunted, Brendan applied for and obtained for Bryson a PhD studentship in Edinburgh Psychology, assuring her she could return to MIT whenever she wanted to. She stayed on longer than originally planned as a result of meeting her future husband as well as becoming engrossed in the fascinating research in Brendan's lab. In the end she discarded the last year of her PhD funding in order to ensure that MIT (which has a strict policy of not accepting students with prior PhDs) would let her finish the PhD there. Although Bryson spent her time in Brendan's lab working with his robot and extending his ideas on behaviour-based robot control, she was fascinated by the lives, minds and experimental outcomes of the monkeys that co-inhabited Brendan's lab with the students and robots, and this experience strongly influenced her future work.

In 1996 Iain Kusel was a psychology undergraduate at Edinburgh, where he attended Brendan's lectures on intelligent systems. Brendan's knowledge ranged from Leonardo da Vinci to Rodney Brooks, and included evolutionary biology, philosophy of mind, developmental psychology and artificial intelligence. Brendan explained how we should understand complex systems by using children, monkeys and robots as 'epistemic tools', and he provided regular visits to the Appleton Tower laboratory to see these 'tools' in action. In early 2007, Brendan agreed to be an external supervisor for Iain's PhD, the aim of which was to model some of the intriguing data profiles that had emerged within his and Maggie's experiments during the 90s. As he began, he re-read his lecture and tutorial notes from 1996, realising how little Brendan had changed and how great his influence as a mentor had been. Iain has continued to work on the model, under Maggie's supervisory guidance, greatly indebted to the path that Brendan set out for him.

In 1996 Brendan was an invited participant in an annual forum sponsored by Kent State's Department of Psychology at a remote resort in Ohio Amish country. The topic for this Forum was 'Animal Cognition and Sequential Behavior' and his contribution was delivered in his distinctive brogue and florid vocabulary. His talk, especially regarding non-human primates, was highly relevant to the work of Bob Treichler, and their initial discussions were soon followed by e-mails. From this time on they were often called upon to review each other's research reports, and these interactions led to an invitation for the McGonigle-Chalmers team to come to Kent State as Distinguished Visitors in late May 2000. That's when the association between the two teams became even closer. After Brendan had delivered his presentation, the faculty thought it might be appropriate to allow graduate students some private meeting time with him. They took him to lunch at a local student bar, and they later reported that he asked every student about their research and then indicated how it could be related to his work. Brendan's ability to mix socialising with work impressed everybody, and led to a warm relationship between the two families. In the fall of 2007, they discussed the prospect of another visit to the Treichlers. Brendan agreed to consider the prospect for 2008, but only if they would 'prepare another breakfast like the one we shared on your terrace a few years ago'. The visit never materialised. Unexpectedly, Brendan died at home on the 29th of November 2007.

Collin Allen writes: 'It is my everlasting disappointment that I never met Brendan McGonigle. We carried on an intermittent correspondence by e-mail that began about a year before his untimely death. Our correspondence began when I received an e-mail message out of the blue in which he reacted to a paper I had written about 'rational' versus 'associative' explanations of transitive inference in animals. I was greeted with "Colin!"...and thus began our discussion of the themes appearing in my contribution to this volume. I sometimes let weeks and months go by between messages during that year. In retrospect, precious time squandered. *Carpe diem.*'

Ulrich Nehmzow

It is with much sadness that we have to report the death of one of the contributors to this volume. Professor Ulrich Nehmzow was a leading light in the UK robotics community and contributed a great deal, both technically but also in terms of collaboration, community service and inspiration. Ulrich Nehmzow was a postdoctoral researcher working for Brendan McGonigle from 1992 to 1994 on the project outlined in Chapter 6. He was very much influenced by Brendan and remembered him with affection and gratitude.

Ulrich was a very positive individual with a clear vision of the research issues that matter and the key value of science. He inspired many colleagues and students with his great enthusiasm for his subject and his keen sense of morality and correctness. He was popular socially and was the founding force and organizer behind the TAROS series of conferences on robotics (originally TIMR) that have been running successfully for ten years.

Ulrich was dedicated to his research and it is significant that he continued working in his final year, delivering his inaugural lecture as Professor of Cognitive Robotics at Ulster and completing his contribution to Chapter 6 only two weeks before he passed away on 15 April 2010, aged only 48. From the early days of his training under Brendan, Ulrich's career blossomed until he became a renowned international authority on the design and operation of intelligent mobile robots. He will be deeply missed by many friends, collaborators and colleagues, but the work he started with Brendan and developed through his own achievements will continue and be built upon by the research community.

Index

absolute discrimination learning, 12–24
absolute hypothesis, 13–16
absolute stimulus learning, 5, 17–18, 23
absolute stimulus value, 17–18, 23
abstract relationships, 147
accommodation, 148, 159, 163
action-based learning, 76
adaptationist hypothesis of language, 248–9, 251
adaptive behaviour, 134–8
agent/world interactions, 144–5, 149–64
agnosticism, 146–7
alternative idealisations, 176–7
altruism, 137
analogy, 170–2
animal cognition, 224
  assessment of, 6
  criteria for judging, 4
  vs. human cognition, 187–96, 225, 233–4
  research on, 3–5, 12–16, 227–8, 230–5
animal communication, 229–30
animal robotics, 99
animals
  see also primates
  number sense of, 80–93
  serial memory in, 26–35
anthropomorphism, 191, 196
a priori equilibrium, 163–4
arbitrary sequence learning, 63, 64
Aristotle, 147
articulatory loop, 189
artificial intelligence
  behaviour-based, 99
  empiricism and, 147–8
  New AI, 101, 103–4
assimilation, 148, 152, 159, 163, 235–8
associationism, 13, 190–1, 224, 227
associative learning, 9, 26, 27, 190–1, 235–8
associative mechanisms, 6, 224, 235–6
attention, selective, 203
attractor networks, 162
augmented reality, 198–201

autonomous agents, 167
autonomous reasoning, 167–81
autonomous systems, 104–8, 116

Bayesian-based models, 144, 150–4
Bayesian belief net, 156–7
Bayesian inference, 69–72
behaviour
  adaptive, 134–8
  culture as source of, 137–8
  hierarchical structure and, 126–38, 266, 269
  learned, 135–7
  modification, 45
  modular structure and, 126–38
  robotic, 100
behavioural sequences, 117
behaviour-based robotics, 103–4, 107–15
behaviourism, 4, 128, 224–5, 236
binary choice, 5
binary codification, 62–3
binary relations, 6, 8–9, 61, 68
biological innovations, 274–9
biolinguistics, 246–51
brain, enlargement in human evolution, 274, see also encephalisation
brightness discrimination, 23
bright-noisy-tasty water experiment, 15

Cauchy's proof, 173–5
causality, 146, 147, 210, 223–5, 236
children
  emergence of linear sequencing in, 55–78
  emergence of new life stages in evolution, 266, 274, 278
  imitation by, 192–3, 228–9, 233
  language learning in, 188–9
  search organisation and performance in, 41–2, 47, 49–50
  seriation by, 206–7
  understanding of strategic deception in, 91–2

Chomsky, Noam, 191, 234, 249, 250, 261
chunking, 50
classical conditioning, 13, 235
clustering, 41–2, 49–50
cognition, 100
  see also animal cognition; human cognition
  animal vs. human, 187–96, 225, 233–4
  comparative, 29–35, 187–96, 234–5
  cultural practices and, 210–13, 216–17
  dynamic theory of, 129–30
  embodied, 99, 103–4, 117, 122–3, 209–10, 226
  externalisation of, 223
  language and, 187–9, 192, 193, 196–205, 215–16, 265, 276
  models of, 148
  search organisation and, 39–40
  simian, 206–15
  species-level differences in, 135–8
  trade-offs, 137–8
cognitive continuity, 26–35, 196
cognitive development, 224, 225, 232–3, 279
cognitive economy, 45–9
cognitive ethology, 100, 225
cognitive evolution, 77–8, 188, 194, 197–219, 243–62, 264–266, 274–279
cognitive growth, 116–17
cognitive robotics, 103–23
cognitive scaffolding, 197–205, 215–17, 223–4, 238
cognitive science, 234–5
cognitive skills
  advanced, 38
  core, 4
cognitivism, 235–8
communication
  see also language
  animal, 229–30
  code model of, 266–8
  discourse model of, 266–8
  phatic, 275
  symbolic, 229–30
  systems, 194, 248–9
comparative cognition, 29–35, 187–96, 234–5
comparative development, 230–4

comparative psychology, 3–4, 13–15, 225–38
comparative study, of working memory, 38–51
complexity, 25, 35, 56
  complex behaviours, 4, 25, 55
  complex decision-making, 5
  in language, 243–62
  in robots, 99–101
comprehension, 4
  relational, 5
computer science, 126–7
concurrent conditional method, 6, 27–8, 31–3
connectionism, 236
constraints, 118–19, 123, 249–50
constructist computational model, 150–4
constructive omega rule, 175
constructivism, 145, 148–50
continuity/discontinuity debate, 7, 29–31, 77–8, 191
control
  modular, 131–8
  robot, 106–10, 115–16
  structured, 127–38
control theory, 103, 108–9
Corsi tapping task, 8, 49
counter-examples, 175–80
counting skills, 80–93
cultural practices, 210–13, 216–17
culture, 137–8, 190
cybernetics, 103

Darwin, Charles, 25
decision-making, complex, 5
declarative representation, 100, 194–5, 273
deixis, 193, 244, 256–61, 273
Descartes, Rene, 146
development
  cognitive, 224, 225, 232–3
  in cognitive robots, 116–23
  in evolution, *see* evo-devo
  comparative, 230–5
developmental psychology, 116–17, 234–5, 238
developmental robotics, 117–22, 123
diagnostic reasoning, 153, 163–4
discontinuities, 191, 192

discourse, 193–6, 256, 265–80
discrimination learning, 5, 12–24
  size discrimination, 16–24
dorsolateral prefrontal cortex (DLPFC), 50
dualism, 146, 271
dynamic intelligence, 128–30

ecological validity, 225
efficient search, 8–9, 41, 43–5
embodied intelligence, 99, 103–4, 117, 122–3, 209–10, 226
embodiment, 104–6
emergence, 129–30
empiricism, 145–50
encephalisation, 274–6
energetic costs, 48–9
environment
  agent/world interactions, 144–5, 149–64, 190
  robot-environment interaction, 110–15
  spatial properties of, 7–8
epistemological access, 149–50
epistemology, 144–64
equilibration, 148, 149
equipotentiality premise, 14–15
equivalence of associability, 14–15
ethological theory, 128
ethology, 225, 234
Euler's theorem, 173–80
evolution, 188, 193
  cognitive, 77–8, 188, 194, 197–219, 243–62
  of language, 243–62, 264–81
evolutionary continuity, 77, 191–2, *see also* saltatory evolution
evolutionary developmental theoretical biology (evo-devo), 101, 195–6, 264
evolutionary discontinuity thesis, 13
executive control, 56
executive functions, 189, 206, 208, *see also* planning

face detection, 120–2
fatigue, 48–9
foraging, 7, 43, 48
functional reference, 229

Gaussian variables, 69
genes, 190, 264

genetic epistemology, 149
geometric knowledge, 119–22
Gestalt psychology, 13
glass ceiling problem, 192, 213
grammar, 193–5, 248–61
  Universal Grammar (UG), 243–4, 251–6
greatest likelihood calculation, 153–64
Great Leap Forward, 243, 244–6

Hebbs' reinforcement rule, 163
hierarchical organisation and structure, 50, 51, 126–38, 266, 269–72, *see also* recursion
homing behaviour, 99
hominids, 244–6
How to solve it (Pólya), 168–72
human cognition, 4
  vs. animal cognition, 187–96, 225, 233–4
  continuity/discontinuity debate, 29–31, 77–8, 191
  cultural practices, 210–13, 216–17
  language and, 187–9, 192, 193, 267
  models of, 148
  uniqueness of, 35
human infants, 196, 201–2, 264, 272–3
human mind, complexity of, 3
human planning, 195, 266, 268–75, *see also* primate planning; thought
Hume, David, 146, 150, 236

imitation, 192–3, 228–9, 233
inductive bias, 148, 224
infants, 196, 201–2, 264, 272–3
information
  processing, 100
  self-generated, 192, 214–15
inhibition, 61
instrumental conditioning, 13
intelligence
  see also artificial intelligence
  dynamic theory of, 128, 129–30
  embodied, 99, 103–4, 117, 122–3, 209–10, 226
  human, 226
  structuring, 126–38
International Society for Adaptive Behaviour, 99
intrinsic motivation, 123

Kant, Immanuel, 148
knowledge
   diagnostic, 153, 163–4
   geometric, 119–22

laboratory experimentation, 229
labelling, 188–9, 199–200
Lakatos, Imre, 167, 173–80
language, 187–9, 192
language and non-linguistic context, 268
   see also communication
   as anchoring thought, 203–5
   biological basis of, 246–51
   cognition and, 193, 196–205, 215–16
   discourse, 193–6, 256, 265–80
   emergence of complex, 243–62
   evolution, 194–5, 264–81
   grammar, 193–5, 248–61
   learning, 235–6
   recursive sentence structure, 256–60, 265–6, 269–70
   semantics, 193, 267
   thought and, 195, 247
learning
   absolute discrimination, 12–24
   action-based, 76
   arbitrary sequence, 63, 64
   associative, 9, 26, 27, 190–1, 235–8
   behaviour, 135–7
   culture and, 190
   discrimination, 5, 12–24
   as evolution, 135–6
   language, 235–6
   machine, 129
   relational, 4–5, 12–24, 189–90, 224
   as remembering, 147
   speed of, 135
   stimulus, 5, 17–18, 23
lemma incorporation, 178
lesion studies, 204
lexical content, 193, 194, 257–8
Lift-Constraint, Act, Saturate (LCAS) approach, 118–19
linear relationships, 6
linear search, 44–5
linear sequencing, in children, 55–78
linguistic evolution, 194
linguistic rehearsal, 201
linguistics, 191, 193–5, 234, 246–51
list linking, 31–5

list memory, 25–35
logic, 191, 193, 256, 269
logico-mathematical structure, 8–10
long-term memory, 192
loopy belief propagation, 164

machine learning, 129
Markov models, 129, 159
mark test, 228, 231
material symbols, 198–9
mathematical methodology, 167, 173–80
mathematical thought, 203–5
mathematics, 146
memory
   long-term, 192
   serial, 25–35, 50–1
   short-term, 26
   spatial, 40–5, 48–50
   working, 38–51, 191, 192
memory demand, 40, 48
mental capacity, 38
mental state attribution, 228
metabolic efficiency, 126
method of proofs and refutations, 179
mind/body dualism, 146
mirror self-recognition task, 192, 228, 231–2
mobile robots, 106–13
model induction, 154
model-refinement process, 150–64
models
   Bayesian-based, 144, 150–4
   connectionist model of seriation, 191, 236
   discourse models, 267
   of human cognition, 148
   Markov, 129, 159
   Narmax, 116
   of robot-environment interaction, 113–15
   of seriation, 60–1
modularity, 101, 126–38
monkeys
   number sense of, 80–1
   relational and absolute discrimination learning in, 12–24
   relational learning in, 189–90
   search organisation and performance in, 40–5, 47, 49–50
   serial memory in, 25–35
   seriation by, 56–7, 206–8, 226

monotonicity, 59
monotonic seriation, 63–8, 73, 76
motor action, 268–9

naive Bayes, 153
Narmax models, 116
narratives, 271–2
nativism, 234
Neanderthals, 244–6
neural innovations, 197, 213–16, 217
neural networks, 30–31, 113–15, 129–30, 148
neuro-imaging, 7
New AI, 101, 103–4
number judgments, 9
numerosities, 9–10
numerosity, 80–93

Occam's razor, 12, 14
ontologies, 167
ontology evolution, 167
OpenCV Intel libraries, 120
ordered-list memory, 25–35
ordering of events, 5–6
ordering operations, 6–7

parsimony, 126
path dependency, 126
pattern discrimination, 23
pattern recognition, 45–8, 121
perception, in mobile robots, 106–8
perceptual bias, 62
persistent agents, 167
philosophy, 145–50
phylogenetic scale, 15
phylogenetic tree, 15
Physical Symbol System Hypothesis, 103
Piaget, Jean, 55, 57, 148, 149, 152, 159, 163, 164, 234
pigeons, serial memory in, 26, 29
planning, 193, 266, 268–75, *see also* communication; human planning; primate planning; thought
plasticity, 134, 137
Plato, 146, 147
pointer, 61
Pólya, George, 167–72, 173
polyhedra, 173–80
Popper, Karl, 173
precondition analysis, 178
prefrontal function, 50–1
primates
    cognitive evolution, 50–1
    communication systems, 194, 264–6, 270–4
    imitation by, 228–9, 233
    limits on, 206–15
    number sense of, 80–93
    planning, 194, 196, 270–1, *see also* human planning; thought
    search organisation and performance in, 40–5
    serial memory in, 25–35
    seriation by, 56–7, 206–8, 226
private codes, 226, 235–8
probabilities, 150–61
problem solving, 101, 167–81
    counter-examples, 175–80
    mathematical methodology, 173–80
    Pólya's method for, 168–72
Production Systems model, 60
proof plan, 170–2
proofs, 173–80
Proofs and Refutations (Lakatos), 173–80
psychology, 193
    comparative, 3–4, 13–15, 225–38
    developmental, 116–17, 234–5, 238
    Gestalt, 13

quantity judgments, 83–93

radial maze studies, 39–40
ranking behaviours, 5–6, 62–3
rationalism, 145–50
rats, search organisation and performance in, 41–2
reasoning
    autonomous, 167–81
    prognostic, 150–61
recapitulationism, 195
recursion, 256–60, 265–6, 269–70
reductionism, 4
reference to absent objects, 272
reincarnation, 146
relational hypothesis, 13–16
relational learning, 4–5, 12–24, 189–90, 224
relational perception, 4
representations, 101, 167–81, 228, 247
reverse contingency task (RCT), 9–10, 84–93
robot-environment interaction, 110–15

robots/robotics, 147–8
  autonomy and embodiment, 104–8, 116
  behaviour-based, 103–4, 107–15
  behaviour of, 100
  cognitive, 103–23
  complexity in, 99–101
  control and perception in, 106–8, 115–16
  developmental, 117–22, 123
  epistemological sophistication in, 101
  mobile, 106–13
  recognition tasks, 119–22
  research, 103–4
  scientific method in, 110–13
rostrolateral prefrontal cortex (RLPFC), 192, 214–15

saltatory evolution 194–5
schemata, 148, 149
search, 55
  efficient, 8–9, 41–5
  linear, 44–5
  organisation, 39–45, 49–50
  patterns, 45–8
  performance, 49–50
  serial, 59
  simplicity in, 45–8
  spatial, 8
  strategies, 39–40
  trial-and-error, 8
selective attention, 203
self-awareness, 192, 228, 231–2
self-generated information, 192, 214–15
self-regulated learning, 9
self-structuring, 210
semantics, 257–8, 273
Sense-Think-Act cycle, 103
sensory-motor development, 117, 118–19
serial list retention, 25–35
serial memory, 7, 25–35
  effect of serial position contrasts, 33–4
  neuronal correlates of, 29–31
  organisation of, 50–1
  structure and, 48–9
serial order, 5–7, 56, 206–7
serial search, 59
serial spatial recall, 48–50

seriation, 8–9, 56–78, 206–8
  monotonic, 63–8, 73, 76
  size, 8, 9, 55, 57–78, 206–8, 226
simian thought, 206–15, 248
simple explanations, 12
simplicity, 45–8
simultaneous chaining method, 6–7, 27–9
size discrimination, 15–24
size-relational learning, 5
size seriation, 8, 9, 55, 57–78, 206–8, 226
snapshot theory, 99
[social group size, 275]
sorting, 55
spatial location, 7–8
spatial memory, 40–5, 48–50
spatial structure, in working memory, 38–51
stimulus, evaluation of, 19–22
strategic deception, 91–2
structured control, 127–38
structured intelligence, 126–38
subsumption, 101
successive chaining technique, 33–4
symbolic communication, 229–30
symbolic distances, 31–2, 189
symbolic representation, 247
symbol systems, 9–10
syntactic theory, 191
syntax, 265

temporal-ordering mechanisms, 6–7
temporal sequencing, 6
theorem generalisation, 179–80
theory of mind, 91, 228–30, 232, 271, 273
thought, 203–5
  language and, 194–5, 203–5, 247
  mathematical, 203–5
  simian, 206–15, 248
3D reconstruction, 120–2
three-stimulus discrimination learning, 18–19
time, order of events in, 5–6
toolmaking, 229, 276–9
touch screens, 225–6
transitivity, 26–7, 29, 224
transitivity training paradigm, 6
transposition, 12–13, 16

trial-and-error search, 8
2D tracking, 120–2
two-stimulus discrimination learning, 12–16, 18–19

Universal Grammar (UG), 243–4, 251–6
urgency variables, 66–7, 69–72, 74–75

value transfer theory, 28
visual cortex, 127

working memory (WM), 7–8, 191, 192
  comparative study of, 38–51
  search organisation and, 39–40